SCHAUM'S OUTLINE OF

THEORY AND PROBLEMS

OF

MATRIX OPERATIONS

•

RICHARD BRONSON, Ph.D.
Professor of Mathematics and Computer Science
Fairleigh Dickinson University

SCHAUM'S OUTLINE SERIES
McGRAW-HILL
New York San Francisco Washington, D.C. Auckland Bogotá Caracus Lisbon London Madrid Mexico City Milan Montreal New Dehli San Juan Singapore Sydney Tokyo Toronto

To Evy

RICHARD BRONSON, who is Professor and Chairman of Mathematics and Computer Science at Fairleigh Dickinson University, received his Ph.D. in applied mathematics from Stevens Institute of Technology in 1968. Dr. Bronson is currently an associate editor of the journal *Simulation*, contributing editor to *SIAM News*, has served as a consultant to Bell Telephone Laboratories, and has published over 25 technical articles and books, the latter including *Schaum's Outline of Modern Introductory Differential Equations* and *Schaum's Outline of Operations Research*.

Schaum's Outline of Theory and Problems of
MATRIX OPERATIONS

11 12 13 14 15 16 17 18 19 20 PRS PRS 5 4 3 2 1 0

ISBN 0-07-007978-1

Sponsoring Editor, David Beckwith
Production Supervisor, Louise Karam
Editing Supervisor, Marthe Grice

Library of Congress Cataloging-in-Publication Data

Bronson, Richard.
Schaum's outline of theory and problems of matrix operations.

(Schaum's outline series)
Includes index.
1. Matrices. I. Title. II. Title: Matrix operations.
QA188.B759 1989 512.9'434 88-8902
ISBN 0-07-007978-1

McGraw-Hill
A Division of The ***McGraw-Hill*** *Companies*

Preface

Perhaps no area of mathematics has changed as dramatically as matrices over the last 25 years. This is due to both the advent of the computer as well as the introduction and acceptance of matrix methods into other applied disciplines. Computers provide an efficient mechanism for doing iterative computations. This, in turn, has revolutionized the methods used for locating eigenvalues and eigenvectors and has altered the usefulness of many classical techniques, such as those for obtaining inverses and solving simultaneous equations. Relatively new fields, such as operations research, lean heavily on matrix algebra, while established fields, such as economics, probability, and differential equations, continue to expand their reliance on matrices for clarifying and simplifying complex concepts.

This book is an algorithmic approach to matrix operations. The more complicated procedures are given as a series of steps which may be coded in a straightforward manner for computer implementation. The emphasis throughout is on computationally efficient methods. These should be of value to anyone who needs to apply matrix methods to his or her own work.

The material in this book is self-contained; all concepts and procedures are stated directly in terms of matrix operations. There are no prerequisites for using most of this book other than a working knowledge of high school algebra. Some of the applications, however, do require additional expertise, but these are self-evident and are limited to short portions of the book. For example, elementary calculus is needed for the material on differential equations.

Each chapter of this book is divided into three sections. The first introduces concepts and methodology. The second section consists of completely worked-out problems which clarify the material presented in the first section and which, on occasion, also expand on that development. Finally, there is a section of problems with answers with which the reader can test his or her mastery of the subject matter.

I wish to thank the many individuals who helped make this book a reality. I warmly acknowledge the contributions of William Anderson, whose comments on coverage and content were particularly valuable. I am also grateful to Howard Karp and Martha Kingsley for their suggestions and assistance. Particular thanks are due Edward Millman for his splendid editing and support, David Beckwith of the Schaum staff for overseeing the entire project, and Marthe Grice for technical editing.

RICHARD BRONSON

Contents

CONTENTS

CONTENTS

Chapter 1

Basic Operations

MATRICES

A *matrix* is a rectangular array of elements arranged in horizontal *rows* and vertical *columns*, and usually enclosed in brackets. In this book, the elements of a matrix will almost always be numbers or functions of the variable t. A matrix is *real-valued* (or, simply, *real*) if all its elements are real numbers or real-valued functions; it is *complex-valued* if at least one element is a complex number or a complex-valued function. If all its elements are numbers, then a matrix is called a *constant* matrix.

Example 1.1

$$\begin{bmatrix} 1 & 2 \\ 3 & 4 \end{bmatrix} \qquad \begin{bmatrix} 0.5 & \sin t & t+1 \\ -6 & 0 & \cos t \end{bmatrix} \qquad \text{and} \qquad [-1.7,\ 2+i6,\ -3i,\ 0]$$

are all matrices. The first two on the left are real-valued, whereas the third is complex-valued (with $i = \sqrt{-1}$); the first and third are constant matrices, but the second is not constant.

Matrices are designated by boldface uppercase letters. A general matrix **A** having r rows and c columns may be written

$$\mathbf{A} = \begin{bmatrix} a_{11} & a_{12} \cdots a_{1c} \\ a_{21} & a_{22} \cdots a_{2c} \\ \cdots & \cdots \\ a_{r1} & a_{r2} \cdots a_{rc} \end{bmatrix}$$

where the elements of the matrix are double subscripted to denote location. By convention, the row index precedes the column index, thus, a_{25} represents the element of **A** appearing in the second row and fifth column, while a_{31} represents the element appearing in the third row and first column. A matrix **A** may also be denoted as $[a_{ij}]$, where a_{ij} denotes the general element of **A** appearing in the ith row and jth column.

A matrix having r rows and c columns has *order* (or size) "r by c," usually written $r \times c$. The matrices in Example 1.1 have order 2×2, 2×3, and 1×4, respectively from left to right. Two matrices are *equal* if they have the same order and their corresponding elements are equal.

The *transpose* of a matrix **A**, denoted as $\mathbf{A}^T$, is obtained by converting the rows of **A** into the columns of $\mathbf{A}^T$ one at a time in sequence. If **A** has order $m \times n$, then $\mathbf{A}^T$ has order $n \times m$.

Example 1.2 If

$$\mathbf{A} = \begin{bmatrix} 1 & 2 \\ 3 & 4 \\ 5 & 6 \end{bmatrix} \qquad \text{then} \qquad \mathbf{A}^T = \begin{bmatrix} 1 & 3 & 5 \\ 2 & 4 & 6 \end{bmatrix}$$

VECTORS AND DOT PRODUCTS

A *vector* is a matrix having either one row or one column. A matrix consisting of a single row is called a *row vector*; a matrix having a single column is called a *column vector*. The *dot product* $\mathbf{A} \cdot \mathbf{B}$ of two vectors of the same order is obtained by multiplying together corresponding elements of **A** and **B** and then summing the results. The dot product is a *scalar*, by which we mean it is of the same general type as the elements themselves. (See Problem 1.1.)

MATRIX ADDITION AND MATRIX SUBTRACTION

The sum $\mathbf{A}+\mathbf{B}$ of two matrices $\mathbf{A}=[a_{ij}]$ and $\mathbf{B}=[b_{ij}]$ having the same order is the matrix obtained by adding corresponding elements of **A** and **B**. That is,

$$\mathbf{A}+\mathbf{B}=[a_{ij}]+[b_{ij}]=[a_{ij}+b_{ij}]$$

Matrix addition is both associative and commutative. Thus,

$$\mathbf{A}+(\mathbf{B}+\mathbf{C})=(\mathbf{A}+\mathbf{B})+\mathbf{C} \qquad \text{and} \qquad \mathbf{A}+\mathbf{B}=\mathbf{B}+\mathbf{A}$$

(See Problem 1.2.)

The matrix subtraction $\mathbf{A}-\mathbf{B}$ is defined similarly: **A** and **B** must have the same order, and the subtractions must be performed on corresponding elements to yield the matrix $[a_{ij}-b_{ij}]$. (See Problem 1.3.)

SCALAR MULTIPLICATION AND MATRIX MULTIPLICATION

For any scalar k (in this book, usually a number or a function of t), the matrix $k\mathbf{A}$ (or, equivalently, $\mathbf{A}k$) is obtained by multiplying every element of **A** by the scalar k. That is, $k\mathbf{A}=k[a_{ij}]=[ka_{ij}]$. (See Problem 1.3.)

Let $\mathbf{A}=[a_{ij}]$ and $\mathbf{B}=[b_{ij}]$ have orders $r\times p$ and $p\times c$, respectively, so that the number of columns of **A** equals the number of rows of **B**. Then the product **AB** is defined to be the matrix $\mathbf{C}=[c_{ij}]$ of order $r\times c$ whose elements are given by

$$c_{ij}=\sum_{k=1}^{p} a_{ik}b_{kj} \qquad (i=1,2,\ldots,r;\ j=1,2,\ldots,c)$$

Each element c_{ij} of **AB** is a dot product; it is obtained by forming the transpose of the ith row of **A** and then taking its dot product with the jth column of **B**. (See Problems 1.4 through 1.7.)

Matrix multiplication is associative and distributes over addition and subtraction; in general, it is not commutative. Thus,

$$\mathbf{A}(\mathbf{BC})=(\mathbf{AB})\mathbf{C} \qquad \mathbf{A}(\mathbf{B}+\mathbf{C})=\mathbf{AB}+\mathbf{AC} \qquad (\mathbf{B}-\mathbf{C})\mathbf{A}=\mathbf{BA}-\mathbf{CA}$$

but, in general, $\mathbf{AB}\neq\mathbf{BA}$. Also,

$$(\mathbf{AB})^T=\mathbf{B}^T\mathbf{A}^T$$

ROW-ECHELON FORM

A zero row in a matrix is a row whose elements are all zero, and a nonzero row is one that contains at least one nonzero element. A matrix is a *zero matrix*, denoted **0**, if it contains only zero rows.

A matrix is in *row-echelon form* if it satisfies four conditions:

(R1): All nonzero rows precede (that is, appear above) zero rows when both types are contained in the matrix.

(R2): The first (leftmost) nonzero element of each nonzero row is unity.

(R3): When the first nonzero element of a row appears in column c, then all elements in column c in succeeding rows are zero.

(R4): The first nonzero element of any nonzero row appears in a later column (further to the right) than the first nonzero element of any preceding row.

Example 1.3 The matrix

$$\begin{bmatrix} 1 & 0 & 3 & 2 & 3 \\ 0 & 0 & 1 & 0 & 2 \\ 0 & 0 & 0 & 0 & 0 \end{bmatrix}$$

satisfies all four conditions and so is in row-echelon form. (See Problems 1.11 to 1.15 and 1.18.)

ELEMENTARY ROW AND COLUMN OPERATIONS

There are three *elementary row operations* which may be used to transform a matrix into row-echelon form. The origins of these operations are discussed in Chapter 2; the operations themselves are:

(E1): Interchange any two rows.

(E2): Multiply the elements of any row by a nonzero scalar.

(E3): Add to any row, element by element, a scalar times the corresponding elements of another row.

Three *elementary column operations* are defined analogously.

An algorithm for using elementary row operations to transform a matrix into row-echelon form is as follows:

STEP 1.1: Let R denote the work row, and initialize $R = 1$ (so that the top row is the first work row).

STEP 1.2: Find the first column containing a nonzero element in either row R or any succeeding row. If no such column exists, stop; the transformation is complete. Otherwise, let C denote this column.

STEP 1.3: Beginning with row R and continuing through successive rows, locate the first row having a nonzero element in column C. If this row is not row R, interchange it with row R (elementary row operation E1). Row R will now have a nonzero element in column C. This element is called the *pivot*; let P denote its value.

STEP 1.4: If P is not 1, multiply the elements of row R by $1/P$ (elementary row operation E2); otherwise continue.

STEP 1.5: Search all rows following row R for one having a nonzero element in column C. If no such row exists, go to Step 1.8; otherwise designate that row as row N, and the value of the nonzero element in row N and column C as V.

STEP 1.6: Add to the elements of row N the scalar $-V$ times the corresponding elements of row R (elementary row operation E3).

STEP 1.7: Return to Step 1.5.

STEP 1.8: Increase R by 1. If this new value of R is larger than the number of rows in the matrix, stop; the transformation is complete. Otherwise, return to Step 1.2.

(See Problems 1.12 through 1.15.)

RANK

The *rank* (or *row rank*) of a matrix is the number of nonzero rows in the matrix after it has been transformed to row-echelon form via elementary row operations. (See Problems 1.16 and 1.17.)

Solved Problems

1.1 Find $\mathbf{A}\cdot\mathbf{B}$ and $\mathbf{B}\cdot\mathbf{C}^T$ for

$$\mathbf{A}=\begin{bmatrix}2\\3\\4\end{bmatrix}\qquad \mathbf{B}=\begin{bmatrix}5\\6\\-7\end{bmatrix}\qquad \mathbf{C}=[7,-8,-9]$$

$$\mathbf{A}\cdot\mathbf{B}=2(5)+3(6)+4(-7)=0,$$

$$\mathbf{B}\cdot\mathbf{C}^T=\begin{bmatrix}5\\6\\-7\end{bmatrix}\cdot\begin{bmatrix}7\\-8\\-9\end{bmatrix}=5(7)+6(-8)+(-7)(-9)=50$$

1.2 Show that $\mathbf{A}+\mathbf{B}=\mathbf{B}+\mathbf{A}$ for

$$\mathbf{A}=\begin{bmatrix}0&1\\2&3\end{bmatrix}\qquad\text{and}\qquad \mathbf{B}=\begin{bmatrix}4&5\\6&-7\end{bmatrix}$$

$$\mathbf{A}+\mathbf{B}=\begin{bmatrix}0&1\\2&3\end{bmatrix}+\begin{bmatrix}4&5\\6&-7\end{bmatrix}=\begin{bmatrix}0+4&1+5\\2+6&3+(-7)\end{bmatrix}=\begin{bmatrix}4&6\\8&-4\end{bmatrix}$$

$$\mathbf{B}+\mathbf{A}=\begin{bmatrix}4&5\\6&-7\end{bmatrix}+\begin{bmatrix}0&1\\2&3\end{bmatrix}=\begin{bmatrix}4+0&5+1\\6+2&-7+3\end{bmatrix}=\begin{bmatrix}4&6\\8&-4\end{bmatrix}$$

Since the resulting matrices have the same order and all corresponding elements are equal, $\mathbf{A}+\mathbf{B}=\mathbf{B}+\mathbf{A}$.

1.3 Find $3\mathbf{A}-0.5\mathbf{B}$ for the matrices of Problem 1.2.

$$3\mathbf{A}-0.5\mathbf{B}=3\begin{bmatrix}0&1\\2&3\end{bmatrix}-0.5\begin{bmatrix}4&5\\6&-7\end{bmatrix}=\begin{bmatrix}3(0)&3(1)\\3(2)&3(3)\end{bmatrix}-\begin{bmatrix}0.5(4)&0.5(5)\\0.5(6)&0.5(-7)\end{bmatrix}$$

$$=\begin{bmatrix}0-2&3-2.5\\6-3&9-(-3.5)\end{bmatrix}=\begin{bmatrix}-2&0.5\\3&12.5\end{bmatrix}$$

1.4 Find $\mathbf{AB}$ and $\mathbf{BA}$ for the matrices of Problem 1.2.

$$\mathbf{AB}=\begin{bmatrix}0&1\\2&3\end{bmatrix}\begin{bmatrix}4&5\\6&-7\end{bmatrix}=\begin{bmatrix}0(4)+1(6)&0(5)+1(-7)\\2(4)+3(6)&2(5)+3(-7)\end{bmatrix}=\begin{bmatrix}6&-7\\26&-11\end{bmatrix}$$

$$\mathbf{BA}=\begin{bmatrix}4&5\\6&-7\end{bmatrix}\begin{bmatrix}0&1\\2&3\end{bmatrix}=\begin{bmatrix}4(0)+5(2)&4(1)+5(3)\\6(0)+(-7)(2)&6(1)+(-7)(3)\end{bmatrix}=\begin{bmatrix}10&19\\-14&-15\end{bmatrix}$$

Note that, for these matrices, $\mathbf{AB}\neq\mathbf{BA}$.

1.5 Find $\mathbf{AB}$ and $\mathbf{BA}$ for

$$\mathbf{A}=\begin{bmatrix}1&2&3\\4&-5&6\end{bmatrix}\qquad\text{and}\qquad \mathbf{B}=\begin{bmatrix}7&8\\0&-9\end{bmatrix}$$

Since $\mathbf{A}$ has three columns while $\mathbf{B}$ has only two rows, the matrix product $\mathbf{AB}$ is not defined. But

$$\mathbf{BA}=\begin{bmatrix}7&8\\0&-9\end{bmatrix}\begin{bmatrix}1&2&3\\4&-5&6\end{bmatrix}=\begin{bmatrix}7(1)+8(4)&7(2)+8(-5)&7(3)+8(6)\\0(1)+(-9)(4)&0(2)+(-9)(-5)&0(3)+(-9)(6)\end{bmatrix}$$

$$=\begin{bmatrix}39&-26&69\\-36&45&-54\end{bmatrix}$$

1.6 Verify that $(\mathbf{BA})^T = \mathbf{A}^T\mathbf{B}^T$ for the matrices of Problem 1.5.

$$\mathbf{A}^T\mathbf{B}^T = \begin{bmatrix} 1 & 4 \\ 2 & -5 \\ 3 & 6 \end{bmatrix}\begin{bmatrix} 7 & 0 \\ 8 & -9 \end{bmatrix} = \begin{bmatrix} 1(7)+4(8) & 1(0)+4(-9) \\ 2(7)+(-5)(8) & 2(0)+(-5)(-9) \\ 3(7)+6(8) & 3(0)+6(-9) \end{bmatrix} = \begin{bmatrix} 39 & -36 \\ -26 & 45 \\ 69 & -54 \end{bmatrix}$$

which is the transpose of the product **BA** found in Problem 1.5.

1.7 Find **AB** and **AC** if

$$\mathbf{A} = \begin{bmatrix} 4 & 2 & 0 \\ 2 & 1 & 0 \\ -2 & -1 & 1 \end{bmatrix} \qquad \mathbf{B} = \begin{bmatrix} 2 & 3 & 1 \\ 2 & -2 & -2 \\ -1 & 2 & 1 \end{bmatrix} \qquad \mathbf{C} = \begin{bmatrix} 3 & 1 & -3 \\ 0 & 2 & 6 \\ -1 & 2 & 1 \end{bmatrix}$$

$$\mathbf{AB} = \begin{bmatrix} 4(2)+2(2)+0(-1) & 4(3)+2(-2)+0(2) & 4(1)+2(-2)+0(1) \\ 2(2)+1(2)+0(-1) & 2(3)+1(-2)+0(2) & 2(1)+1(-2)+0(1) \\ -2(2)+(-1)(2)+1(-1) & -2(3)+(-1)(-2)+1(2) & -2(1)+(-1)(-2)+1(1) \end{bmatrix}$$

$$= \begin{bmatrix} 12 & 8 & 0 \\ 6 & 4 & 0 \\ -7 & -2 & 1 \end{bmatrix}$$

$$\mathbf{AC} = \begin{bmatrix} 4(3)+2(0)+0(-1) & 4(1)+2(2)+0(2) & 4(-3)+2(6)+0(1) \\ 2(3)+1(0)+0(-1) & 2(1)+1(2)+0(2) & 2(-3)+1(6)+0(1) \\ -2(3)+(-1)(0)+1(-1) & -2(1)+(-1)(2)+1(2) & -2(-3)+(-1)(6)+1(1) \end{bmatrix} = \begin{bmatrix} 12 & 8 & 0 \\ 6 & 4 & 0 \\ -7 & -2 & 1 \end{bmatrix}$$

Note that, for these matrices, $\mathbf{AB} = \mathbf{AC}$ and yet $\mathbf{B} \neq \mathbf{C}$. This shows that the cancellation law is not valid for matrix multiplication.

1.8 A matrix is *partitioned* if it is divided into smaller matrices by horizontal or vertical lines drawn between entire rows and columns. Determine three partitionings of the matrix

$$\mathbf{A} = \begin{bmatrix} 1 & 2 & 3 & 4 \\ 0 & 0 & 5 & 6 \\ 7 & 8 & -1 & -2 \end{bmatrix}$$

There are $2^5 - 1 = 31$ different ways in which **A** can be partitioned with at least one partitioning line. By placing a line between each two rows and each two columns, we divide **A** into twelve 1×1 matrices, obtaining

$$\mathbf{A} = \left[\begin{array}{c:c:c:c} 1 & 2 & 3 & 4 \\ \hdashline 0 & 0 & 5 & 6 \\ \hdashline 7 & 8 & -1 & -2 \end{array}\right]$$

By placing one line between the first and second rows and another line between the second and third columns, we construct the partitioning

$$\mathbf{A} = \left[\begin{array}{cc:cc} 1 & 2 & 3 & 4 \\ \hdashline 0 & 0 & 5 & 6 \\ 7 & 8 & -1 & -2 \end{array}\right] = \begin{bmatrix} \mathbf{B} & \mathbf{C} \\ \mathbf{E} & \mathbf{F} \end{bmatrix}$$

where

$$\mathbf{B} = [1, 2] \qquad \mathbf{C} = [3, 4] \qquad \mathbf{E} = \begin{bmatrix} 0 & 0 \\ 7 & 8 \end{bmatrix} \qquad \mathbf{F} = \begin{bmatrix} 5 & 6 \\ -1 & -2 \end{bmatrix}$$

A third partitioning can be constructed by placing a single line between the third and fourth columns of **A**. Then $\mathbf{A} = [\mathbf{G}, \mathbf{H}]$, where

$$\mathbf{G}=\begin{bmatrix}1 & 2 & 3\\ 0 & 0 & 5\\ 7 & 8 & -1\end{bmatrix} \quad \text{and} \quad \mathbf{H}=\begin{bmatrix}4\\ 6\\ -2\end{bmatrix}$$

A partitioned matrix can be viewed as a matrix whose elements are themselves matrices.

1.9 The arithmetic operations defined above for matrices having scalar elements apply as well to partitioned matrices. Determine **AB** and **A** − **B** if

$$\mathbf{A}=\begin{bmatrix}\mathbf{C} & \mathbf{D}\\ \mathbf{E} & \mathbf{C}\end{bmatrix} \quad \text{and} \quad \mathbf{B}=\begin{bmatrix}\mathbf{F} & \mathbf{G}\\ \mathbf{F} & \mathbf{E}\end{bmatrix}$$

where $\mathbf{C}=\begin{bmatrix}1 & 2\\ 3 & 4\end{bmatrix}$ $\mathbf{D}=\begin{bmatrix}0 & 0\\ 0 & 0\end{bmatrix}$ $\mathbf{E}=\begin{bmatrix}5 & 6\\ 7 & 8\end{bmatrix}$ $\mathbf{F}=\begin{bmatrix}-1 & 0\\ 2 & 1\end{bmatrix}$ $\mathbf{G}=\begin{bmatrix}-2 & -3\\ 1 & 1\end{bmatrix}$

$$\mathbf{AB}=\begin{bmatrix}\mathbf{CF}+\mathbf{DF} & \mathbf{CG}+\mathbf{DE}\\ \mathbf{EF}+\mathbf{CF} & \mathbf{EG}+\mathbf{CE}\end{bmatrix}$$

$$=\left[\begin{array}{c:c}\begin{bmatrix}1 & 2\\ 3 & 4\end{bmatrix}\begin{bmatrix}-1 & 0\\ 2 & 1\end{bmatrix}+\begin{bmatrix}0 & 0\\ 0 & 0\end{bmatrix}\begin{bmatrix}-1 & 0\\ 2 & 1\end{bmatrix} & \begin{bmatrix}1 & 2\\ 3 & 4\end{bmatrix}\begin{bmatrix}-2 & -3\\ 1 & 1\end{bmatrix}+\begin{bmatrix}0 & 0\\ 0 & 0\end{bmatrix}\begin{bmatrix}5 & 6\\ 7 & 8\end{bmatrix}\\ \hdashline \begin{bmatrix}5 & 6\\ 7 & 8\end{bmatrix}\begin{bmatrix}-1 & 0\\ 2 & 1\end{bmatrix}+\begin{bmatrix}1 & 2\\ 3 & 4\end{bmatrix}\begin{bmatrix}-1 & 0\\ 2 & 1\end{bmatrix} & \begin{bmatrix}5 & 6\\ 7 & 8\end{bmatrix}\begin{bmatrix}-2 & -3\\ 1 & 1\end{bmatrix}+\begin{bmatrix}1 & 2\\ 3 & 4\end{bmatrix}\begin{bmatrix}5 & 6\\ 7 & 8\end{bmatrix}\end{array}\right]$$

$$=\left[\begin{array}{c:c}\begin{bmatrix}3 & 2\\ 5 & 4\end{bmatrix}+\begin{bmatrix}0 & 0\\ 0 & 0\end{bmatrix} & \begin{bmatrix}0 & -1\\ -2 & -5\end{bmatrix}+\begin{bmatrix}0 & 0\\ 0 & 0\end{bmatrix}\\ \hdashline \begin{bmatrix}7 & 6\\ 9 & 8\end{bmatrix}+\begin{bmatrix}3 & 2\\ 5 & 4\end{bmatrix} & \begin{bmatrix}-4 & -9\\ -6 & -13\end{bmatrix}+\begin{bmatrix}19 & 22\\ 43 & 50\end{bmatrix}\end{array}\right]$$

$$=\left[\begin{array}{c:c}\begin{bmatrix}3 & 2\\ 5 & 4\end{bmatrix} & \begin{bmatrix}0 & -1\\ -2 & -5\end{bmatrix}\\ \hdashline \begin{bmatrix}10 & 8\\ 14 & 12\end{bmatrix} & \begin{bmatrix}15 & 13\\ 37 & 37\end{bmatrix}\end{array}\right]=\begin{bmatrix}3 & 2 & 0 & -1\\ 5 & 4 & -2 & -5\\ 10 & 8 & 15 & 13\\ 14 & 12 & 37 & 37\end{bmatrix}$$

$$\mathbf{A}-\mathbf{B}=\begin{bmatrix}\mathbf{C}-\mathbf{F} & \mathbf{D}-\mathbf{G}\\ \mathbf{E}-\mathbf{F} & \mathbf{C}-\mathbf{E}\end{bmatrix}=\left[\begin{array}{c:c}\begin{bmatrix}1 & 2\\ 3 & 4\end{bmatrix}-\begin{bmatrix}-1 & 0\\ 2 & 1\end{bmatrix} & \begin{bmatrix}0 & 0\\ 0 & 0\end{bmatrix}-\begin{bmatrix}-2 & -3\\ 1 & 1\end{bmatrix}\\ \hdashline \begin{bmatrix}5 & 6\\ 7 & 8\end{bmatrix}-\begin{bmatrix}-1 & 0\\ 2 & 1\end{bmatrix} & \begin{bmatrix}1 & 2\\ 3 & 4\end{bmatrix}-\begin{bmatrix}5 & 6\\ 7 & 8\end{bmatrix}\end{array}\right]$$

$$=\left[\begin{array}{c:c}\begin{bmatrix}2 & 2\\ 1 & 3\end{bmatrix} & \begin{bmatrix}2 & 3\\ -1 & -1\end{bmatrix}\\ \hdashline \begin{bmatrix}6 & 6\\ 5 & 7\end{bmatrix} & \begin{bmatrix}-4 & -4\\ -4 & -4\end{bmatrix}\end{array}\right]=\begin{bmatrix}2 & 2 & 2 & 3\\ 1 & 3 & -1 & -1\\ 6 & 6 & -4 & -4\\ 5 & 7 & -4 & -4\end{bmatrix}$$

1.10 Partitioning can be used to check matrix multiplication. Suppose the product **AB** is to be found and checked. Then **A** and **B** are replaced by two larger partitioned matrices, such that their product is

$$\begin{bmatrix}\mathbf{A}\\ \mathbf{R}\end{bmatrix}[\mathbf{B}\ \mathbf{C}]=\begin{bmatrix}\mathbf{AB} & \mathbf{AC}\\ \mathbf{RB} & \mathbf{RC}\end{bmatrix}$$

where **R** is a new row consisting of the column sums of **A**, and **C** is a new column consisting of the row sums of **B**. The resulting matrix has the original product **AB** in the upper left partition. If no errors have been made, **AC** consists of the row sums of **AB**; **RB** consists of the column sums of **AB**; and **RC** is the sum of the elements of **AC** as well as the sum of the elements of **RB**. Use this procedure to obtain the product

$$\begin{bmatrix}1 & 3\\ 2 & -1\end{bmatrix}\begin{bmatrix}3 & -1 & 1\\ 3 & 2 & 2\end{bmatrix}$$

We form the partitioned matrices and find their product:

$$\left[\begin{array}{cc} 1 & 3 \\ 2 & -1 \\ \hdashline 3 & 2 \end{array}\right]\left[\begin{array}{ccc:c} 3 & -1 & 1 & 3 \\ 3 & 2 & 2 & 7 \end{array}\right] = \left[\begin{array}{c:c} \begin{bmatrix} 1 & 3 \\ 2 & -1 \end{bmatrix}\begin{bmatrix} 3 & -1 & 1 \\ 3 & 2 & 2 \end{bmatrix} & \begin{bmatrix} 1 & 3 \\ 2 & -1 \end{bmatrix}\begin{bmatrix} 3 \\ 7 \end{bmatrix} \\ \hdashline [3 \quad 2]\begin{bmatrix} 3 & -1 & 1 \\ 3 & 2 & 2 \end{bmatrix} & [3 \quad 2]\begin{bmatrix} 3 \\ 7 \end{bmatrix} \end{array}\right]$$

$$= \left[\begin{array}{ccc:c} 12 & 5 & 7 & 24 \\ 3 & -4 & 0 & -1 \\ \hdashline 15 & 1 & 7 & 23 \end{array}\right]$$

The product **AB** is the upper left part of the resulting matrix. Since the row sums of this product, the column sums, and their sums are correctly given in the matrix, the multiplication checks.

1.11 Determine which of the following matrices are in row-echelon form:

$$\mathbf{A} = \begin{bmatrix} 1 & 2 & -1 & 0 \\ 0 & 0 & 1 & 4 \\ 0 & 0 & 0 & 0 \end{bmatrix} \qquad \mathbf{B} = \begin{bmatrix} 0 & 1 & 4 \\ 1 & 2 & 3 \end{bmatrix} \qquad \mathbf{C} = \begin{bmatrix} 0 & 2 & 4 \\ 0 & 0 & 1 \end{bmatrix}$$

$$\mathbf{D} = \begin{bmatrix} 0 & 1 & 0 & 4 \\ 0 & 0 & 1 & 3 \\ 0 & 0 & 0 & 1 \end{bmatrix} \qquad \mathbf{E} = \begin{bmatrix} 1 & 2 & 3 \\ 4 & 9 & 7 \end{bmatrix}$$

Only **A** and **D** are in row-echelon form. **B** is not, because the first (leftmost) nonzero element in the second row is further left than the first nonzero element in the top row, violating condition R4. Condition R2 is violated in the first row of **C**. Matrix **E** violates condition R3, because the first nonzero element in the lower row appears in the same column as the first nonzero element of the upper row.

1.12 Use elementary row operations to transform matrices **B**, **C**, and **E** of Problem 1.11 into row-echelon form.

We follow Steps 1.1 through 1.8 in each case, but for simplicity list only those steps that result in a matrix manipulation. For **B**, with $R = 1$ (Step 1.1) and $C = 1$ (Step 1.2), we apply Step 1.3 and interchange rows 1 and 2, obtaining.

$$\begin{bmatrix} 1 & 2 & 3 \\ 0 & 1 & 4 \end{bmatrix}$$

which is in row-echelon form. For matrix **C**, with $R = 1$ (Step 1.1), $C = 2$ (Step 1.2), and $P = 2$ (Step 1.3), we apply Step 1.4 and multiply all elements in the first row by $1/2$, obtaining

$$\begin{bmatrix} 0 & 1 & 2 \\ 0 & 0 & 1 \end{bmatrix}$$

which is in row-echelon form. For matrix **E**, with $R = 1$ (Step 1.1), $C = 1$ (Step 1.2), and $N = 2$ and $V = 4$ (Step 1.5), we apply Step 1.6 by adding, to each element in row 2, -4 times the corresponding element in row 1; the result is

$$\begin{bmatrix} 1 & 2 & 3 \\ 4+(-4)(1) & 9+(-4)(2) & 7+(-4)(3) \end{bmatrix} = \begin{bmatrix} 1 & 2 & 3 \\ 0 & 1 & -5 \end{bmatrix}$$

which is in row-echelon form.

1.13 Transform the following matrix into row-echelon form:

$$\begin{bmatrix} 1 & 2 & -1 & 6 \\ 3 & 8 & 9 & 10 \\ 2 & -1 & 2 & -2 \end{bmatrix}$$

Here (and in later problems) we shall use an arrow to indicate the row that results from each elementary row operation.

$$\begin{matrix} \\ \rightarrow \\ \\ \end{matrix}\begin{bmatrix} 1 & 2 & -1 & 6 \\ 0 & 2 & 12 & -8 \\ 2 & -1 & 2 & -2 \end{bmatrix}$$

Step 1.6 with $R = 1$, $C = 1$, $N = 2$, and $V = 3$: Add -3 times the first row to the second row.

$$\begin{matrix} \\ \\ \rightarrow \end{matrix}\begin{bmatrix} 1 & 2 & -1 & 6 \\ 0 & 2 & 12 & -8 \\ 0 & -5 & 4 & -14 \end{bmatrix}$$

Step 1.6 with $R = 1$, $C = 1$, $N = 3$, and $V = 2$: Add -2 times the first row to the third row.

$$\begin{matrix} \\ \rightarrow \\ \\ \end{matrix}\begin{bmatrix} 1 & 2 & -1 & 6 \\ 0 & 1 & 6 & -4 \\ 0 & -5 & 4 & -14 \end{bmatrix}$$

Step 1.4 with $R = 2$, $C = 2$, and $P = 2$: Multiply the second row by $1/2$.

$$\begin{matrix} \\ \\ \rightarrow \end{matrix}\begin{bmatrix} 1 & 2 & -1 & 6 \\ 0 & 1 & 6 & -4 \\ 0 & 0 & 34 & -34 \end{bmatrix}$$

Step 1.6 with $R = 2$, $C = 2$, $N = 3$, and $V = -5$: Add 5 times the second row to the third row.

$$\begin{matrix} \\ \\ \rightarrow \end{matrix}\begin{bmatrix} 1 & 2 & -1 & 6 \\ 0 & 1 & 6 & -4 \\ 0 & 0 & 1 & -1 \end{bmatrix}$$

Step 1.4 with $R = 3$, $C = 3$, and $P = 34$: Multiply the third row by $1/34$.

1.14 Transform the following matrix into row-echelon form:

$$\begin{bmatrix} 2 & 1 & 0 & 5 \\ 3 & 6 & 1 & 1 \\ 5 & 7 & 1 & 8 \end{bmatrix}$$

$$\begin{matrix} \rightarrow \\ \\ \\ \end{matrix}\begin{bmatrix} 1 & 1/2 & 0 & 5/2 \\ 3 & 6 & 1 & 1 \\ 5 & 7 & 1 & 8 \end{bmatrix}$$

Step 1.4 with $R = 1$, $C = 1$, and $P = 2$: Multiply the first row by $1/2$.

$$\begin{matrix} \\ \rightarrow \\ \\ \end{matrix}\begin{bmatrix} 1 & 1/2 & 0 & 5/2 \\ 0 & 9/2 & 1 & -13/2 \\ 5 & 7 & 1 & 8 \end{bmatrix}$$

Step 1.6 with $R = 1$, $C = 1$, $N = 2$, and $V = 3$: Add -3 times the first row to the second row.

$$\begin{matrix} \\ \\ \rightarrow \end{matrix}\begin{bmatrix} 1 & 1/2 & 0 & 5/2 \\ 0 & 9/2 & 1 & -13/2 \\ 0 & 9/2 & 1 & -9/2 \end{bmatrix}$$

Step 1.6 with $R = 1$, $C = 1$, $N = 3$, and $V = 5$: Add -5 times the first row to the third row.

$$\begin{matrix} \\ \rightarrow \\ \\ \end{matrix}\begin{bmatrix} 1 & 1/2 & 0 & 5/2 \\ 0 & 1 & 2/9 & -13/9 \\ 0 & 9/2 & 1 & -9/2 \end{bmatrix}$$

Step 1.4 with $R = 2$, $C = 2$, and $P = 9/2$: Multiply the second row by $2/9$.

$$\begin{matrix} \\ \\ \rightarrow \end{matrix}\begin{bmatrix} 1 & 1/2 & 0 & 5/2 \\ 0 & 1 & 2/9 & -13/9 \\ 0 & 0 & 0 & 2 \end{bmatrix}$$

Step 1.6 with $R = 2$, $C = 2$, $N = 3$, and $V = 9/2$: Add $-9/2$ times the second row to the third row.

$$\begin{matrix} \\ \\ \rightarrow \end{matrix}\begin{bmatrix} 1 & 1/2 & 0 & 5/2 \\ 0 & 1 & 2/9 & -13/9 \\ 0 & 0 & 0 & 1 \end{bmatrix}$$

Step 1.4 with $R = 3$, $C = 4$, and $P = 2$: Multiply the third row by $1/2$.

1.15 Transform the following matrix into row-echelon form:

$$\begin{bmatrix} 3 & 2 & 1 & -4 & 1 \\ 2 & 3 & 0 & -1 & -1 \\ 1 & -6 & 3 & -8 & 7 \end{bmatrix}$$

$$\rightarrow \begin{bmatrix} 1 & 2/3 & 1/3 & -4/3 & 1/3 \\ 2 & 3 & 0 & -1 & -1 \\ 1 & -6 & 3 & -8 & 7 \end{bmatrix}$$ Step 1.4: Multiply the first row by 1/3.

$$\begin{bmatrix} 1 & 2/3 & 1/3 & -4/3 & 1/3 \\ 0 & 5/3 & -2/3 & 5/3 & -5/3 \\ 1 & -6 & 3 & -8 & 7 \end{bmatrix}$$ Step 1.6: Add −2 times the first row to the second row.

$$\begin{bmatrix} 1 & 2/3 & 1/3 & -4/3 & 1/3 \\ 0 & 5/3 & -2/3 & 5/3 & -5/3 \\ 0 & -20/3 & 8/3 & -20/3 & 20/3 \end{bmatrix}$$ Step 1.6: Add −1 times the first row to the third row.

$$\begin{bmatrix} 1 & 2/3 & 1/3 & -4/3 & 1/3 \\ 0 & 1 & -2/5 & 1 & -1 \\ 0 & -20/3 & 8/3 & -20/3 & 20/3 \end{bmatrix}$$ Step 1.4: Multiply the second row by 3/5.

$$\begin{bmatrix} 1 & 2/3 & 1/3 & -4/3 & 1/3 \\ 0 & 1 & -2/5 & 1 & -1 \\ 0 & 0 & 0 & 0 & 0 \end{bmatrix}$$ Step 1.6: Add 20/3 times the second row to the third row.

1.16 Determine the rank of the matrix of Problem 1.14.

Because the row-echelon form of this matrix has three nonzero rows, the rank of the original matrix is 3.

1.17 Determine the rank of the matrix of Problem 1.15.

Because the row-echelon form of this matrix has two nonzero rows, the rank of the original matrix is 2.

1.18 Show that row-echelon form is not unique if a matrix has rank 2 or greater.

Such a matrix has at least two nonzero rows after it is transformed into row-echelon form. By adding the second row to the first row, a different row-echelon-form matrix is produced. As an example, if we add the second row to the first row of the row-echelon-form matrix obtained in Problem 1.14, we obtain

$$\begin{bmatrix} 1 & 3/2 & 2/9 & 19/18 \\ 0 & 1 & 2/9 & -13/9 \\ 0 & 0 & 0 & 1 \end{bmatrix}$$

which is also in row-echelon form.

Supplementary Problems

In Problems 1.19 through 1.32 let

$$\mathbf{A}=\begin{bmatrix}1 & 2\\ 3 & -4\end{bmatrix}\quad \mathbf{B}=\begin{bmatrix}0 & 2\\ 1 & 2\end{bmatrix}\quad \mathbf{C}=\begin{bmatrix}3 & 5 & 0\\ -2 & -3 & 1\\ 1 & 2 & 1\end{bmatrix}\quad \mathbf{D}=\begin{bmatrix}2 & 2 & 1\\ 4 & 4 & 2\\ 6 & 6 & 3\end{bmatrix}\quad \mathbf{E}=\begin{bmatrix}1\\ 2\\ 3\end{bmatrix}\quad \mathbf{F}=[1,\ 2,\ 3]$$

1.19 Find (*a*) $\mathbf{A}+\mathbf{B}$; (*b*) $3\mathbf{A}$; (*c*) $2\mathbf{A}-3\mathbf{B}$; (*d*) $\mathbf{C}-\mathbf{D}$; and (*e*) $\mathbf{A}+\mathbf{F}$.

1.20 Designate the columns of **A** as $\mathbf{A}_1$ and $\mathbf{A}_2$, and the columns of **C** as $\mathbf{C}_1$, $\mathbf{C}_2$, and $\mathbf{C}_3$, from left to right. Then calculate (*a*) $\mathbf{A}_1 \cdot \mathbf{A}_2$; (*b*) $\mathbf{C}_1 \cdot \mathbf{C}_2$; and (*c*) $\mathbf{C}_1 \cdot \mathbf{C}_3$.

1.21 Find (*a*) **AB**; (*b*) **BA**; (*c*) $(\mathbf{AB})^T$; (*d*) $\mathbf{B}^T\mathbf{A}^T$; and (*e*) $\mathbf{A}^T\mathbf{B}^T$.

1.22 Find (*a*) **CD** and (*b*) **DC**.

1.23 Find $\mathbf{A}(\mathbf{A}+\mathbf{B})$.

1.24 Find (*a*) **CE** and (*b*) **EC**.

1.25 Find (*a*) **CF** and (*b*) **FC**.

1.26 Find (*a*) **EF** and (*b*) **FE**.

1.27 Transform **A** to row-echelon form.

1.28 Transform **B** to row-echelon form.

1.29 Transform **C** to row-echelon form.

1.30 Transform **D** to row-echelon form.

1.31 Transform **E** to row-echelon form.

1.32 Find the rank of (*a*) **A**; (*b*) **B**; (*c*) **C**; (*d*) **D**; and (*e*) **E**.

1.33 Find two matrices, neither of which is a zero matrix, whose product is a zero matrix.

1.34 The price schedule for a New York to Miami flight is given by the vector $\mathbf{P}=[240, 180, 89]$, where the elements denote the costs of first class, business class, and tourist class tickets, respectively. The number of tickets of each class purchased for a particular flight is given by the vector $\mathbf{N}=[8, 21, 115]$. What is the significance of $\mathbf{P}\cdot\mathbf{N}$?

1.35 The inventory of computers at each outlet of a three-store chain is given by the matrix

$$\mathbf{N}=\begin{bmatrix}9 & 12\\ 15 & 4\\ 7 & 0\end{bmatrix}$$

where the rows pertain to the different stores and the columns denote the number of brand X and brand Y computers, respectively, in each store. The wholesale costs of these computers are given by the vector $\mathbf{D}=[700, 1200]^T$. Calculate **ND** and state its significance.

Chapter 2

Simultaneous Linear Equations

CONSISTENCY

A system of simultaneous linear equations is a set of equations of the form

$$\begin{aligned}
a_{11}x_1 + a_{12}x_2 + a_{13}x_3 + \cdots + a_{1n}x_n &= b_1 \\
a_{21}x_1 + a_{22}x_2 + a_{23}x_3 + \cdots + a_{2n}x_n &= b_2 \\
\cdots\cdots\cdots\cdots\cdots\cdots\cdots\cdots\cdots \\
a_{m1}x_1 + a_{m2}x_2 + a_{m3}x_3 + \cdots + a_{mn}x_n &= b_m
\end{aligned} \tag{2.1}$$

The coefficients a_{ij} ($i = 1, 2, \ldots, m$; $j = 1, 2, \ldots, n$) and the quantities b_i ($i = 1, 2, \ldots, m$) are known constants. The x_j ($j = 1, 2, \ldots, n$) are the unknowns whose values are sought.

A *solution* for system (*2.1*) is a set of values, one for each unknown, that, when substituted in the system, renders all its equations valid. (See Problem 2.1.) A system of simultaneous linear equations may possess no solutions, exactly one solution, or more than one solution.

Example 2.1 The system

$$\begin{aligned} x_1 + x_2 &= 1 \\ x_1 + x_2 &= 0 \end{aligned}$$

has no solutions, because there are no values for x_1 and x_2 that sum to 1 and 0 simultaneously. The system

$$\begin{aligned} x_1 + x_2 &= 1 \\ x_1 + 2x_2 &= 2 \end{aligned}$$

has the single solution $x_1 = 0$, $x_2 = 1$; and

$$\begin{aligned} x_1 - x_2 &= 0 \\ 2x_1 - 2x_2 &= 0 \end{aligned}$$

has a solution, $x_1 = x_2$ for every value of x_2.

A set of simultaneous equations is *consistent* if it possesses at least one solution; otherwise it is *inconsistent*.

MATRIX NOTATION

System (*2.1*) is algebraically equivalent to the matrix equation

$$\mathbf{AX} = \mathbf{B} \tag{2.2}$$

where

$$\mathbf{A} = \begin{bmatrix} a_{11} & a_{12} & a_{13} \cdots a_{1n} \\ a_{21} & a_{22} & a_{23} \cdots a_{2n} \\ \cdots & \cdots & \cdots \\ a_{m1} & a_{m2} & a_{m3} \cdots a_{mn} \end{bmatrix} \qquad \mathbf{X} = \begin{bmatrix} x_1 \\ x_2 \\ \cdots \\ x_n \end{bmatrix} \qquad \mathbf{B} = \begin{bmatrix} b_1 \\ b_2 \\ \cdots \\ b_m \end{bmatrix}$$

The matrix **A** is called the *coefficient matrix*, because it contains the coefficients of the unknowns. The ith row of **A** ($i = 1, 2, \ldots, m$) corresponds to the ith equation in system (*2.1*), while the jth column of **A** ($j = 1, 2, \ldots, n$) contains all the coefficients of x_j, one coefficient for each equation.

The *augmented matrix* corresponding to system (*2.1*) is the partitioned matrix $[\mathbf{A} \mid \mathbf{B}]$. (See Problems 2.2 through 2.4.)

THEORY OF SOLUTIONS

Theorem 2.1: The system $\mathbf{AX} = \mathbf{B}$ is consistent if and only if the rank of $\mathbf{A}$ equals the rank of $[\mathbf{A} \mid \mathbf{B}]$.

Theorem 2.2: Denote the rank of $\mathbf{A}$ as k, and the number of unknowns as n. If the system $\mathbf{AX} = \mathbf{B}$ is consistent, then the solution contains $n - k$ arbitrary scalars.

(See Problems 2.5 to 2.7.)

System (*2.1*) is said to be *homogeneous* if $\mathbf{B} = \mathbf{0}$; that is, if $b_1 = b_2 = \cdots = b_m = 0$. If $\mathbf{B} \neq \mathbf{0}$ [i.e., if at least one b_i $(i = 1, 2, \ldots, m)$ is not zero], the system is *nonhomogeneous*. Homogeneous systems are consistent and admit the solution $x_1 = x_2 \cdots = x_n = 0$, which is called the *trivial solution*; a *nontrivial solution* is one that contains at least one nonzero value.

Theorem 2.3: Denote the rank of $\mathbf{A}$ as k, and the number of unknowns as n. The homogeneous system $\mathbf{AX} = \mathbf{0}$ has a nontrivial solution if and only if $n \neq k$. (See Problem 2.7.)

SIMPLIFYING OPERATIONS

Three operations that alter the form of a system of simultaneous linear equations but do not alter its solution set are:

(O1): Interchanging the sequence of two equations.

(O2): Multiplying an equation by a nonzero scalar.

(O3): Adding to one equation a scalar times another equation.

Applying operations O1, O2, and O3 to system (*2.1*) is equivalent to applying the elementary row operations E1, E2, and E3 (see Chapter 1) to the augmented matrix associated with that system. *Gaussian elimination* is an algorithm for applying these operations systematically, to obtain a set of equations that is easy to analyze for consistency and easy to solve if it is consistent.

GAUSSIAN ELIMINATION ALGORITHM

STEP 2.1: Form the augmented matrix $[\mathbf{A} \mid \mathbf{B}]$ associated with the given system of equations.

STEP 2.2: Use elementary row operations to transform $[\mathbf{A} \mid \mathbf{B}]$ into row-echelon form (see Chapter 1). Denote the result as $[\mathbf{C} \mid \mathbf{D}]$.

STEP 2.3: Determine the ranks of $\mathbf{C}$ and $[\mathbf{C} \mid \mathbf{D}]$. If these ranks are equal, comtinue; the system is consistent (by Theorem 2.1). If not, stop; the original system has no solution.

STEP 2.4: Consider the system of equations corresponding to $[\mathbf{C} \mid \mathbf{D}]$, discarding any identically zero equations. (If the rank of $\mathbf{C}$ is k and the number of unknowns is n, there will be $n - k$ such equations.) Solve each equation for its first (lowest indexed) variable having a nonzero coefficient.

STEP 2.5: Any variable not appearing on the left side of any equation is arbitrary. All other variables can be determined uniquely in terms of the arbitrary variables by back substitution.

(See Problems 2.5 through 2.8.) Other solution procedures are discussed in Chapters 3, 4, 5, and 21.

PIVOTING STRATEGIES

Errors due to rounding can become a problem in Gaussian elimination. To minimize the effect of roundoff errors, a variety of pivoting strategies have been proposed, each modifying Step 1.3 of the algorithm given in Chapter 1. Pivoting strategies are merely criteria for choosing the pivot element.

Partial pivoting involves searching the work column of the augmented matrix for the largest element in absolute value appearing in the current work row or a succeeding row. That element becomes the new pivot. To use partial pivoting, replace Step 1.3 of the algorithm for transforming a matrix to row-echelon form with the following:

STEP 1.3′: Beginning with row R and continuing through successive rows, locate the largest element in absolute value appearing in work column C. Denote the first row in which this element appears as row I. If I is different from R, interchange rows I and R (elementary row operation E1). Row R will now have, in column C, the largest nonzero element in absolute value appearing in column C of row R or any row succeeding it. This element in row R and column C is called the *pivot*; let P denote its value.

(See Problems 2.9 and 2.10.)

Two other pivoting strategies are described in Problems 2.11 and 2.12; they are successively more powerful but require additional computations. Since the goal is to avoid significant roundoff error, it is not necessary to find the best pivot element at each stage, but rather to avoid bad ones. Thus, partial pivoting is the strategy most often implemented.

Solved Problems

2.1 Determine whether $x_1 = 2$, $x_2 = 1$, and $x_3 = -11$ is a solution set for the system

$$\begin{aligned} 2x_1 + x_2 &= 5 \\ 3x_1 + 6x_2 + x_3 &= 1 \\ 5x_1 + 7x_2 + x_3 &= 8 \end{aligned}$$

Substituting the proposed values for the unknowns into the left side of each equation gives

$$\begin{aligned} 2(2) + 1(1) &= 5 \\ 3(2) + 6(1) + 1(-11) &= 1 \\ 5(2) + 7(1) + 1(-11) &= 6 \end{aligned}$$

The last equation does not yield 8 as required; hence the proposed values do not constitute a solution set.

2.2 Write the system of equations given in Problem 2.1 as a matrix equation, and then determine its associated augmented matrix.

$$\mathbf{A} = \begin{bmatrix} 2 & 1 & 0 \\ 3 & 6 & 1 \\ 5 & 7 & 1 \end{bmatrix} \qquad \mathbf{X} = \begin{bmatrix} x_1 \\ x_2 \\ x_3 \end{bmatrix} \qquad \mathbf{B} = \begin{bmatrix} 5 \\ 1 \\ 8 \end{bmatrix}$$

The original system can be written as $\mathbf{AX} = \mathbf{B}$; its augmented matrix is

$$[\mathbf{A} \mid \mathbf{B}] = \left[\begin{array}{ccc|c} 2 & 1 & 0 & 5 \\ 3 & 6 & 1 & 1 \\ 5 & 7 & 1 & 8 \end{array}\right]$$

2.3 Write the following system of equations in matrix form, and then determine its augmented matrix:

$$\begin{aligned} 3x_1 + 2x_2 + \ x_3 - 4x_4 &= \ \ 1 \\ 2x_1 + 3x_2 \qquad\quad - x_4 &= -1 \\ x_1 - 6x_2 + 3x_3 - 8x_4 &= \ \ 7 \end{aligned}$$

This system is equivalent to the matrix equation

$$\begin{bmatrix} 3 & 2 & 1 & -4 \\ 2 & 3 & 0 & -1 \\ 1 & -6 & 3 & -8 \end{bmatrix} \begin{bmatrix} x_1 \\ x_2 \\ x_3 \\ x_4 \end{bmatrix} = \begin{bmatrix} 1 \\ -1 \\ 7 \end{bmatrix}$$

The associated augmented matrix is

$$[\mathbf{A} \mid \mathbf{B}] = \left[\begin{array}{cccc|c} 3 & 2 & 1 & -4 & 1 \\ 2 & 3 & 0 & -1 & -1 \\ 1 & -6 & 3 & -8 & 7 \end{array}\right]$$

Observe that in both **A** and [**A** | **B**], the zero in the second row and third column corresponds to the zero coefficient of x_3 in the second equation of the original system.

2.4 Write the set of simultaneous equations that corresponds to the augmented matrix

$$[\mathbf{A} \mid \mathbf{B}] = \left[\begin{array}{cccc|c} 1 & 2/3 & 1/3 & -4/3 & 1/3 \\ 0 & 1 & -2/5 & 1 & -1 \\ 0 & 0 & 0 & 0 & 0 \end{array}\right]$$

The corresponding set of equations is

$$\begin{aligned} x_1 + \tfrac{2}{3}x_2 + \tfrac{1}{3}x_3 - \tfrac{4}{3}x_4 &= \tfrac{1}{3} \\ x_2 - \tfrac{2}{5}x_3 + \ x_4 &= -1 \end{aligned}$$

The third equation reduces to the tautology $0 = 0$ and is not written. Nor do we write any variable having a zero coefficient.

2.5 Solve the set of equations given in Problem 2.1 by Gaussian elimination.

The augmented matrix for this system was determined in Problem 2.2 to be

$$[\mathbf{A} \mid \mathbf{B}] = \left[\begin{array}{ccc|c} 2 & 1 & 0 & 5 \\ 3 & 6 & 1 & 1 \\ 5 & 7 & 1 & 8 \end{array}\right]$$

Using the results of Problem 1.14, we transform this matrix into the row-echelon form

$$[\mathbf{C} \mid \mathbf{D}] = \left[\begin{array}{ccc|c} 1 & 1/2 & 0 & 5/2 \\ 0 & 1 & 2/9 & -13/9 \\ 0 & 0 & 0 & 1 \end{array}\right]$$

It follows from Problem 1.16 that the rank of [**C** | **D**] is 3. Submatrix **C** is also in row-echelon form and has rank 2. Since the rank of **C** does not equal the rank of [**C** | **D**], the original set of equations is inconsistent. The problem is the last equation associated with [**C** | **D**], which is

$$0x_1 + 0x_2 + 0x_3 = 1$$

and which clearly has no solution.

2.6 Solve the set of equations given in Problem 2.3 by Gaussian elimination.

The augmented matrix for this system was determined in Problem 2.3 to be

$$[\mathbf{A} \mid \mathbf{B}] = \left[\begin{array}{cccc|c} 3 & 2 & 1 & -4 & 1 \\ 2 & 3 & 0 & -1 & -1 \\ 1 & -6 & 3 & -8 & 7 \end{array}\right]$$

Using the results of Problem 1.15, we transform this matrix into the row-echelon form

$$[\mathbf{C} \mid \mathbf{D}] = \left[\begin{array}{cccc|c} 1 & 2/3 & 1/3 & -4/3 & 1/3 \\ 0 & 1 & -2/5 & 1 & -1 \\ 0 & 0 & 0 & 0 & 0 \end{array}\right]$$

It follows from Problem 1.17 that the rank of $[\mathbf{C} \mid \mathbf{D}]$ is 2. Submatrix $\mathbf{C}$ is also in row-echelon form, and it also has rank 2. Thus, the original set of equations is consistent.

Now, using the results of Problem 2.4, we write

$$\begin{aligned} x_1 + \tfrac{2}{3}x_2 + \tfrac{1}{3}x_3 - \tfrac{4}{3}x_4 &= \tfrac{1}{3} \\ x_2 - \tfrac{2}{5}x_3 + \phantom{\tfrac{4}{3}} x_4 &= -1 \end{aligned}$$

as the set of equations associated with $[\mathbf{C} \mid \mathbf{D}]$. Solving the first equation for x_1 and the second for x_2, we get

$$\begin{aligned} x_1 &= \tfrac{1}{3} - \tfrac{2}{3}x_2 - \tfrac{1}{3}x_3 + \tfrac{4}{3}x_4 \\ x_2 &= -1 + \tfrac{2}{5}x_3 - x_4 \end{aligned}$$

Since x_3 and x_4 do not appear on the left side of any equation, they are arbitrary. The unknown x_2 is completely determined in terms of the arbitrary unknowns. Substituting it into the first equation, we calculate

$$\begin{aligned} x_1 &= \tfrac{1}{3} - \tfrac{2}{3}(-1 + \tfrac{2}{5}x_3 - x_4) - \tfrac{1}{3}x_3 + \tfrac{4}{3}x_4 \\ &= 1 - \tfrac{3}{5}x_3 + 2x_4 \end{aligned}$$

The complete solution to the original set of equations is

$$\begin{aligned} x_1 &= 1 - \tfrac{3}{5}x_3 + 2x_4 \\ x_2 &= -1 + \tfrac{2}{5}x_3 - x_4 \end{aligned}$$

with x_3 and x_4 arbitrary.

2.7 Solve the following set of homogeneous equations by Gaussian elimination:

$$\begin{aligned} 7x_2 + 9x_3 &= 0 \\ 2x_1 + x_2 - x_3 &= 0 \\ 5x_1 + 6x_2 + 2x_3 &= 0 \end{aligned}$$

By converting this system to augmented-matrix form and then transforming the matrix into row-echelon form (Steps 1.1 through 1.8), we get

$$[\mathbf{A} \mid \mathbf{B}] = \left[\begin{array}{ccc|c} 0 & 7 & 9 & 0 \\ 2 & 1 & -1 & 0 \\ 5 & 6 & 2 & 0 \end{array}\right]$$

$$[\mathbf{C} \mid \mathbf{D}] = \left[\begin{array}{ccc|c} 1 & 1/2 & -1/2 & 0 \\ 0 & 1 & 9/7 & 0 \\ 0 & 0 & 0 & 0 \end{array}\right]$$

The rank of the coefficient matrix **A** is thus 2, and because there are three unknowns in the original set of equations, the system has nontrivial solutions. The set of equations associated with the augmented matrix [**C** | **D**] is

$$\begin{aligned} x_1 + \tfrac{1}{2}x_2 - \tfrac{1}{2}x_3 &= 0 \\ x_2 + \tfrac{9}{7}x_3 &= 0 \\ 0 &= 0 \end{aligned}$$

Solving for the first variable in each equation with a nonzero coefficient, we obtain

$$\begin{aligned} x_1 &= -\tfrac{1}{2}x_2 + \tfrac{1}{2}x_3 \\ x_2 &= -\tfrac{9}{7}x_3 \end{aligned}$$

Therefore, x_3 is arbitrary. Solving for x_1 and x_2 in terms of x_3 by back substitution, we find

$$\begin{aligned} x_2 &= -\tfrac{9}{7}x_3 \\ x_1 &= -\tfrac{1}{2}(-\tfrac{9}{7}x_3) + \tfrac{1}{2}x_3 = \tfrac{8}{7}x_3 \end{aligned}$$

2.8 Solve the following set of equations:

$$\begin{aligned} x_1 + 2x_2 - x_3 &= 6 \\ 3x_1 + 8x_2 + 9x_3 &= 10 \\ 2x_1 - x_2 + 2x_3 &= -2 \end{aligned}$$

The augmented matrix associated with this system is

$$[\mathbf{A} \mid \mathbf{B}] = \left[\begin{array}{ccc:c} 1 & 2 & -1 & 6 \\ 3 & 8 & 9 & 10 \\ 2 & -1 & 2 & -2 \end{array}\right]$$

which, in Problem 1.13, was transformed into the row-echelon form

$$[\mathbf{C} \mid \mathbf{D}] = \left[\begin{array}{ccc:c} 1 & 2 & -1 & 6 \\ 0 & 1 & 6 & -4 \\ 0 & 0 & 1 & -1 \end{array}\right]$$

Both **C** and [**C** | **D**] have rank three, so the system is consistent. The set of equations associated with this augmented matrix is

$$\begin{aligned} x_1 + 2x_2 - x_3 &= 6 \\ x_2 + 6x_3 &= -4 \\ x_3 &= -1 \end{aligned}$$

Solving each equation for the first variable with a nonzero coefficient, we obtain the system

$$\begin{aligned} x_1 &= 6 - 2x_2 + x_3 \\ x_2 &= -4 - 6x_3 \\ x_3 &= -1 \end{aligned}$$

which can be solved easily by back substitution beginning with the last equation. The solution to this system and to the original set of equations is $x_1 = 1$, $x_2 = 2$, and $x_3 = -1$.

2.9 Solve the following set of equations by (*a*) standard Gaussian elimination and (*b*) Gaussian elimination with partial pivoting, *rounding all computations to four significant figures*:

$$\begin{aligned} 0.00001x_1 + x_2 &= 1.00001 \\ x_1 + x_2 &= 2 \end{aligned}$$

(*a*) We write the system in matrix form, rounding 1.00001 to 1.000. Then we transform the augmented matrix into row-echelon form using the algorithm of Chapter 1, in the following steps:

$$\left[\begin{array}{cc|c} 0.00001 & 1 & 1.000 \\ 1 & 1 & 2 \end{array}\right]$$

$$\rightarrow \left[\begin{array}{cc|c} 1 & 100000 & 100000 \\ 1 & 1 & 2 \end{array}\right]$$

$$\rightarrow \left[\begin{array}{cc|c} 1 & 100000 & 100000 \\ 0 & -100000 & -100000 \end{array}\right]$$

$$\rightarrow \left[\begin{array}{cc|c} 1 & 100000 & 100000 \\ 0 & 1 & 1 \end{array}\right]$$

(Note that we round to $-100\,000$ twice in the next-to-last step.) The resulting augmented matrix shows that the system is consistent. The equations associated with this matrix are

$$\begin{aligned} x_1 + 100000x_2 &= 100000 \\ x_2 &= 1 \end{aligned}$$

which have the solution $x_1 = 0$ and $x_2 = 1$. However, substitution into the original equations shows that this is not the solution to the original system.

(*b*) Transforming the augmented matrix into row-echelon form using partial pivoting yields

$$\left[\begin{array}{cc|c} 0.00001 & 1 & 1.000 \\ 1 & 1 & 2 \end{array}\right]$$

$$\rightarrow \left[\begin{array}{cc|c} 1 & 1 & 2 \\ 0.00001 & 1 & 1.000 \end{array}\right]$$ Rows 1 and 2 are interchanged because row 2 has the largest element in column 1, the current work column.

$$\rightarrow \left[\begin{array}{cc|c} 1 & 1 & 2 \\ 0 & 1 & 1 \end{array}\right]$$ Rounding to four significant figures.

The system of equations associated with the last augmented matrix is consistent and is

$$\begin{aligned} x_1 + x_2 &= 2 \\ x_2 &= 1 \end{aligned}$$

Its solution is $x_1 = x_2 = 1$, which is also the solution to the original set of equations.

All computers round to a number of significant figures k that depends on the machine being used. Then an equation of the form

$$10^{-(k+1)}x_1 + x_2 = 1 + 10^{-(k+1)}$$

will generate results like that of part *a* unless some pivoting strategy is used. (We had $k = 4$ in part *a*.) as a rule, dividing by very small numbers can lead to significant roundoff error and should be avoided when possible.

2.10 Solve the following set of equations using partial pivoting:

$$\begin{aligned} x_1 + 2x_2 + 3x_3 &= 18 \\ 2x_1 + x_2 - 4x_3 &= -30 \\ -5x_1 + 8x_2 + 17x_3 &= 96 \end{aligned}$$

The augmented matrix for this system is

$$\left[\begin{array}{ccc|c} 1 & 2 & 3 & 18 \\ 2 & 1 & -4 & -30 \\ -5 & 8 & 17 & 96 \end{array}\right]$$

In transforming this matrix, we need to use Step 1.3′ immediately, with $R = 1$ and $C = 1$. The largest element in absolute value in column 1 is -5, appearing in row 3. We interchange the first and third rows, and then continue the transformation to row-echelon form:

$$\begin{array}{l}\rightarrow \\ \\ \rightarrow\end{array}\left[\begin{array}{rrr|r} -5 & 8 & 17 & 96 \\ 2 & 1 & -4 & -30 \\ 1 & 2 & 3 & 18 \end{array}\right]$$

$$\begin{array}{l}\rightarrow \\ \\ \\ \end{array}\left[\begin{array}{rrr|r} 1 & -1.6 & -3.4 & -19.2 \\ 2 & 1 & -4 & -30 \\ 1 & 2 & 3 & 18 \end{array}\right]$$

$$\begin{array}{l} \\ \rightarrow \\ \\ \end{array}\left[\begin{array}{rrr|r} 1 & -1.6 & -3.4 & -19.2 \\ 0 & 4.2 & 2.8 & 8.4 \\ 1 & 2 & 3 & 18 \end{array}\right]$$

$$\begin{array}{l} \\ \\ \rightarrow \end{array}\left[\begin{array}{rrr|r} 1 & -1.6 & -3.4 & -19.2 \\ 0 & 4.2 & 2.8 & 8.4 \\ 0 & 3.6 & 6.4 & 37.2 \end{array}\right]$$

We next apply Step 1.3′ with $R = 2$ and $C = 2$. Considering only rows 2 and 3, we find that largest element in absolute value in column 2 is 4.2, so $I = 2$ and no row interchange is required. Continuing with the Gaussian elimination, we calculate

$$\begin{array}{l} \\ \rightarrow \\ \\ \end{array}\left[\begin{array}{rrr|r} 1 & -1.6 & -3.4 & -19.2 \\ 0 & 1 & 0.666667 & 2 \\ 0 & 3.6 & 6.4 & 37.2 \end{array}\right]$$

$$\begin{array}{l} \\ \\ \rightarrow \end{array}\left[\begin{array}{rrr|r} 1 & -1.6 & -3.4 & -19.2 \\ 0 & 1 & 0.666667 & 2 \\ 0 & 0 & 4 & 30 \end{array}\right]$$

$$\begin{array}{l} \\ \\ \rightarrow \end{array}\left[\begin{array}{rrr|r} 1 & -1.6 & -3.4 & -19.2 \\ 0 & 1 & 0.666667 & 2 \\ 0 & 0 & 1 & 7.5 \end{array}\right]$$

The system of equations associated with the last augmented matrix is consistent and is

$$\begin{aligned} x_1 - 1.6x_2 - \quad 3.4x_3 &= -19.2 \\ x_2 + 0.666667x_3 &= 2 \\ x_3 &= 7.5 \end{aligned}$$

Using back substitution (beginning with x_3), we obtain, as the solution to this set of equations as well as the original system, $x_1 = 1.5$, $x_2 = -3$, and $x_3 = 7.5$.

2.11 To use *scaled pivoting*, we first define, as the scale factor for each row of the coefficient matrix **A**, the largest element in absolute value appearing in that row. The scale factors are computed once and only once and, for easy reference, are added onto the augmented matrix [**A** | **B**] as another partitioned column. Then Step 1.3 of Chapter 1 is replaced with the following:

Divide the absolute value of each nonzero element that is in the work column and on or below the work row by the scale factor for its row. The element yielding the largest quotient is the new pivot; denote its row as row I. If row I is different from the current work row (row R), then interchange rows I and R. Row interchanges are the only elementary row operations that are performed on the scale factors; all other steps in the Gaussian elimination are limited to **A** and **B**.

Solve Problem 2.10 using scaled pivoting.

The scale factors for the system of Problem 2.10 are

$$\begin{aligned} s_1 &= \max\{1, 2, 3\} = 3 \\ s_2 &= \max\{2, 1, |-4|\} = 4 \\ s_3 &= \max\{|-5|, 8, 17\} = 17 \end{aligned}$$

We add a column consisting of these scale factors to the augmented matrix for the system, and then transforming it to row-echelon form as follows:

$$\left[\begin{array}{rrr:r|r} 1 & 2 & 3 & 18 & 3 \\ 2 & 1 & -4 & -30 & 4 \\ -5 & 8 & 17 & 96 & 17 \end{array}\right]$$

The scale-factor quotients for the elements in column 1 are $1/3 = 0.333$, $2/4 = 0.500$, and $5/17 = 0.294$.

$$\rightarrow \left[\begin{array}{rrr:r|r} 2 & 1 & -4 & -30 & 4 \\ 1 & 2 & 3 & 18 & 3 \\ -5 & 8 & 17 & 96 & 17 \end{array}\right]$$

The largest quotient is 0.500, so the pivot is 2, which appears in row 2. Since $I = 2$ and $R = 1$, the first and second rows are interchanged.

$$\rightarrow \left[\begin{array}{rrr:r|r} 1 & 0.5 & -2 & -15 & 4 \\ 1 & 2 & 3 & 18 & 3 \\ -5 & 8 & 17 & 96 & 17 \end{array}\right]$$

$$\rightarrow \left[\begin{array}{rrr:r|r} 1 & 0.5 & -2 & -15 & 4 \\ 0 & 1.5 & 5 & 33 & 3 \\ -5 & 8 & 17 & 96 & 17 \end{array}\right]$$

$$\rightarrow \left[\begin{array}{lrr:r|r} 1 & 0.5 & -2 & -15 & 4 \\ 0 & 1.5 & 5 & 33 & 3 \\ 0 & 10.5 & 7 & 21 & 17 \end{array}\right]$$

Now work row is 2, and the work column is 2. The quotients are $1.5/3 = 0.500$ and $10.5/17 = 0.618$.

$$\rightarrow \left[\begin{array}{lrr:r|r} 1 & 0.5 & -2 & -15 & 4 \\ 0 & 10.5 & 7 & 21 & 17 \\ 0 & 1.5 & 5 & 33 & 3 \end{array}\right]$$

The largest quotient is 0.618, so the pivot is 10.5, which appears in row 3. The second and third rows are interchanged.

$$\rightarrow \left[\begin{array}{lll:r|r} 1 & 0.5 & -2 & -15 & 4 \\ 0 & 1 & 0.66667 & 2 & 17 \\ 0 & 1.5 & 5 & 33 & 3 \end{array}\right]$$

$$\rightarrow \left[\begin{array}{lll:r|r} 1 & 0.5 & -2 & -15 & 4 \\ 0 & 1 & 0.66667 & 2 & 17 \\ 0 & 0 & 4 & 30 & 3 \end{array}\right]$$

$$\rightarrow \left[\begin{array}{lll:r|r} 1 & 0.5 & -2 & -15 & 4 \\ 0 & 1 & 0.66667 & 2 & 17 \\ 0 & 0 & 1 & 7.5 & 3 \end{array}\right]$$

Writing the set of equations associated with this augmented matrix (ignoring the column of scale factors) and solving them by back substitution, we obtain the solution $x_1 = 1.5$, $x_2 = -3$, $x_3 = 7.5$.

2.12 To use *complete pivoting*, we replace Step 1.3 of Chapter 1 with the following steps, which involve both row and column interchanges: Let the current work row be R, and the current work column C. Scan all the elements of submatrix **A** of the augmented matrix that are on or below row R and on or to the right of column C, to determine which is largest in absolute value. Denote the row and column in which this element appears as row I and column J. If $I \neq R$, interchange rows I and R; if $J \neq C$, interchange rows J and C. Because column interchanges change the order of the unknowns, a bookkeeping mechanism for associating columns with unknowns must be implemented. To do so, add a new partitioned row, row 0, above the usual agumented matrix. Its elements, which are initially in the order $1, 2, \ldots, n$ to denote the subscripts on the unknowns, will designate which unknown is associated with each column.

Solve the system of Problem 2.10 using complete pivoting. We add the bookkeeping row 0 to the augmented matrix of Problem 2.10. Then, beginning with row 1, we transform the remaining rows into row-echelon form.

$$\left[\begin{array}{rrr|r} \mathit{1} & \mathit{2} & \mathit{3} & — \\ \hline 1 & 2 & 3 & 18 \\ 2 & 1 & -4 & -30 \\ -5 & 8 & 17 & 96 \end{array}\right]$$

$R = 1$ and $C = 1$. The largest element in absolute value in the lower left submatrix is 17, in row 3 and column 3. We first interchange rows 1 and 3, and then columns 1 and 3.

$$\rightarrow \left[\begin{array}{rrr|r} \mathit{3} & \mathit{2} & \mathit{1} & — \\ \hline 17 & 8 & -5 & 96 \\ -4 & 1 & 2 & -30 \\ 3 & 2 & 1 & 18 \end{array}\right]$$

$$\rightarrow \left[\begin{array}{rrr|r} \mathit{3} & \mathit{2} & \mathit{1} & — \\ \hline 1 & 0.470588 & -0.294118 & 5.64706 \\ -4 & 1 & 2 & -30 \\ 3 & 2 & 1 & 18 \end{array}\right]$$

$$\rightarrow \left[\begin{array}{rrr|r} \mathit{3} & \mathit{2} & \mathit{1} & — \\ \hline 1 & 0.470588 & -0.294118 & 5.64706 \\ 0 & 2.88235 & 0.823528 & -7.41176 \\ 3 & 2 & 1 & 18 \end{array}\right]$$

$$\rightarrow \left[\begin{array}{rrr|r} \mathit{3} & \mathit{2} & \mathit{1} & — \\ \hline 1 & 0.470588 & -0.294118 & 5.64706 \\ 0 & 2.88235 & 0.823528 & -7.41176 \\ 0 & 0.588236 & 1.88235 & 1.05882 \end{array}\right]$$

$$\rightarrow \left[\begin{array}{rrr|r} \mathit{3} & \mathit{2} & \mathit{1} & — \\ \hline 1 & 0.470588 & -0.294118 & 5.64706 \\ 0 & 1 & 0.285714 & -2.57143 \\ 0 & 0.588236 & 1.88235 & 1.05882 \end{array}\right]$$

The work row and work column are now $R = 2$ and $C = 2$. The largest element in absolute value of the four under consideration is 2.88235, for which $I = 2$ and $J = 2$. Since $I = R$ and $J = C$, no interchange is required.

$$\rightarrow \left[\begin{array}{rrr|r} \mathit{3} & \mathit{2} & \mathit{1} & — \\ \hline 1 & 0.470588 & -0.294118 & 5.64706 \\ 0 & 1 & 0.285714 & -2.57143 \\ 0 & 0 & 1.71428 & 2.57143 \end{array}\right]$$

$$\rightarrow \left[\begin{array}{rrr|r} \mathit{3} & \mathit{2} & \mathit{1} & — \\ \hline 1 & 0.470588 & -0.294118 & 5.64706 \\ 0 & 1 & 0.285714 & -2.57143 \\ 0 & 0 & 1 & 1.50001 \end{array}\right]$$

The first column of the resulting row-echelon matrix corresponds to x_3, and the third column to x_1, so the associated set of equations is

$$\begin{aligned} x_3 + 0.470588x_2 - 0.294118x_1 &= 5.64706 \\ x_2 + 0.285714x_1 &= -2.57143 \\ x_1 &= 1.50001 \end{aligned}$$

Solving each equation for the first variable with a nonzero coefficient, we obtain

$$\begin{aligned} x_3 &= 5.64706 - 0.470588x_2 + 0.294118x_1 \\ x_2 &= -2.57143 - 0.285714x_1 \\ x_1 &= 1.50001 \end{aligned}$$

which, when solved by back substitution, yields the solution $x_1 = 1.50001$, $x_2 = -3.00000$, and $x_3 = 7.50001$.

2.13 *Gauss-Jordan elimination* adds a step between Steps 2.3 and 2.4 of the algorithm for Gaussian elimination. Once the augmented matrix has been reduced to row-echelon form, it is then reduced still further. Beginning with the last pivot element and continuing sequentially backward to the first, each pivot element is used to transform all other elements in its column to zero.

Use Gauss-Jordan elimination to solve Problem 2.8.

The first two steps of the Gaussian elimination algorithm are used to reduce the augmented matrix to row-echelon form as in Problems 1.13 and 2.8:

$$\left[\begin{array}{ccc|c} 1 & 2 & -1 & 6 \\ 0 & 1 & 6 & -4 \\ 0 & 0 & 1 & -1 \end{array}\right]$$

Then the matrix is reduced further, as follows:

$$\rightarrow \left[\begin{array}{ccc|c} 1 & 2 & -1 & 6 \\ 0 & 1 & 0 & 2 \\ 0 & 0 & 1 & -1 \end{array}\right]$$ Add -6 times the third row to the second row.

$$\rightarrow \left[\begin{array}{ccc|c} 1 & 2 & 0 & 5 \\ 0 & 1 & 0 & 2 \\ 0 & 0 & 1 & -1 \end{array}\right]$$ Add the third row to the first row.

$$\rightarrow \left[\begin{array}{ccc|c} 1 & 0 & 0 & 1 \\ 0 & 1 & 0 & 2 \\ 0 & 0 & 1 & -1 \end{array}\right]$$ Add -2 times the second row to the first row.

The set of equations associated with this augmented matrix is $x_1 = 1$, $x_2 = 2$, and $x_3 = -1$, which is the solution set for the original system (no back substitution is required).

2.14 Use Gauss-Jordan elimination to solve the system of Problem 2.7.

The first two steps of the Gaussian elimination algorithm provide the augmented row-echelon-form matrix

$$[\mathbf{C} \mid \mathbf{D}] = \left[\begin{array}{ccc|c} 1 & 1/2 & -1/2 & 0 \\ 0 & 1 & 9/7 & 0 \\ 0 & 0 & 0 & 0 \end{array}\right]$$

as in Problem 2.7. This matrix is reduced further by using the pivot in the (2,2) position to place a zero in the (1,2) position:

$$\rightarrow \left[\begin{array}{ccc|c} 1 & 0 & -8/7 & 0 \\ 0 & 1 & 9/7 & 0 \\ 0 & 0 & 0 & 0 \end{array}\right]$$ Add $-1/2$ times the second row to the first row.

The set of equations associated with this augmented matrix is

$$\begin{aligned} x_1 - \tfrac{8}{7}x_3 &= 0 \\ x_2 + \tfrac{9}{7}x_3 &= 0 \\ 0 &= 0 \end{aligned}$$

Solving for the first variable in each equation with a nonzero coefficient, we obtain $x_1 = \frac{8}{7}x_3$ and $x_2 = -\frac{9}{7}x_3$, which is the solution (no back substitution is required) with x_3 aribitrary.

Supplementary Problems

2.15 Which of

(*a*) $x_1 = x_2 = x_3 = 1$ (*b*) $x_1 = 8, x_2 = -1, x_3 = 0$
(*c*) $x_1 = 12, x_2 = -3, x_3 = 2$ (*d*) $x_1 = 2, x_2 = -2, x_3 = 9$

are solutions to the system

$$\begin{aligned} x_1 + 3x_2 + x_3 &= 5 \\ 2x_1 + x_2 - 3x_3 &= 15 \\ x_1 + 7x_2 + 5x_3 &= 1 \end{aligned}$$

2.16 Write the augmented matrix for the system given in Problem 2.15.

2.17 Write the augmented matrix for system

$$\begin{aligned} 2x_1 - 4x_2 + 7x_3 + 6x_4 - 4x_5 &= 17 \\ 6x_2 - 3x_3 - 4x_4 - 5x_5 &= 2 \\ 2x_1 + 8x_2 + x_3 - 2x_4 - 14x_5 &= 10 \end{aligned}$$

2.18 Solve the set of equations associated with each of the following augmented matrices:

$$(a)\ \left[\begin{array}{ccc|c} 1 & -2 & 3 & 17 \\ 0 & 1 & 2 & -3 \\ 0 & 0 & 1 & -4 \end{array}\right] \qquad (b)\ \left[\begin{array}{cccc|c} 1 & 3 & 2 & 1 & 3 \\ 0 & 0 & 1 & 2 & 5 \\ 0 & 0 & 0 & 1 & 0 \end{array}\right]$$

2.19 Solve the system given in Problem 2.15.

2.20 Solve the system given in Problem 2.17.

In Problems 2.21 through 2.27, solve for the unknowns in the given system.

2.21
$$\begin{aligned} x_1 + 2x_2 - x_3 &= 0 \\ 2x_1 - 2x_2 + 3x_3 &= 0 \\ 3x_1 + x_2 + 2x_3 &= 0 \end{aligned}$$

2.22
$$\begin{aligned} x_1 + 2x_2 + 3x_3 &= 4 \\ 4x_1 + 5x_2 + 6x_3 &= 16 \\ 7x_1 + 8x_2 + 9x_3 &= 28 \end{aligned}$$

2.23
$$\begin{aligned} 2x_1 - x_2 + 4x_3 &= 6 \\ x_1 + 3x_2 + 3x_3 &= -7 \\ -x_1 + 2x_2 \qquad &= -12 \end{aligned}$$

2.24
$$\begin{aligned} x_1 + 2x_2 + 3x_3 + 4x_4 &= 8 \\ 2x_1 - 2x_2 - x_3 + x_4 &= -3 \\ x_1 - 3x_2 + 4x_3 - 4x_4 &= 8 \\ 2x_1 + 2x_2 - 3x_3 + 4x_4 &= -2 \end{aligned}$$

2.25
$$\begin{aligned} \tfrac{1}{2}x_1 + \tfrac{1}{3}x_2 + \tfrac{1}{4}x_3 + \tfrac{1}{5}x_4 &= 10 \\ \tfrac{1}{3}x_1 + \tfrac{1}{4}x_2 + \tfrac{1}{5}x_3 + \tfrac{1}{6}x_4 &= 11 \\ \tfrac{1}{4}x_1 + \tfrac{1}{5}x_2 + \tfrac{1}{6}x_3 + \tfrac{1}{7}x_4 &= 12 \\ \tfrac{1}{5}x_1 + \tfrac{1}{6}x_2 + \tfrac{1}{7}x_3 + \tfrac{1}{8}x_4 &= 13 \end{aligned}$$

2.26
$$1.0001x_1 + 2.0000x_2 + 3.0000x_3 + 4.0000x_4 = 5$$
$$1.0000x_1 + 2.0001x_2 + 3.0000x_3 + 4.0000x_4 = 6$$
$$1.0000x_1 + 2.0000x_2 + 3.0001x_3 + 4.0000x_4 = 7$$
$$1.0000x_1 + 2.0000x_2 + 3.0000x_3 + 4.0001x_4 = 8$$

What would be the result of solving this system by working with only four significant figures?

2.27
$$\begin{aligned} 0.00001x_1 + x_2 + 0.00001x_3 &= 0.00002 \\ x_1 + 2x_2 + x_3 &= 1 \\ 0.00001x_1 + x_2 - 0.00001x_3 &= 0.00001 \end{aligned}$$

2.28 Use Gaussian elimination to determine values of k for which solutions exist to the following systems, and then find the solutions:

(*a*)
$$\begin{aligned} x_1 + 2x_2 - x_3 &= 4 \\ 2x_1 - x_2 + 3x_3 &= 3 \\ 3x_1 + x_2 + 2x_3 &= k \end{aligned}$$

(*b*)
$$\begin{aligned} x_1 - 3x_2 &= -4 \\ 2x_1 + x_2 &= 6 \\ 3x_1 - 2x_2 &= k \end{aligned}$$

2.29 A manufacturer produces three types of desks: custom, deluxe, and regular. Each custom desk c requires 12 worker hours to cut and assemble, and 5 worker hours to finish. Each deluxe desk d requires 10 hours to cut and assemble, and 3 hours to finish; each regular desk r requires 6 hours to cut and assemble, and 1 hour to finish. On a daily basis, the manufacturer has available 440 worker hours for cutting and assembling, and 120 worker hours for finishing. Show that the problem of determining how many desks of each type to produce so that all workpower is used is equivalent to solving two equations in the three unknowns c, d, and r. How many solutions are there?

2.30 The end-of-the-year employee bonus b is 3 percent of taxable income i after city and state taxes are deducted. The city tax c is 1 percent of taxable income, while the state tax s is 4 percent of taxable income with credit allowed for the city tax as a pretax deduction. Show that the problem of determining the bonus is equivalent to solving three equations in the four unknowns b, i, c, and s.

2.31 Prove that if $\mathbf{Y}$ and $\mathbf{Z}$ are two solutions of the linear system $\mathbf{AX} = \mathbf{B}$, then $\mathbf{Y} - \mathbf{Z}$ is a solution of the homogeneous system $\mathbf{AX} = \mathbf{0}$.

2.32 Prove that if $\mathbf{Y}$ and $\mathbf{Z}$ are two solutions of the linear system $\mathbf{AX} = \mathbf{B}$, then $\mathbf{Y} = \mathbf{Z} + \mathbf{H}$, where $\mathbf{H}$ is a solution of the homogeneous system $\mathbf{AX} = \mathbf{0}$.

Chapter 3

Square Matrices

DIAGONALS

A matrix is *square* if it has the same number of rows and columns. Its general form is then

$$\mathbf{A} = \begin{bmatrix} a_{11} & a_{12} & a_{13} & \cdots & a_{1n} \\ a_{21} & a_{22} & a_{23} & \cdots & a_{2n} \\ a_{31} & a_{32} & a_{33} & \cdots & a_{3n} \\ \cdots & \cdots & \cdots & \cdots & \cdots \\ a_{n1} & a_{n2} & a_{n3} & \cdots & a_{nn} \end{bmatrix}$$

The elements $a_{11}, a_{22}, a_{33}, \ldots, a_{nn}$ lie on and form the *diagonal*, also called the *main diagonal* or *principal diagonal*. The elements $a_{12}, a_{23}, \ldots, a_{n-1,n}$ immediately above the diagonal elements form the *superdiagonal*, and the elements $a_{21}, a_{32}, \ldots, a_{n,n-1}$ immediately below the diagonal elements constitute the *subdiagonal*.

A *diagonal matrix* is a square matrix in which all elements not on the main diagonal are equal to zero; the diagonal elements may have any values. An *identity matrix* **I** is a diagonal matrix in which all of the diagonal elements are equal to unity. The 2×2 and 4×4 identity matrices are

$$\begin{bmatrix} 1 & 0 \\ 0 & 1 \end{bmatrix} \quad \text{and} \quad \begin{bmatrix} 1 & 0 & 0 & 0 \\ 0 & 1 & 0 & 0 \\ 0 & 0 & 1 & 0 \\ 0 & 0 & 0 & 1 \end{bmatrix}$$

Identity matrices play the same role in matrix arithmetic as the number 1 plays in real-number arithmetic. In particular, for any matrix **A**, $\mathbf{AI} = \mathbf{A}$ and $\mathbf{IA} = \mathbf{A}$ provided, in each case, that **I** is of the appropriate order for the indicated multiplication.

ELEMENTARY MATRICES

An *elementary matrix* **E** is a square matrix that generates an elementary row operation on a given matrix **A** under the multiplication **EA**. The order of **E** is dictated by the order of **A**, such that the multiplication is defined. There are three general kinds of elementary matrices, corresponding to the three different elementary row operations (see Chapter 1). A specific elementary matrix is obtained by applying the desired elementary row operation to an identity matrix of the appropriate order. (See Problems 3.1 and 3.2.)

LU DECOMPOSITION

A square matrix is *upper triangular* if all elements below the main diagonal are zero; it is *lower triangular* if all elements above the main diagonal are zero. The elements on or above the diagonal in an upper triangular matrix (and on or below the diagonal in a lower triangular matrix) may have any values, including zero.

In most cases, a square matrix **A** can be written as the product of a lower triangular matrix **L** and an upper triangular matrix **U**, where **L** and **U** have the same order as **A**. This factorization, when it exists, is unique if the elements on the main diagonal of **U** are all 1s. That is,

$$\mathbf{A} = \mathbf{LU} \tag{3.1}$$

where
$$\mathbf{L} = \begin{bmatrix} l_{11} & 0 & 0 & \cdots & 0 \\ l_{21} & l_{22} & 0 & \cdots & 0 \\ l_{31} & l_{32} & l_{33} & \cdots & 0 \\ \cdots & \cdots & \cdots & \cdots & \cdots \\ l_{n1} & l_{n2} & l_{n3} & \cdots & l_{nn} \end{bmatrix} \quad \text{and} \quad \mathbf{U} = \begin{bmatrix} 1 & u_{12} & u_{13} & \cdots & u_{1n} \\ 0 & 1 & u_{23} & \cdots & u_{2n} \\ 0 & 0 & 1 & \cdots & u_{3n} \\ \cdots & \cdots & \cdots & \cdots & \cdots \\ 0 & 0 & 0 & \cdots & 1 \end{bmatrix}$$

Example 3.1
$$\begin{bmatrix} 2 & 1 & 1 \\ 2 & -1 & 1 \\ 4 & 1 & 1 \end{bmatrix} = \begin{bmatrix} 2 & 0 & 0 \\ 2 & -2 & 0 \\ 4 & -1 & -1 \end{bmatrix} \begin{bmatrix} 1 & 1/2 & 1/2 \\ 0 & 1 & 0 \\ 0 & 0 & 1 \end{bmatrix}$$

Crout's reduction is an algorithm for calculating the elements of **L** and **U**. In this procedure, the first column of **L** is determined first, then the first row of **U**, the second column of **L**, the second row of **U**, the third column of **L**, the third row of **U**, and so on until all elements have been found. The order of **L** and **U** is the same as that of **A**, which we here assume is $n \times n$.

STEP 3.1: *Initialization*: If $a_{11} = 0$, stop; factorization is not possible. Otherwise, the first column of **L** is the first column of **A**; remaining elements of the first row of **L** are zero. The first row of **U** is the first row of **A** divided by $l_{11} = a_{11}$; remaining elements of the first column of **U** are zero. Set a counter at $N = 2$.

STEP 3.2: For $i = N, N+1, \ldots, n$, set $\mathbf{L}'_i$ equal to that portion of the ith row of **L** that has already been determined. That is, $\mathbf{L}'_i$ consists of the first $N-1$ elements of the ith row of **L**.

STEP 3.3: For $j = N, N+1, \ldots, n$, set $\mathbf{U}'_j$ equal to that portion of the jth column of **U** that has already been determined. That is, $\mathbf{U}'_j$ consists of the first $N-1$ elements of the jth column of **U**.

STEP 3.4: Compute the Nth column of **L**. For each element of that column on or below the main diagonal, compute

$$l_{iN} = a_{iN} - (\mathbf{L}'_i)^T \cdot \mathbf{U}'_N \qquad (i = N, N+1, \ldots, n)$$

If any $l_{NN} = 0$ when $N \neq n$, stop; the factorization is not possible. Otherwise, set the remaining elements of the Nth row of **L** equal to zero.

STEP 3.5: Set $u_{NN} = 1$. If $N = n$, stop; the factorization is complete. Otherwise, set the remaining elements of the Nth column of **U** equal to zero and compute the Nth row of **U**. For each element of that row to the right of the main diagonal, compute

$$u_{Nj} = \frac{a_{Nj} - (\mathbf{L}'_N)^T \cdot U'_j}{l_{NN}} \qquad (j = N+1, N+2, \ldots, n)$$

STEP 3.6: Increase N by 1, and return to Step 3.2.

(See Problems 3.4 through 3.6.)

Partial pivoting (see Chapter 2) is recommended when exact arithmetic is not used and roundoff error is possible. Prior to Steps 3.1 and 3.2 (for $N = 2, 3, \ldots, n$), scan the Nth column of **A** for the largest element in absolute value appearing in that column and on or below the main diagonal. If this element is in row p, with $p \neq N$, then interchange the pth and Nth rows of **A**, as well as the pth and Nth rows **L** up to the Nth column (which represents the parts of those two rows in **L** that have already been determined).

SIMULTANEOUS LINEAR EQUATIONS

LU decompositions are useful for solving systems of simultaneous linear equations when the number of unknowns is equal to the number of equations. The matrix form of such a system is

$\mathbf{AX} = \mathbf{B}$, which, in light of Eq. (3.1), may be rewritten as $\mathbf{L}(\mathbf{UX}) = \mathbf{B}$. To obtain $\mathbf{X}$, we first decompose $\mathbf{A}$ and then solve the system associated with

$$\mathbf{LY} = \mathbf{B} \tag{3.2}$$

for $\mathbf{Y}$. Then, once $\mathbf{Y}$ is known, we solve the system associated with

$$\mathbf{UX} = \mathbf{Y} \tag{3.3}$$

for $\mathbf{X}$. Both (*3.2*) and (*3.3*) are easy to solve—the first by forward substitution, and the second by backward substitution. (See Problem 3.7.)

When $\mathbf{A}$ is a square matrix, **LU** factorization and Gaussian elimination are equally efficient for solving a single set of equations. **LU** factorization is superior when the system $\mathbf{AX} = \mathbf{B}$ must be solved repeatedly with different right sides, because the same **LU** factorization of $\mathbf{A}$ is used for all $\mathbf{B}$. (See Problem 3.8.) A drawback with **LU** factorization is that the factorization does not exist when a pivot element is zero. However, this rarely occurs in practice, and the problem can usually be eliminated by reordering the equations. Gaussian elimination is applicable to all systems, and for that reason is often the preferred algorithm.

POWERS OF A MATRIX

If n is a positive integer and $\mathbf{A}$ is a square matrix, then

$$\mathbf{A}^n = \underbrace{\mathbf{AA}\cdots\mathbf{A}}_{n \text{ times}}$$

In particular, $\mathbf{A}^2 = \mathbf{AA}$ and $\mathbf{A}^3 = \mathbf{AAA}$. By definition $\mathbf{A}^0 = \mathbf{I}$. (See Problems 3.10 and 3.11.)

Solved Problems

3.1 Find elementary matrices that when multiplied on the right by any 3×3 matrix $\mathbf{A}$ will (*a*) interchange the first and third rows of $\mathbf{A}$; (*b*) multiply the second row of $\mathbf{A}$ by $1/2$; and (*c*) add -4 times the second row of $\mathbf{A}$ to the third row of $\mathbf{A}$.

Since an elementary matrix is constructed by performing the desired elementary row operation on an identity matrix of the appropriate size, in this case the 3×3 identity, we have

$$(a)\quad \mathbf{E} = \begin{bmatrix} 0 & 0 & 1 \\ 0 & 1 & 0 \\ 1 & 0 & 0 \end{bmatrix} \qquad (b)\quad \mathbf{E} = \begin{bmatrix} 1 & 0 & 0 \\ 0 & 1/2 & 0 \\ 0 & 0 & 1 \end{bmatrix} \qquad (c)\quad \mathbf{E} = \begin{bmatrix} 1 & 0 & 0 \\ 0 & 1 & 0 \\ 0 & -4 & 1 \end{bmatrix}$$

3.2 Find elementary matrices that when multiplied on the right by any 4×4 matrix $\mathbf{A}$ will (*a*) interchange the second and fourth rows of $\mathbf{A}$; (*b*) multiply the third row of $\mathbf{A}$ by -6; and (*c*) add 8 times the first row of $\mathbf{A}$ to the fourth row of $\mathbf{A}$.

$$(a)\quad \mathbf{E} = \begin{bmatrix} 1 & 0 & 0 & 0 \\ 0 & 0 & 0 & 1 \\ 0 & 0 & 1 & 0 \\ 0 & 1 & 0 & 0 \end{bmatrix} \qquad (b)\quad \mathbf{E} = \begin{bmatrix} 1 & 0 & 0 & 0 \\ 0 & 1 & 0 & 0 \\ 0 & 0 & -6 & 0 \\ 0 & 0 & 0 & 1 \end{bmatrix} \qquad (c)\quad \mathbf{E} = \begin{bmatrix} 1 & 0 & 0 & 0 \\ 0 & 1 & 0 & 0 \\ 0 & 0 & 1 & 0 \\ 8 & 0 & 0 & 1 \end{bmatrix}$$

3.3 Find a matrix **P** such that **PA** is in row-echelon form when

$$\mathbf{A} = \begin{bmatrix} 1 & 2 & -1 \\ 3 & 8 & 9 \\ 2 & -1 & 2 \end{bmatrix}$$

The matrix **A** consists of the first three columns of the matrix considered in Problem 1.13, so the same sequence of elementary row operations utilized in that problem will convert this matrix to row-echelon form. The elementary matrices corresponding to those operations are, sequentially,

$$\mathbf{E}_1 = \begin{bmatrix} 1 & 0 & 0 \\ -3 & 1 & 0 \\ 0 & 0 & 1 \end{bmatrix} \quad \mathbf{E}_2 = \begin{bmatrix} 1 & 0 & 0 \\ 0 & 1 & 0 \\ -2 & 0 & 1 \end{bmatrix} \quad \mathbf{E}_3 = \begin{bmatrix} 1 & 0 & 0 \\ 0 & 1/2 & 0 \\ 0 & 0 & 1 \end{bmatrix}$$

$$\mathbf{E}_4 = \begin{bmatrix} 1 & 0 & 0 \\ 0 & 1 & 0 \\ 0 & 5 & 1 \end{bmatrix} \quad \mathbf{E}_5 = \begin{bmatrix} 1 & 0 & 0 \\ 0 & 1 & 0 \\ 0 & 0 & 1/34 \end{bmatrix}$$

Then
$$\mathbf{P} = \mathbf{E}_5\mathbf{E}_4\mathbf{E}_3\mathbf{E}_2\mathbf{E}_1 = \begin{bmatrix} 1 & 0 & 0 \\ -3/2 & 1/2 & 0 \\ -19/68 & 5/68 & 2/68 \end{bmatrix}$$

and
$$\mathbf{PA} = \begin{bmatrix} 1 & 0 & 0 \\ -3/2 & 1/2 & 0 \\ -19/68 & 5/68 & 2/68 \end{bmatrix}\begin{bmatrix} 1 & 2 & -1 \\ 3 & 8 & 9 \\ 2 & -1 & 2 \end{bmatrix} = \begin{bmatrix} 1 & 2 & -1 \\ 0 & 1 & 6 \\ 0 & 0 & 1 \end{bmatrix}$$

3.4 Factor the following matrix into an upper triangular matrix and a lower triangular matrix:

$$\mathbf{A} = \begin{bmatrix} 1 & 2 & -2 & 3 \\ -1 & 1 & 0 & 2 \\ 3 & -3 & 4 & 1 \\ 2 & 1 & 1 & -2 \end{bmatrix}$$

Using Crout's reduction, we have

STEP 3.1:

$$\mathbf{L} = \begin{bmatrix} 1 & 0 & 0 & 0 \\ -1 & - & - & - \\ 3 & - & - & - \\ 2 & - & - & - \end{bmatrix} \quad \mathbf{U} = \begin{bmatrix} 1 & 2 & -2 & 3 \\ 0 & - & - & - \\ 0 & - & - & - \\ 0 & - & - & - \end{bmatrix} \quad N = 2$$

STEP 3.2: $\mathbf{L}_2' = [-1]$, $\mathbf{L}_3' = [3]$, and $\mathbf{L}_4' = [2]$.

STEP 3.3: $\mathbf{U}_2' = [2]$, $\mathbf{U}_3' = [-2]$, and $\mathbf{U}_4' = [3]$.

STEP 3.4:

$$l_{22} = a_{22} - (\mathbf{L}_2')^T \cdot \mathbf{U}_2' = 1 - [-1] \cdot [2] = 1 - (-2) = 3$$

$$l_{32} = a_{32} - (\mathbf{L}_3')^T \cdot \mathbf{U}_2' = -3 - [3] \cdot [2] = -3 - 6 = -9$$

$$l_{42} = a_{42} - (\mathbf{L}_4')^T \cdot \mathbf{U}_2' = 1 - [2] \cdot [2] = 1 - 4 = -3$$

STEP 3.5:

$$u_{22} = 1$$

$$u_{23} = \frac{a_{23} - (\mathbf{L}_2')^T \cdot \mathbf{U}_3'}{l_{22}} = \frac{0 - [-1] \cdot [-2]}{3} = -\frac{2}{3}$$

$$u_{24} = \frac{a_{24} - (\mathbf{L}_2')^T \cdot \mathbf{U}_4'}{l_{22}} = \frac{2 - [-1] \cdot [3]}{3} = \frac{5}{3}$$

STEP 3.6: To this point we have

$$\mathbf{L}=\begin{bmatrix} 1 & 0 & 0 & 0 \\ -1 & 3 & 0 & 0 \\ 3 & -9 & - & - \\ 2 & -3 & - & - \end{bmatrix} \quad \text{and} \quad \mathbf{U}=\begin{bmatrix} 1 & 2 & -2 & 3 \\ 0 & 1 & -2/3 & 5/3 \\ 0 & 0 & - & - \\ 0 & 0 & - & - \end{bmatrix}$$

Since $N=2$ and $n=4$, we increase N by 1 to $N=3$.

STEP 3.2: $\mathbf{L}_3'=[3,-9]$ and $\mathbf{L}_4'=[2,-3]$.

STEP 3.3:

$$\mathbf{U}_3'=\begin{bmatrix} -2 \\ -2/3 \end{bmatrix} \quad \text{and} \quad \mathbf{U}_4'=\begin{bmatrix} 3 \\ 5/3 \end{bmatrix}$$

STEP 3.4:

$$l_{33}=a_{33}-(\mathbf{L}_3')^T\cdot\mathbf{U}_3'=4-\begin{bmatrix} 3 \\ -9 \end{bmatrix}\cdot\begin{bmatrix} -2 \\ -2/3 \end{bmatrix}=4-0=4$$

$$l_{43}=a_{43}-(\mathbf{L}_4')^T\cdot\mathbf{U}_3'=1-\begin{bmatrix} 2 \\ -3 \end{bmatrix}\cdot\begin{bmatrix} -2 \\ -2/3 \end{bmatrix}=1-(-2)=3$$

STEP 3.5:

$$u_{33}=1$$

$$u_{34}=\frac{a_{34}-(\mathbf{L}_3')^T\cdot\mathbf{U}_4'}{l_{33}}=\frac{1}{4}\left(1-\begin{bmatrix} 3 \\ -9 \end{bmatrix}\cdot\begin{bmatrix} 3 \\ 5/3 \end{bmatrix}\right)=\frac{1-(-6)}{4}=\frac{7}{4}$$

STEP 3.6: To this point we have

$$\mathbf{L}=\begin{bmatrix} 1 & 0 & 0 & 0 \\ -1 & 3 & 0 & 0 \\ 3 & -9 & 4 & 0 \\ 2 & -3 & 3 & - \end{bmatrix} \quad \text{and} \quad \mathbf{U}=\begin{bmatrix} 1 & 2 & -2 & 3 \\ 0 & 1 & -2/3 & 5/3 \\ 0 & 0 & 1 & 7/4 \\ 0 & 0 & 0 & - \end{bmatrix}$$

Since $N=3$ and $n=4$, we increase N by 1 to $N=4$.

STEP 3.2: $\mathbf{L}_4'=[2,-3,3]$.

STEP 3.3:

$$\mathbf{U}_4'=\begin{bmatrix} 3 \\ 5/3 \\ 7/4 \end{bmatrix}$$

STEP 3.4:

$$l_{44}=a_{44}-(\mathbf{L}_4')^T\cdot\mathbf{U}_4'=-2-\begin{bmatrix} 2 \\ -3 \\ 3 \end{bmatrix}\cdot\begin{bmatrix} 3 \\ 5/3 \\ 7/4 \end{bmatrix}=-2-\frac{25}{4}=-\frac{33}{4}$$

STEP 3.5: $u_{44}=1$. Since $N=4=n$, the factorization is done. We have $\mathbf{A}=\mathbf{LU}$, with

$$\mathbf{L}=\begin{bmatrix} 1 & 0 & 0 & 0 \\ -1 & 3 & 0 & 0 \\ 3 & -9 & 4 & 0 \\ 2 & -3 & 3 & -33/4 \end{bmatrix} \quad \text{and} \quad \mathbf{U}=\begin{bmatrix} 1 & 2 & -2 & 3 \\ 0 & 1 & -2/3 & 5/3 \\ 0 & 0 & 1 & 7/4 \\ 0 & 0 & 0 & 1 \end{bmatrix}$$

3.5 Factor the following matrix into an upper triangular matrix and a lower triangular matrix:

$$\mathbf{A}=\begin{bmatrix} 1 & 2 & -2 & 3 \\ -1 & -2 & 0 & 2 \\ 3 & -3 & 0 & 1 \\ 2 & 1 & 1 & -2 \end{bmatrix}$$

The first three steps of Grout's reduction here are identical to those in Problem 3.4. Then:

STEP 3.4:

$$l_{22}=a_{22}-(\mathbf{L}_2')^T\cdot\mathbf{U}_2'=-2-[-1]\cdot[2]=-2-(-2)=0$$

Since $l_{22}=0$ but $N\neq n$, the original matrix cannot be factored as an $\mathbf{LU}$ product.

3.6 Factor the following matrix into an upper triangular matrix and a lower triangular matrix:

$$\mathbf{A}=\begin{bmatrix} 2 & 2 & 1 & 0 \\ 3 & 0 & -1 & 1 \\ 0 & 1 & 0 & 5 \\ -1 & 1 & 0 & 0 \end{bmatrix}$$

Using Crout's reduction, we have

STEP 3.1:

$$\mathbf{L}=\begin{bmatrix} 2 & 0 & 0 & 0 \\ 3 & - & - & - \\ 0 & - & - & - \\ -1 & - & - & - \end{bmatrix} \qquad \mathbf{U}=\begin{bmatrix} 1 & 1 & 1/2 & 0 \\ 0 & - & - & - \\ 0 & - & - & - \\ 0 & - & - & - \end{bmatrix} \qquad N=2$$

STEP 3.2: $\mathbf{L}_2'=[3], \mathbf{L}_3'=[0]$, and $\mathbf{L}_4'=[-1]$.

STEP 3.3: $\mathbf{U}_2'=[1], \mathbf{U}_3'=[1/2]$, and $\mathbf{U}_4'=[0]$.

STEP 3.4:

$$l_{22}=a_{22}-(\mathbf{L}_2')^T\cdot\mathbf{U}_2'=0-[3]\cdot[1]=0-(3)=-3$$

$$l_{32}=a_{32}-(\mathbf{L}_3')^T\cdot\mathbf{U}_2'=1-[0]\cdot[1]=1-0=1$$

$$l_{42}=a_{42}-(\mathbf{L}_4')^T\cdot\mathbf{U}_2'=1-[-1]\cdot[1]=1-(-1)=2$$

STEP 3.5:

$$u_{22}=1$$

$$u_{23}=\frac{a_{23}-(\mathbf{L}_2')^T\cdot\mathbf{U}_3'}{l_{22}}=\frac{-1-[3]\cdot[1/2]}{-3}=\frac{5}{6}$$

$$u_{24}=\frac{a_{24}-(\mathbf{L}_2')^T\cdot\mathbf{U}_4'}{l_{22}}=\frac{1-[3]\cdot[0]}{-3}=-\frac{1}{3}$$

STEP 3.6: To this point we have

$$\mathbf{L}=\begin{bmatrix} 2 & 0 & 0 & 0 \\ 3 & -3 & 0 & 0 \\ 0 & 1 & - & - \\ -1 & 2 & - & - \end{bmatrix} \quad \text{and} \quad \mathbf{U}=\begin{bmatrix} 1 & 1 & 1/2 & 0 \\ 0 & 1 & 5/6 & -1/3 \\ 0 & 0 & - & - \\ 0 & 0 & - & - \end{bmatrix}$$

Since $N=2$ and $n=4$, we increase N by 1 to $N=3$.

STEP 3.2: $\mathbf{L}_3'=[0,1]$ and $\mathbf{L}_4'=[-1,2]$.

STEP 3.3:

$$\mathbf{U}_3'=\begin{bmatrix} 1/2 \\ 5/6 \end{bmatrix} \quad \text{and} \quad \mathbf{U}_4'=\begin{bmatrix} 0 \\ -1/3 \end{bmatrix}$$

STEP 3.4:

$$l_{33}=a_{33}-(\mathbf{L}_3')^T\cdot\mathbf{U}_3'=0-\begin{bmatrix} 0 \\ 1 \end{bmatrix}\cdot\begin{bmatrix} 1/2 \\ 5/6 \end{bmatrix}=0-\frac{5}{6}=-\frac{5}{6}$$

$$l_{43}=a_{43}-(\mathbf{L}_4')^T\cdot\mathbf{U}_3'=0-\begin{bmatrix} -1 \\ 2 \end{bmatrix}\cdot\begin{bmatrix} 1/2 \\ 5/6 \end{bmatrix}=0-\frac{7}{6}=-\frac{7}{6}$$

STEP 3.5:

$$u_{33}=1$$

$$u_{34}=\frac{a_{34}-(\mathbf{L}_3')^T\cdot\mathbf{U}_4'}{l_{33}}=\left(5-\begin{bmatrix} 0 \\ 1 \end{bmatrix}\begin{bmatrix} 0 \\ -1/3 \end{bmatrix}\right)\Big/(-5/6)=\frac{5-(-1/3)}{-5/6}=-\frac{32}{5}$$

STEP 3.6: To this point we have

$$\mathbf{L} = \begin{bmatrix} 2 & 0 & 0 & 0 \\ 3 & -3 & 0 & 0 \\ 0 & 1 & -5/6 & 0 \\ -1 & 2 & -7/6 & - \end{bmatrix} \quad \text{and} \quad \mathbf{U} = \begin{bmatrix} 1 & 1 & 1/2 & 0 \\ 0 & 1 & 5/6 & -1/3 \\ 0 & 0 & 1 & -32/5 \\ 0 & 0 & 0 & - \end{bmatrix}$$

Since $N = 3$ and $n = 4$, we increase N by 1 to $N = 4$.

STEP 3.2: $\mathbf{L}'_4 = [-1, 2, -7/6]$.

STEP 3.3:

$$\mathbf{U}'_4 = \begin{bmatrix} 0 \\ -1/3 \\ -32/5 \end{bmatrix}$$

STEP 3.4:

$$l_{44} = a_{44} - (\mathbf{L}'_4)^T \cdot \mathbf{U}'_4 = 0 - \begin{bmatrix} -1 \\ 2 \\ -7/6 \end{bmatrix} \cdot \begin{bmatrix} 0 \\ -1/3 \\ -32/5 \end{bmatrix} = \frac{0 - 34}{5} = -\frac{34}{5}$$

STEP 3.5: $u_{44} = 1$. Since $N = 4 = n$, the factorization is done. We have $\mathbf{A} = \mathbf{LU}$, with

$$\mathbf{L} = \begin{bmatrix} 2 & 0 & 0 & 0 \\ 3 & -3 & 0 & 0 \\ 0 & 1 & -5/6 & 0 \\ -1 & 2 & -7/6 & -34/5 \end{bmatrix} \quad \text{and} \quad \mathbf{U} = \begin{bmatrix} 1 & 1 & 1/2 & 0 \\ 0 & 1 & 5/6 & -1/3 \\ 0 & 0 & 1 & -32/5 \\ 0 & 0 & 0 & 1 \end{bmatrix}$$

3.7 Solve the system of equations

$$\begin{aligned} 2x_1 + 2x_2 + x_3 &= 10 \\ 3x_1 - x_3 + x_4 &= -11 \\ x_2 + 5x_4 &= 5 \\ -x_1 + x_2 &= 14 \end{aligned}$$

The coefficient matrix for this system is matrix $\mathbf{A}$ of Problem 3.6. Using $\mathbf{L}$ as determined in that problem, we write the system corresponding to $\mathbf{LY} = \mathbf{B}$ as

$$\begin{aligned} 2y_1 &= 10 \\ 3y_1 - 3y_2 &= -11 \\ y_2 - 5/6y_3 &= 5 \\ -y_1 + 2y_2 - 7/6y_3 - 34/5y_4 &= 14 \end{aligned}$$

Solving this system sequentially from top to bottom, we obtain $y_1 = 5$, $y_2 = 26/3$, $y_3 = 22/5$, and $y_4 = -1$.

With these values and $\mathbf{U}$ as given in Problem 3.6, we can write the system corresponding to $\mathbf{UX} = \mathbf{Y}$:

$$\begin{aligned} x_1 + x_2 + \tfrac{1}{2}x_3 \phantom{- \tfrac{1}{3}x_4} &= 5 \\ x_2 + \tfrac{5}{6}x_3 - \tfrac{1}{3}x_4 &= \tfrac{26}{3} \\ x_3 - \tfrac{32}{5}x_4 &= \tfrac{22}{5} \\ x_4 &= -1 \end{aligned}$$

Solving this system sequentially from bottom to top, we obtain the solution to the original system: $x_1 = -4$, $x_2 = 10$, $x_3 = -2$, and $x_4 = -1$.

3.8 Solve the system of equations given in Problem 3.7 if the right side of the second equation is changed from -11 to 11.

The coefficient matrix **A** is unchanged, so both **L** and **U** are as they were. From (*3.2*),

$$\begin{aligned} 2y_1 \qquad\qquad\qquad\qquad &= 10 \\ 3y_1 - 3y_2 \qquad\qquad\quad &= 11 \\ y_2 - \tfrac{5}{6}y_3 \qquad\quad &= 5 \\ -y_1 + 2y_2 - \tfrac{7}{6}y_3 - \tfrac{34}{5}y_4 &= 14 \end{aligned}$$

Solving this system sequentially from top to bottom, we obtain $y_1 = 5$, $y_2 = 4/3$, $y_3 = -22/5$, and $y_4 = -28/17$. With these values and **U** as given in Problem 3.6, (*3.3*) becomes

$$\begin{aligned} x_1 + x_2 + \tfrac{1}{2}x_3 \qquad\quad &= 5 \\ x_2 + \tfrac{5}{6}x_3 - \tfrac{1}{3}x_4 &= \tfrac{4}{3} \\ x_3 - \tfrac{32}{5}x_4 &= -\tfrac{22}{5} \\ x_4 &= -\tfrac{28}{17} \end{aligned}$$

Solving this system sequentially from bottom to top, we obtain the solution to the system of interest: $x_1 = -13/17$, $x_2 = 225/17$, $x_3 = -254/17$, and $x_4 = -28/17$.

3.9 Verify Crout's algorithm for 3×3 matrices.

For an arbitrary 3×3 matrix **A**, we seek a factorization of the form

$$\begin{bmatrix} a_{11} & a_{12} & a_{13} \\ a_{21} & a_{22} & a_{23} \\ a_{31} & a_{32} & a_{33} \end{bmatrix} = \begin{bmatrix} l_{11} & 0 & 0 \\ l_{21} & l_{22} & 0 \\ l_{31} & l_{32} & l_{33} \end{bmatrix} \begin{bmatrix} 1 & u_{12} & u_{13} \\ 0 & 1 & u_{23} \\ 0 & 0 & 1 \end{bmatrix} = \begin{bmatrix} l_{11} & l_{11}u_{12} & l_{11}u_{13} \\ l_{21} & l_{21}u_{12} + l_{22} & l_{21}u_{13} + l_{22}u_{23} \\ l_{31} & l_{31}u_{12} + l_{32} & l_{31}u_{13} + l_{32}u_{23} + l_{33} \end{bmatrix}$$

By equating corresponding coefficients in the order of first column, remaining first row, remaining second column, remaining second row, and remaining third column, and then solving successively for the single unknown in each equation, we would obtain the formulas of the Crout reduction algorithm.

3.10 Find $\mathbf{A}^2$ and $\mathbf{A}^4$ when

$$\mathbf{A} = \begin{bmatrix} 2 & -1 \\ 4 & 3 \end{bmatrix}$$

$$\mathbf{A}^2 = \mathbf{AA} = \begin{bmatrix} 2 & -1 \\ 4 & 3 \end{bmatrix}\begin{bmatrix} 2 & -1 \\ 4 & 3 \end{bmatrix} = \begin{bmatrix} 0 & -5 \\ 20 & 5 \end{bmatrix}$$

$$\mathbf{A}^4 = \mathbf{A}^2\mathbf{A}^2 = \begin{bmatrix} 0 & -5 \\ 20 & 5 \end{bmatrix}\begin{bmatrix} 0 & -5 \\ 20 & 5 \end{bmatrix} = \begin{bmatrix} -100 & -25 \\ 100 & -75 \end{bmatrix}$$

3.11 Show that $\mathbf{A}^3 - 9\mathbf{A} + 10\mathbf{I} = \mathbf{0}$ when

$$\mathbf{A} = \begin{bmatrix} 1 & -2 & 2 \\ 0 & 2 & 0 \\ 1 & -1 & -3 \end{bmatrix}$$

We have

$$\mathbf{A}^2 = \mathbf{AA} = \begin{bmatrix} 1 & -2 & 2 \\ 0 & 2 & 0 \\ 1 & -1 & -3 \end{bmatrix}\begin{bmatrix} 1 & -2 & 2 \\ 0 & 2 & 0 \\ 1 & -1 & -3 \end{bmatrix} = \begin{bmatrix} 3 & -8 & -4 \\ 0 & 4 & 0 \\ -2 & -1 & 11 \end{bmatrix}$$

and

$$\mathbf{A}^3 = \mathbf{A}^2\mathbf{A} = \begin{bmatrix} 3 & -8 & -4 \\ 0 & 4 & 0 \\ -2 & -1 & 11 \end{bmatrix}\begin{bmatrix} 1 & -2 & 2 \\ 0 & 2 & 0 \\ 1 & -1 & -3 \end{bmatrix} = \begin{bmatrix} -1 & -18 & 18 \\ 0 & 8 & 0 \\ 9 & -9 & -37 \end{bmatrix}$$

Then

$$\mathbf{A}^3 - 9\mathbf{A} + 10\mathbf{I} = \begin{bmatrix} -1 & -18 & 18 \\ 0 & 8 & 0 \\ 9 & -9 & -37 \end{bmatrix} - 9\begin{bmatrix} 1 & -2 & 2 \\ 0 & 2 & 0 \\ 1 & -1 & -3 \end{bmatrix} + 10\begin{bmatrix} 1 & 0 & 0 \\ 0 & 1 & 0 \\ 0 & 0 & 1 \end{bmatrix} = \begin{bmatrix} 0 & 0 & 0 \\ 0 & 0 & 0 \\ 0 & 0 & 0 \end{bmatrix}$$

3.12 A square matrix A is said to be *nilpotent* if $\mathbf{A}^p = \mathbf{0}$ for some positive integer p. If p is the least positive integer for which $\mathbf{A}^p = \mathbf{0}$, then $\mathbf{A}$ is said to be nilpotent *of index p*. Show that

$$\mathbf{A} = \begin{bmatrix} 1 & 5 & -2 \\ 1 & 2 & -1 \\ 3 & 6 & -3 \end{bmatrix}$$

is nilpotent of index 3.

That is indeed the case, because

$$\mathbf{A}^2 = \begin{bmatrix} 1 & 5 & -2 \\ 1 & 2 & -1 \\ 3 & 6 & -3 \end{bmatrix}\begin{bmatrix} 1 & 5 & -2 \\ 1 & 2 & -1 \\ 3 & 6 & -3 \end{bmatrix} = \begin{bmatrix} 0 & 3 & -1 \\ 0 & 3 & -1 \\ 0 & 9 & -3 \end{bmatrix}$$

and

$$\mathbf{A}^3 = \mathbf{A}^2\mathbf{A} = \begin{bmatrix} 0 & 3 & -1 \\ 0 & 3 & -1 \\ 0 & 9 & -3 \end{bmatrix}\begin{bmatrix} 1 & 5 & -2 \\ 1 & 2 & -1 \\ 3 & 6 & -3 \end{bmatrix} = \begin{bmatrix} 0 & 0 & 0 \\ 0 & 0 & 0 \\ 0 & 0 & 0 \end{bmatrix}$$

Supplementary Problems

3.13 Find elementary matrices that, when multiplied on the right by any 3×3 matrix $\mathbf{A}$ (*a*) will interchange the second and third rows of $\mathbf{A}$; (*b*) will multiply the first row of $\mathbf{A}$ by 7; and (*c*) will add -3 times the first row of $\mathbf{A}$ to the second row of $\mathbf{A}$.

3.14 Find elementary matrices that when multiplied on the right by any 4×4 matrix $\mathbf{A}$ (*a*) will interchange the second and third rows of $\mathbf{A}$; (*b*) will add -3 times the first row of $\mathbf{A}$ to the fourth row of $\mathbf{A}$; and (*c*) will add 5 times the third row of $\mathbf{A}$ to the first row of $\mathbf{A}$.

3.15 Find (*a*) a matrix $\mathbf{P}$ such that $\mathbf{PA}$ is in row-echelon form and (*b*) a matrix $\mathbf{Q}$ such that $\mathbf{QA} = \mathbf{I}$ when

$$\mathbf{A} = \begin{bmatrix} 1 & 2 \\ 3 & 4 \end{bmatrix}$$

3.16 Use elementary matrices to find a matrix $\mathbf{P}$ such that $\mathbf{PA} = \mathbf{I}$ when

$$\mathbf{A} = \begin{bmatrix} 1 & 0 & 2 \\ 0 & 1 & 2 \\ 3 & 2 & 5 \end{bmatrix}$$

3.17 Prove that the product of two lower triangular matrices of the same order is itself lower triangular.

In Problems 3.18 through 3.23, write each of the given matrices as the product of a lower triangular matrix and an upper triangular matrix.

3.18 $\begin{bmatrix} 1 & 2 & 3 \\ 4 & 5 & 6 \\ 7 & 8 & 9 \end{bmatrix}$ **3.19** $\begin{bmatrix} 2 & -1 & 4 \\ 1 & 3 & 3 \\ -1 & 2 & 0 \end{bmatrix}$ **3.20** $\begin{bmatrix} 1 & 0 & 0 & 2 \\ 2 & 1 & 0 & 1 \\ 0 & 2 & 2 & 0 \\ 0 & 3 & 0 & 1 \end{bmatrix}$

3.21 $\begin{bmatrix} 1 & 2 & 3 & 4 \\ 2 & 4 & -1 & 1 \\ 1 & -3 & 4 & -4 \\ 2 & 2 & -3 & 4 \end{bmatrix}$ **3.22** $\begin{bmatrix} 1 & 2 & 3 & 4 \\ 2 & -2 & -1 & 1 \\ 1 & -3 & 4 & -4 \\ 2 & 2 & -3 & 4 \end{bmatrix}$

In Problems 3.23 through 3.29, use **LU** factorization to solve for the unknowns.

3.23
$$\begin{aligned} 2x_1 + 2x_2 + x_3 \quad\quad &= 6 \\ 3x_1 \quad - x_3 + x_4 &= -1 \\ x_2 \quad + 5x_4 &= -9 \\ -x_1 + x_2 \quad\quad &= 0 \end{aligned}$$

(*Hint:* See Problems 3.7 and 3.8.)

3.24
$$\begin{aligned} x_1 + 2x_2 - 2x_3 + 3x_4 &= 2 \\ -x_1 + x_2 \quad + 2x_4 &= -4 \\ 3x_1 - 3x_2 + 4x_3 + x_4 &= 16 \\ 2x_1 + x_2 + x_3 - 2x_4 &= 9 \end{aligned}$$

(*Hint:* See Problem 3.4.)

3.25 Repeat Problem 3.24, but with $\mathbf{B} = [-3, -1, 0, 4,]^T$.

3.26
$$\begin{aligned} x_1 + 2x_2 + 3x_3 &= 4 \\ 4x_1 + 5x_2 + 6x_3 &= 16 \\ 7x_1 + 8x_2 + 9x_3 &= 28 \end{aligned}$$

(*Hint:* See Problem 3.18.)

3.27 Repeat Problem 3.26, but with $\mathbf{B} = [6, -7, -12]^T$.

3.28
$$\begin{aligned} 2x_1 - x_2 + 4x_3 &= 6 \\ x_1 + 3x_2 + 3x_3 &= -7 \\ -x_1 + 2x_2 \quad &= -12 \end{aligned}$$

(*Hint:* See Problem 3.19.)

3.29
$$\begin{aligned} x_1 + 2x_2 + 3x_3 + 4x_4 &= 8 \\ 2x_1 - 2x_2 - x_3 + x_4 &= -3 \\ x_1 - 3x_2 + 4x_3 - 4x_4 &= 8 \\ 2x_1 + 2x_2 - 3x_3 + 4x_4 &= -2 \end{aligned}$$

(*Hint:* See Problem 3.21.)

3.30 Find $\mathbf{A}^2$ and $\mathbf{A}^3$ for the matrix given in Problem 3.15.

3.31 Find $\mathbf{A}^5$ for

$$\mathbf{A} = \begin{bmatrix} 2 & 0 & 0 \\ 0 & 1 & 0 \\ 0 & 0 & -1 \end{bmatrix}$$

3.32 What does $\mathbf{A}^p$ look like when $\mathbf{A}$ is a diagonal matrix?

3.33 A square matrix is said to be *idempotent* if $\mathbf{A}^2 = \mathbf{A}$. Show that the following matrix is idempotent:

$$\mathbf{A} = \begin{bmatrix} 2 & -2 & -4 \\ -1 & 3 & 4 \\ 1 & -2 & -3 \end{bmatrix}$$

3.34 Prove that if $\mathbf{A}$ is idempotent, then so too is $\mathbf{I} - \mathbf{A}$.

3.35 Prove that $(\mathbf{A}^p)^T = (\mathbf{A}^T)^p$.

Chapter 4

Matrix Inversion

THE INVERSE

Matrix **B** is the *inverse* of a square matrix **A** if

$$\mathbf{AB} = \mathbf{BA} = \mathbf{I} \tag{4.1}$$

For both products to be defined simultaneously, **A** and **B** must be square matrices of the same order.

Example 4.1

$$\begin{bmatrix} -2 & 1 \\ 3/2 & -1/2 \end{bmatrix} \quad \text{is the inverse of} \quad \begin{bmatrix} 1 & 2 \\ 3 & 4 \end{bmatrix}$$

because $$\begin{bmatrix} 1 & 2 \\ 3 & 4 \end{bmatrix}\begin{bmatrix} -2 & 1 \\ 3/2 & -1/2 \end{bmatrix} = \begin{bmatrix} -2 & 1 \\ 3/2 & -1/2 \end{bmatrix}\begin{bmatrix} 1 & 2 \\ 3 & 4 \end{bmatrix} = \begin{bmatrix} 1 & 0 \\ 0 & 1 \end{bmatrix}$$

A square matrix is said to be *singular* if it does not have an inverse; a matrix that has an inverse is called *nonsingular* or *invertible*. The inverse of **A**, when it exists, is denoted as $\mathbf{A}^{-1}$.

SIMPLE INVERSES

Elementary matrices corresponding to elementary row operations (see Chapter 3) are invertible. An elementary matrix *of the first kind*, one that corresponds to an interchange of two rows, is its own inverse. The inverse of an elementary matrix *of the second kind*, one that corresponds to multiplying one row of a matrix by a nonzero scalar k, is obtained simply by replacing the value of k in the elementary matrix with its reciprocal $1/k$. The inverse of an elementary matrix *of the third kind*, which corresponds to adding to one row a constant k times another row, is obtained by replacing the value k in the elementary matrix with its additive inverse $-k$. (See Problem 4.2.)

The inverse of an upper triangular matrix is itself upper triangular, while that of a lower triangular matrix is lower triangular (see Problem 4.13), provided none of the diagonal elements is zero. If at least one diagonal element is zero, then no inverse exists. The inverses of triangular matrices are constructed iteratively, one column at a time, using Eq. (*4.1*). (See Problems 4.3 and 4.4.)

CALCULATING INVERSES

Inverses may be found through the use of elementary row operations (see Chapter 1). This procedure not only yields the inverse when it exists, but also indicates when the inverse does not exist. An algorithm for finding the inverse of a matrix **A** is as follows:

STEP 4.1: Form the partitioned matrix $[\mathbf{A} \mid \mathbf{I}]$, where **I** is the identity matrix having the same order as **A**.

STEP 4.2: Using elementary row operations, transform **A** into row-echelon form (see Chapter 1), applying each row operation to the entire partitioned matrix formed in Step 1. Denote the result as $[\mathbf{C} \mid \mathbf{D}]$, where **C** is in row-echelon form.

STEP 4.3: If **C** has a zero row, stop; the original matrix **A** is singular and does not have an inverse. Otherwise continue; the original matrix is invertible.

STEP 4.4:　　Beginning with the last column of **C** and progressing backward iteratively through the second column, use elementary row operation E3 to transform all elements above the diagonal of **C** to zero. Apply each operation, however, to the entire matrix [**C** | **D**]. Denote the result as [**I** | **B**]. The matrix **B** is the inverse of the original matrix **A**.

(See Problems 4.5 through 4.7.) If exact arithmetic is not used in Step 4.2, then a pivoting strategy (see Chapter 2) should be employed. No pivoting strategy is used in Step 4.4; the pivot is always one of the unity elements on the diagonal of **C**. Interchanging any rows after Step 4.2 has been completed will undo the work of that step and, therefore, is not allowed.

SIMULTANEOUS LINEAR EQUATIONS

A set of linear equations in the matrix form

$$\mathbf{AX} = \mathbf{B} \tag{4.2}$$

can be solved easily if **A** is invertible and its inverse is known. Multiplying each side of this matrix equation by $\mathbf{A}^{-1}$ yields $\mathbf{A}^{-1}\mathbf{AX} = \mathbf{A}^{-1}\mathbf{B}$, which simplifies to *Answer to system*

$$\mathbf{X} = \mathbf{A}^{-1}\mathbf{B} \tag{4.3}$$

(See Problems 4.8 and 4.9.) Equation (*4.3*) is most useful as a theoretical representation of the solution to (*4.2*). The methods given in Chapter 2 for solving simultaneous linear equations generally require fewer computations than the method indicated in (*4.3*) when $\mathbf{A}^{-1}$ is not known.

PROPERTIES OF THE INVERSE

4.0: A left inverse is a right inverse

Property 4.1:　The inverse of a nonsingular matrix is unique.

Property 4.2:　If **A** is nonsingular, then $(\mathbf{A}^{-1})^{-1} = \mathbf{A}$.

Property 4.3:　If **A** and **B** are nonsingular, then $(\mathbf{AB})^{-1} = \mathbf{B}^{-1}\mathbf{A}^{-1}$.

Property 4.4:　If **A** is nonsingular, then so too is $\mathbf{A}^T$. Further, $(\mathbf{A}^T)^{-1} = (\mathbf{A}^{-1})^T$.

(See Problems 4.10 to 4.12 and 4.30.)

Solved Problems

4.1　　Determine whether

$$\mathbf{G} = \begin{bmatrix} 0 & 0.5 \\ -0.25 & 0.25 \end{bmatrix}$$

is the inverse of any of the following matrices:

$$\mathbf{A} = \begin{bmatrix} 4 & -8 \\ 4 & 0 \end{bmatrix} \qquad \mathbf{B} = \begin{bmatrix} 1 & 2 & 3 \\ 4 & 5 & 6 \end{bmatrix} \qquad \mathbf{C} = \begin{bmatrix} 2 & -4 \\ 2 & 0 \end{bmatrix} \qquad \mathbf{D} = \begin{bmatrix} 2 & -4 & 0 \\ 2 & 0 & 0 \\ 0 & 0 & 1 \end{bmatrix}$$

We consider each of the given matrices in turn. Since

$$\mathbf{AG} = \begin{bmatrix} 4 & -8 \\ 4 & 0 \end{bmatrix}\begin{bmatrix} 0 & 0.5 \\ -0.25 & 0.25 \end{bmatrix} = \begin{bmatrix} 2 & 0 \\ 0 & 2 \end{bmatrix}$$

is not the identity matrix, **G** is not the inverse of **A**.

B is not square, so it has no inverse. In particular, the product **BG** is not defined.

For **C**, matrix multiplication gives

$$\mathbf{CG} = \begin{bmatrix} 2 & -4 \\ 2 & 0 \end{bmatrix} \begin{bmatrix} 0 & 0.5 \\ -0.25 & 0.25 \end{bmatrix} = \begin{bmatrix} 1 & 0 \\ 0 & 1 \end{bmatrix}$$

and

$$\mathbf{GC} = \begin{bmatrix} 0 & 0.5 \\ -0.25 & 0.25 \end{bmatrix} \begin{bmatrix} 2 & -4 \\ 2 & 0 \end{bmatrix} = \begin{bmatrix} 1 & 0 \\ 0 & 1 \end{bmatrix}$$

so **G** is the inverse of **C**.

G and **D** do not have the same order, so they cannot be inverses of one another.

4.2 Determine the inverses of the following elementary matrices:

$$\mathbf{A} = \begin{bmatrix} 0 & 0 & 1 \\ 0 & 1 & 0 \\ 1 & 0 & 0 \end{bmatrix} \qquad \mathbf{B} = \begin{bmatrix} 1 & 0 & 0 \\ 0 & 0 & 1 \\ 0 & 1 & 0 \end{bmatrix} \qquad \mathbf{C} = \begin{bmatrix} 4 & 0 & 0 \\ 0 & 1 & 0 \\ 0 & 0 & 1 \end{bmatrix}$$

$$\mathbf{D} = \begin{bmatrix} 1 & 0 & 0 \\ 0 & -2 & 0 \\ 0 & 0 & 1 \end{bmatrix} \qquad \mathbf{E} = \begin{bmatrix} 1 & 0 & 0 \\ 0 & 1 & 0 \\ 0 & 2 & 1 \end{bmatrix} \qquad \mathbf{F} = \begin{bmatrix} 1 & 0 & 0 \\ -3 & 1 & 0 \\ 0 & 0 & 1 \end{bmatrix}$$

Both **A** and **B** are elementary matrices of the first kind; thus, $\mathbf{A}^{-1} = \mathbf{A}$ and $\mathbf{B}^{-1} = \mathbf{B}$. Matrices **C** and **D** are elementary matrices of the second kind. Their inverses are

$$\mathbf{C}^{-1} = \begin{bmatrix} 1/4 & 0 & 0 \\ 0 & 1 & 0 \\ 0 & 0 & 1 \end{bmatrix} \qquad \text{and} \qquad \mathbf{D}^{-1} = \begin{bmatrix} 1 & 0 & 0 \\ 0 & -1/2 & 0 \\ 0 & 0 & 1 \end{bmatrix}$$

Matrices **E** and **F** are elementary matrices of the third kind. Their inverses are

$$\mathbf{E}^{-1} = \begin{bmatrix} 1 & 0 & 0 \\ 0 & 1 & 0 \\ 0 & -2 & 1 \end{bmatrix} \qquad \text{and} \qquad \mathbf{F}^{-1} = \begin{bmatrix} 1 & 0 & 0 \\ 3 & 1 & 0 \\ 0 & 0 & 1 \end{bmatrix}$$

4.3 Determine the inverse of

$$\mathbf{A} = \begin{bmatrix} 2 & 1 & 3 \\ 0 & 1 & 2 \\ 0 & 0 & 3 \end{bmatrix}$$

Since **A** is upper triangular with no zero elements on its diagonal, it has an inverse and the inverse is upper triangular. Furthermore, since $\mathbf{A}^{-1}\mathbf{A} = \mathbf{I}$, we may write

$$\begin{bmatrix} a & b & c \\ 0 & d & e \\ 0 & 0 & f \end{bmatrix} \begin{bmatrix} 2 & 1 & 3 \\ 0 & 1 & 2 \\ 0 & 0 & 3 \end{bmatrix} = \begin{bmatrix} 1 & 0 & 0 \\ 0 & 1 & 0 \\ 0 & 0 & 1 \end{bmatrix}$$

with the first matrix on the left representing $\mathbf{A}^{-1}$. We perform the indicated matrix multiplication and equate corresponding elements on and above the diagonal. Beginning with the leftmost column and sequentially moving through successive columns, we determine that

$$\begin{aligned} a(2) + b(0) + c(0) &= 1 & \text{so} \quad a &= \tfrac{1}{2} \\ \tfrac{1}{2}(1) + b(1) + c(0) &= 0 & \text{so} \quad b &= -\tfrac{1}{2} \\ 0(1) + d(1) + e(0) &= 1 & \text{so} \quad d &= 1 \\ \tfrac{1}{2}(3) + (-\tfrac{1}{2})(2) + c(3) &= 0 & \text{so} \quad c &= -\tfrac{1}{6} \\ 0(3) + 1(2) + e(3) &= 0 & \text{so} \quad e &= -\tfrac{2}{3} \\ 0(3) + 0(2) + f(3) &= 1 & \text{so} \quad f &= \tfrac{1}{3} \end{aligned}$$

Thus,

$$\mathbf{A}^{-1} = \begin{bmatrix} 1/2 & -1/2 & -1/6 \\ 0 & 1 & -2/3 \\ 0 & 0 & 1/3 \end{bmatrix}$$

4.4 Determine the inverses of

$$\mathbf{A} = \begin{bmatrix} 3 & 0 & 0 & 0 \\ 1 & -2 & 0 & 0 \\ 2 & 4 & 1 & 0 \\ 1 & 3 & -1 & 0 \end{bmatrix} \quad \text{and} \quad \mathbf{B} = \begin{bmatrix} -1 & 0 & 0 & 0 \\ 2 & -2 & 0 & 0 \\ 3 & 1 & -2 & 0 \\ 1 & -1 & 3 & 3 \end{bmatrix}$$

Both matrices are lower triangular. Since **A** has a zero element on its main diagonal, it does not have an inverse. In contrast, all the elements on the main diagonal of **B** are nonzero so it has an inverse which itself must be lower triangular. Since $\mathbf{BB}^{-1} = \mathbf{I}$, we may write

$$\begin{bmatrix} -1 & 0 & 0 & 0 \\ 2 & -2 & 0 & 0 \\ 3 & 1 & -2 & 0 \\ 1 & -1 & 3 & 3 \end{bmatrix} \begin{bmatrix} a & 0 & 0 & 0 \\ b & c & 0 & 0 \\ d & e & f & 0 \\ g & h & i & j \end{bmatrix} = \begin{bmatrix} 1 & 0 & 0 & 0 \\ 0 & 1 & 0 & 0 \\ 0 & 0 & 1 & 0 \\ 0 & 0 & 0 & 1 \end{bmatrix}$$

with the second matrix on the left representing $\mathbf{B}^{-1}$. We perform the indicated matrix multiplication and equate corresponding elements on and below the diagonal. Beginning with the leftmost column and sequentially moving through successive columns, we determine that

$$\begin{aligned}
-1a + 0b + 0d + 0g &= 1 \quad &\text{so} \quad a &= -1 \\
2(-1) + (-2)b + 0d + 0g &= 0 \quad &\text{so} \quad b &= -1 \\
3(-1) + 1(-1) + (-2)d + 0g &= 0 \quad &\text{so} \quad d &= -2 \\
1(-1) + (-1)(-1) + 3(-2) + 3g &= 0 \quad &\text{so} \quad g &= 2 \\
2(0) + (-2)c + 0e + 0h &= 1 \quad &\text{so} \quad c &= -1/2 \\
3(0) + 1(-1/2) + (-2)e + 0h &= 0 \quad &\text{so} \quad e &= -1/4 \\
1(0) + (-1)(-1/2) + 3(-1/4) + 3h &= 0 \quad &\text{so} \quad h &= 1/12 \\
3(0) + 1(0) + (-2)f + 0i &= 1 \quad &\text{so} \quad f &= -1/2 \\
1(0) + (-1)(0) + 3(-1/2) + 3i &= 0 \quad &\text{so} \quad i &= 1/2 \\
1(0) + (-1)(0) + 3(0) + 3j &= 1 \quad &\text{so} \quad j &= 1/3
\end{aligned}$$

Therefore

$$\mathbf{B}^{-1} = \begin{bmatrix} -1 & 0 & 0 & 0 \\ -1 & -1/2 & 0 & 0 \\ -2 & -1/4 & -1/2 & 0 \\ 2 & 1/12 & 1/2 & 1/3 \end{bmatrix}$$

4.5 Determine the inverse of

$$\mathbf{A} = \begin{bmatrix} 5 & 3 \\ 2 & 1 \end{bmatrix}$$

We follow Steps 4.1 through 4.4, beginning with $[\mathbf{A} \mid \mathbf{I}]$:

$$\left[\begin{array}{cc|cc} 5 & 3 & 1 & 0 \\ 2 & 1 & 0 & 1 \end{array}\right]$$

$$\rightarrow \begin{bmatrix} 1 & 0.6 & 0.2 & 0 \\ 2 & 1 & 0 & 1 \end{bmatrix} \quad \text{Multiplying the first row by } 1/5$$

$$\rightarrow \left[\begin{array}{cc|cc} 1 & 0.6 & 0.2 & 0 \\ 0 & -0.2 & -0.4 & 1 \end{array}\right] \quad \text{Adding } -2 \text{ times the first row to the second row}$$

$$\rightarrow \left[\begin{array}{cc|cc} 1 & 0.6 & 0.2 & 0 \\ 0 & 1 & 2 & -5 \end{array}\right] \quad \text{Multiplying the second row by } -1/0.2$$

The left side of this partitioned matrix is in row-echelon form. Since it contains no zero rows, the original matrix has an inverse. Applying Step 4.4 to the second column, we obtain

$$\rightarrow \left[\begin{array}{cc|cc} 1 & 0 & -1 & 3 \\ 0 & 1 & 2 & -5 \end{array}\right] \quad \text{Adding } -0.6 \text{ times the second row to the first row}$$

Therefore,

$$\mathbf{A}^{-1} = \begin{bmatrix} -1 & 3 \\ 2 & -5 \end{bmatrix}$$

4.6 Determine the inverse of

$$\mathbf{A} = \begin{bmatrix} 1 & 2 & 3 \\ 4 & 5 & 6 \\ 7 & 8 & 9 \end{bmatrix}$$

$$[\mathbf{A} \mid \mathbf{I}] = \left[\begin{array}{ccc|ccc} 1 & 2 & 3 & 1 & 0 & 0 \\ 4 & 5 & 6 & 0 & 1 & 0 \\ 7 & 8 & 9 & 0 & 0 & 1 \end{array}\right]$$

$$\left[\begin{array}{ccc|ccc} 1 & 2 & 3 & 1 & 0 & 0 \\ 0 & -3 & -6 & -4 & 1 & 0 \\ 7 & 8 & 9 & 0 & 0 & 1 \end{array}\right] \quad \text{Adding } -4 \text{ times the first row to the second row}$$

$$\left[\begin{array}{ccc|ccc} 1 & 2 & 3 & 1 & 0 & 0 \\ 0 & -3 & -6 & -4 & 1 & 0 \\ 0 & -6 & -12 & -7 & 0 & 1 \end{array}\right] \quad \text{Adding } -7 \text{ times the first row to the third row}$$

$$\left[\begin{array}{ccc|ccc} 1 & 2 & 3 & 1 & 0 & 0 \\ 0 & 1 & 2 & 4/3 & -1/3 & 0 \\ 0 & -6 & -12 & -7 & 0 & 1 \end{array}\right] \quad \text{Multiplying the second row by } -1/3$$

$$\left[\begin{array}{ccc|ccc} 1 & 2 & 3 & 1 & 0 & 0 \\ 0 & 1 & 2 & 4/3 & -1/3 & 0 \\ 0 & 0 & 0 & 1 & -2 & 1 \end{array}\right] \quad \text{Adding 6 times the second row to the third row}$$

The left side of this partitioned matrix is in row-echelon form. Since its third row is zero, the original matrix does not have an inverse.

4.7 Determine the inverse of

$$\mathbf{A} = \begin{bmatrix} 0 & 1 & 1 \\ 5 & 1 & -1 \\ 2 & -3 & -3 \end{bmatrix}$$

$$[\mathbf{A} \mid \mathbf{I}] = \left[\begin{array}{ccc|ccc} 0 & 1 & 1 & 1 & 0 & 0 \\ 5 & 1 & -1 & 0 & 1 & 0 \\ 2 & -3 & -3 & 0 & 0 & 1 \end{array}\right]$$

$$\left[\begin{array}{ccc|ccc} 5 & 1 & -1 & 0 & 1 & 0 \\ 0 & 1 & 1 & 1 & 0 & 0 \\ 2 & -3 & -3 & 0 & 0 & 1 \end{array}\right] \quad \text{Interchanging the first and second rows}$$

$$\left[\begin{array}{ccc|ccc} 1 & 1/5 & -1/5 & 0 & 1/5 & 0 \\ 0 & 1 & 1 & 1 & 0 & 0 \\ 2 & -3 & -3 & 0 & 0 & 1 \end{array}\right] \quad \text{Multiplying the first row by } 1/5$$

$$\left[\begin{array}{ccc|ccc} 1 & 1/5 & -1/5 & 0 & 1/5 & 0 \\ 0 & 1 & 1 & 1 & 0 & 0 \\ 0 & -17/5 & -13/5 & 0 & -2/5 & 1 \end{array}\right] \quad \text{Adding } -2 \text{ times the first row to the third row}$$

$$\rightarrow \left[\begin{array}{ccc|ccc} 1 & 1/5 & -1/5 & 0 & 1/5 & 0 \\ 0 & 1 & 1 & 1 & 0 & 0 \\ 0 & 0 & 4/5 & 17/5 & -2/5 & 1 \end{array}\right] \quad \text{Adding 17/5 times the second row to the third row}$$

$$\rightarrow \left[\begin{array}{ccc|ccc} 1 & 1/5 & -1/5 & 0 & 1/5 & 0 \\ 0 & 1 & 1 & 1 & 0 & 0 \\ 0 & 0 & 1 & 17/4 & -2/4 & 5/4 \end{array}\right] \quad \text{Multiplying the third row by 5/4}$$

The left side of this partitioned matrix is in row-echelon form and contains no zero rows; thus, the original matrix has an inverse. Applying Step 4.4, we obtain

$$\rightarrow \left[\begin{array}{ccc|ccc} 1 & 1/5 & -1/5 & 0 & 1/5 & 0 \\ 0 & 1 & 0 & -13/4 & 2/4 & -5/4 \\ 0 & 0 & 1 & 17/4 & -2/4 & 5/4 \end{array}\right] \quad \text{Adding } -1 \text{ times the third row to the second row}$$

$$\rightarrow \left[\begin{array}{ccc|ccc} 1 & 1/5 & 0 & 17/20 & 1/10 & 1/4 \\ 0 & 1 & 0 & -13/4 & 2/4 & -5/4 \\ 0 & 0 & 1 & 17/4 & -2/4 & 5/4 \end{array}\right] \quad \text{Adding 1/5 times the third row to the first row}$$

$$\rightarrow \left[\begin{array}{ccc|ccc} 1 & 0 & 0 & 6/4 & 0 & 2/4 \\ 0 & 1 & 0 & -13/4 & 2/4 & -5/4 \\ 0 & 0 & 1 & 17/4 & -2/4 & 5/4 \end{array}\right] \quad \text{Adding } -1/5 \text{ times the second row to the first row}$$

Thus

$$\mathbf{A}^{-1} = \begin{bmatrix} 6/4 & 0 & 2/4 \\ -13/4 & 2/4 & -5/4 \\ 17/4 & -2/4 & 5/4 \end{bmatrix} = \frac{1}{4}\begin{bmatrix} 6 & 0 & 2 \\ -13 & 2 & -5 \\ 17 & -2 & 5 \end{bmatrix}$$

4.8 Solve the system

$$\begin{aligned} 5x_1 + 3x_2 &= 8 \\ 2x_1 + x_2 &= -1 \end{aligned}$$

This system can be written in the matrix form

$$\begin{bmatrix} 5 & 3 \\ 2 & 1 \end{bmatrix}\begin{bmatrix} x_1 \\ x_2 \end{bmatrix} = \begin{bmatrix} 8 \\ -1 \end{bmatrix}$$

Using the result of Problem 4.5 with Eq. (*4.3*), we have

$$\begin{bmatrix} x_1 \\ x_2 \end{bmatrix} = \begin{bmatrix} -1 & 3 \\ 2 & -5 \end{bmatrix}\begin{bmatrix} 8 \\ -1 \end{bmatrix} = \begin{bmatrix} -11 \\ 21 \end{bmatrix}$$

The solution is $x_1 = -11$, $x_2 = 21$.

4.9 Solve the system

$$\begin{aligned} x_2 + x_3 &= 2 \\ 5x_1 + x_2 - x_3 &= 3 \\ 2x_1 - 3x_2 - 3x_3 &= -6 \end{aligned}$$

This system can be written in the matrix form

$$\begin{bmatrix} 0 & 1 & 1 \\ 5 & 1 & -1 \\ 2 & -3 & -3 \end{bmatrix}\begin{bmatrix} x_1 \\ x_2 \\ x_3 \end{bmatrix} = \begin{bmatrix} 2 \\ 3 \\ -6 \end{bmatrix}$$

Using the result of Problem 4.7 in Eq. (*4.3*), we have

$$\begin{bmatrix} x_1 \\ x_2 \\ x_3 \end{bmatrix} = \frac{1}{4}\begin{bmatrix} 6 & 0 & 2 \\ -13 & 2 & -5 \\ 17 & -2 & 5 \end{bmatrix}\begin{bmatrix} 2 \\ 3 \\ -6 \end{bmatrix} = \begin{bmatrix} 0 \\ 5/2 \\ -1/2 \end{bmatrix}$$

The solution is $x_1 = 0$, $x_2 = 5/2$, $x_3 = -1/2$.

4.10 Prove that the inverse is unique when it exists.

Assume that $\mathbf{A}$ has two inverses, $\mathbf{B}$ and $\mathbf{C}$. Then $\mathbf{AB} = \mathbf{I}$ and $\mathbf{CA} = \mathbf{I}$. It follows that

$$\mathbf{C} = \mathbf{CI} = \mathbf{C}(\mathbf{AB}) = (\mathbf{CA})\mathbf{B} = \mathbf{IB} = \mathbf{B}$$

4.11 Prove that $(\mathbf{A}^{-1})^{-1} = \mathbf{A}$ when $\mathbf{A}$ is nonsingular.

$(\mathbf{A}^{-1})^{-1}$ is, by definition, the inverse of $\mathbf{A}^{-1}$. $\mathbf{A}$ also is the inverse of $\mathbf{A}^{-1}$. These inverses must be equal as a consequence of Problem 4.10.

4.12 Prove that $(\mathbf{AB})^{-1} = \mathbf{B}^{-1}\mathbf{A}^{-1}$ if both $\mathbf{A}$ and $\mathbf{B}$ are invertible.

$(\mathbf{AB})^{-1}$ is, by definition, the inverse of $\mathbf{AB}$. Furthermore,

$$(\mathbf{B}^{-1}\mathbf{A}^{-1})(\mathbf{AB}) = \mathbf{B}^{-1}(\mathbf{A}^{-1}\mathbf{A})\mathbf{B} = \mathbf{B}^{-1}\mathbf{IB} = \mathbf{B}^{-1}\mathbf{B} = \mathbf{I}$$

and

$$(\mathbf{AB})(\mathbf{B}^{-1}\mathbf{A}^{-1}) = \mathbf{A}(\mathbf{BB}^{-1})\mathbf{A}^{-1} = \mathbf{AIA}^{-1} = \mathbf{AA}^{-1} = \mathbf{I}$$

so $\mathbf{B}^{-1}\mathbf{A}^{-1}$ is also an inverse of $\mathbf{AB}$. These inverses must be equal as a consequence of Problem 4.10.

4.13 Prove that the inverse of a lower triangular matrix $\mathbf{A}$ with nonzero diagonal elements is itself lower triangular.

The proof is inductive on the rows of $\mathbf{A}^{-1}$. Denote the inverse of $\mathbf{A} = [a_{ij}]$ as $\mathbf{A}^{-1} = [\alpha_{ij}]$. Since the product $\mathbf{AA}^{-1}$ is the identity matrix, the element in the ith row and jth column of this product must be zero when $i \neq j$. In particular, the element in the first row and jth column of $\mathbf{AA}^{-1}$, with $j > 1$, is zero. We may write that element as

$$0 = \sum_{k=1}^{n} a_{1k}\alpha_{kj} = a_{11}\alpha_{1j} + \sum_{k=2}^{n} a_{1k}\alpha_{kj} = a_{11}\alpha_{1j} + \sum_{k=2}^{n} (0)(\alpha_{kj}) = a_{11}\alpha_{1j}$$

We are given $a_{11} \neq 0$, which implies that $\alpha_{1j} = 0$ for $j > 1$.

Now assume that $\alpha_{ij} = 0$ for $j > i$ and all $i \leq p - 1$; compute the pth row of $\mathbf{AA}^{-1}$. Since $\mathbf{AA}^{-1}$ is the identity matrix, the element in its pth row and jth column, for $j > p$, must satisfy

$$\begin{aligned} 0 = \sum_{k=1}^{n} a_{pk}\alpha_{kj} &= \sum_{k=1}^{p-1} a_{pk}\alpha_{kj} + a_{pp}\alpha_{pj} + \sum_{k=p+1}^{n} a_{pk}\alpha_{kj} \\ &= \sum_{k=1}^{p-1} (a_{pk})(0) + a_{pp}\alpha_{pj} + \sum_{k=p+1}^{n} (0)(\alpha_{kj}) = a_{pp}\alpha_{pj} \end{aligned}$$

Since $a_{pp} \neq 0$, it follows that $\alpha_{pj} = 0$ when $j > p$.

4.14 Prove that any square matrix that can be reduced to row-echelon form without rearranging any rows can be factored into a lower triangular matrix $\mathbf{L}$ times an upper triangular matrix $\mathbf{U}$.

The reduction of a matrix $\mathbf{A}$ to row-echelon form can be expressed as the product of a sequence of elementary matrices, one for each elementary row operation in the reduction process, multiplied by $\mathbf{A}$. If $\mathbf{U}$ is the resulting row-echelon form of $\mathbf{A}$, then $\mathbf{U}$ is upper triangular and

$$(\mathbf{E}_k\mathbf{E}_{k-1}\cdots\mathbf{E}_2\mathbf{E}_1)\mathbf{A} = \mathbf{U} \tag{1}$$

Each $\mathbf{E}_i$ is an elementary matrix of either the second or third kind, so each is lower triangular and invertible. It follows from Problem 4.12 that if

$$\mathbf{P} = \mathbf{E}_k\mathbf{E}_{k-1}\cdots\mathbf{E}_2\mathbf{E}_1 \tag{2}$$

then

$$\mathbf{P}^{-1} = (\mathbf{E}_k\mathbf{E}_{k-1}\cdots\mathbf{E}_2\mathbf{E}_1)^{-1} = \mathbf{E}_1^{-1}\mathbf{E}_2^{-1}\cdots\mathbf{E}_{k-1}^{-1}\mathbf{E}_k^{-1}$$

$\mathbf{P}^{-1}$ is thus the product of lower triangular matrices and is itself lower triangular (Problem 3.17). From *(1)*, $\mathbf{PA} = \mathbf{U}$ whereupon

$$\mathbf{A} = \mathbf{IA} = (\mathbf{P}^{-1}\mathbf{P})\mathbf{A} = \mathbf{P}^{-1}(\mathbf{PA}) = \mathbf{P}^{-1}\mathbf{U}$$

Supplementary Problems

In Problems 4.15 through 4.26 find the inverse of the given matrix if it exists.

4.15 (*a*) $\begin{bmatrix} 0 & 0 & 0 & 1 \\ 0 & 1 & 0 & 0 \\ 0 & 0 & 1 & 0 \\ 1 & 0 & 0 & 0 \end{bmatrix}$ (*b*) $\begin{bmatrix} 1 & 0 & 0 & 0 \\ 0 & 7 & 0 & 0 \\ 0 & 0 & 1 & 0 \\ 0 & 0 & 0 & 1 \end{bmatrix}$ (*c*) $\begin{bmatrix} 1 & 0 & 0 & 0 \\ 0 & 1 & 0 & 0 \\ 0 & 4 & 1 & 0 \\ 0 & 0 & 0 & 1 \end{bmatrix}$ (*d*) $\begin{bmatrix} 1 & 0 & 0 & 0 \\ 0 & 1 & 0 & -3 \\ 0 & 0 & 1 & 0 \\ 0 & 0 & 0 & 1 \end{bmatrix}$

4.16 $\begin{bmatrix} 2 & 0 \\ 1 & 1 \end{bmatrix}$ **4.17** $\begin{bmatrix} 1 & 3 \\ 2 & 5 \end{bmatrix}$ **4.18** $\begin{bmatrix} 2 & 3 \\ 4 & 6 \end{bmatrix}$

4.19 $\begin{bmatrix} -3 & 4 \\ 2 & 1 \end{bmatrix}$ **4.20** $\begin{bmatrix} 1 & 0 & 0 \\ 2 & 2 & 0 \\ 1 & 0 & 3 \end{bmatrix}$ **4.21** $\begin{bmatrix} 1 & -1 & 2 \\ 0 & -1 & 3 \\ 0 & 0 & 5 \end{bmatrix}$

4.22 $\begin{bmatrix} 3 & 0 & 0 \\ -1 & 2 & 0 \\ 1 & 1 & 0 \end{bmatrix}$ **4.23** $\begin{bmatrix} 2 & 1 & 3 \\ 4 & 2 & -1 \\ 2 & -1 & 1 \end{bmatrix}$ **4.24** $\begin{bmatrix} 1 & -2 & 3 \\ 3 & 5 & 1 \\ 6 & 4 & 2 \end{bmatrix}$

4.25 $\begin{bmatrix} 1 & 2 & 1 \\ 1 & 1 & 1 \\ 3 & -1 & 1 \end{bmatrix}$ **4.26** $\begin{bmatrix} 2 & 2 & 3 & 3 \\ 2 & 3 & 3 & 2 \\ 5 & 3 & 7 & 9 \\ 3 & 2 & 4 & 7 \end{bmatrix}$

In Problems 4.27 through 4.29, use matrix inversion to solve for the unknowns:

4.27

$$\begin{aligned} 2x_1 + x_2 + 3x_3 &= -3 \\ 4x_1 + 2x_2 - x_3 &= 5 \\ 2x_1 - x_2 + x_3 &= 2 \end{aligned}$$

(*Hint:* See Problem 4.23.)

4.28

$$\begin{aligned} x_1 + 2x_2 + x_3 &= 0 \\ x_1 + x_2 + x_3 &= 0 \\ 3x_1 - x_2 + x_3 &= 6 \end{aligned}$$

(*Hint:* See Problem 4.25.)

4.29

$$\begin{aligned} 2x_1 + 2x_2 + 3x_3 + 3x_4 &= 1 \\ 2x_1 + 3x_2 + 3x_3 + 2x_4 &= 1 \\ 5x_1 + 3x_2 + 7x_3 + 9x_4 &= 1 \\ 3x_1 + 2x_2 + 4x_3 + 7x_4 &= 1 \end{aligned}$$

(*Hint*: See Problem 4.26.)

4.30 Prove that $(\mathbf{A}^{-1})^T = (\mathbf{A}^T)^{-1}$.

4.31 Prove that if the commutative property of multiplication holds for nonsingular matrices **A** and **B**, then it also holds for the following pairs of matrices: (*a*) $\mathbf{A}^{-1}$ and $\mathbf{B}^{-1}$; (*b*) $\mathbf{A}^{-1}$ and **B**; (*c*) **A** and $\mathbf{B}^{-1}$.

4.32 Let A be a non-square matrix

$B = A(A'A)^{-1}A'$

$B = B^2$

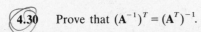

Chapter 5

Determinants

EXPANSION BY COFACTORS

The *determinant* of a square matrix $\mathbf{A}$, denoted $\det \mathbf{A}$ or $|\mathbf{A}|$, is a scalar. If the matrix is written out as an array of elements, then its determinant is indicated by replacing the brackets with vertical lines. For 1×1 matrices,

$$\det \mathbf{A} = |a_{11}| = a_{11}$$

For 2×2 matrices,

$$\det \mathbf{A} = \begin{vmatrix} a_{11} & a_{12} \\ a_{21} & a_{22} \end{vmatrix} = a_{11}a_{22} - a_{12}a_{21}$$

Determinants for $n \times n$ matrices with $n > 2$ are calculated through a process of reduction and expansion utilizing minors and cofactors, as follows.

A *minor* M_{ij} of an $n \times n$ matrix $\mathbf{A}$ is the determinant of the $(n-1) \times (n-1)$ submatrix that remains after the entire ith row and jth column have been deleted from $\mathbf{A}$.

Example 5.1 For

$$\mathbf{A} = \begin{bmatrix} 0 & 1 & 2 \\ 3 & 4 & 5 \\ 6 & 7 & 8 \end{bmatrix}$$

$$M_{11} = \begin{vmatrix} 4 & 5 \\ 7 & 8 \end{vmatrix} = 4(8) - 5(7) = -3$$

$$M_{23} = \begin{vmatrix} 0 & 1 \\ 6 & 7 \end{vmatrix} = 0(7) - 1(6) = -6$$

$$M_{31} = \begin{vmatrix} 1 & 2 \\ 4 & 5 \end{vmatrix} = 1(5) - 2(4) = -3$$

A *cofactor* A_{ij} of an $n \times n$ matrix $\mathbf{A}$ is defined in terms of its associated minor as

$$A_{ij} = (-1)^{i+j} M_{ij}$$

Now for any i or j $(i, j = 1, 2, \ldots, n)$,

$$\det \mathbf{A} = \sum_{k=1}^{n} a_{ik}A_{ik} = \sum_{k=1}^{n} a_{kj}A_{kj} \tag{5.1}$$

For each i, the first sum in (*5.1*) represents an expansion along the ith row of $\mathbf{A}$; for each j, the second sum represents an expansion along the jth column of $\mathbf{A}$. Choosing to expand along a row or column having many zeros, if it exists, greatly reduces the number of calculations required to compute $\det \mathbf{A}$. (See Problems 5.2 through 5.4.)

PROPERTIES OF DETERMINANTS

Property 5.1: If $\mathbf{A}$ and $\mathbf{B}$ are square matrices of the same order, then $\det \mathbf{AB} = \det \mathbf{A} \det \mathbf{B}$.

Property 5.2: The determinant of an upper or lower triangular square matrix is the product of the diagonal elements.

Property 5.3: If **B** is formed from a square matrix **A** by interchanging two rows or two columns of **A**, then $\det \mathbf{A} = -\det \mathbf{B}$.

Property 5.4: If **B** is formed from a square matrix **A** by multiplying every element of a row or column of **A** by a scalar k, then $\det \mathbf{A} = (1/k) \det \mathbf{B}$.

Property 5.5: If **B** is formed from a square matrix **A** by adding a constant times one row (or column) of **A** to another row (or column) of **A**, then $\det \mathbf{A} = \det \mathbf{B}$.

Property 5.6: If one row or one column of a square matrix is zero, its determinant is zero.

Property 5.7: $\det \mathbf{A}^T = \det \mathbf{A}$, provided **A** is a square matrix.

Property 5.8: If two rows of a square matrix are equal, its determinant is zero.

Property 5.9: A matrix **A** (not necessarily square) has rank k if and only if it possesses at least one $k \times k$ submatrix with a nonzero determinant while all square submatrices of larger order have zero determinants.

Property 5.10: If **A** has an inverse, then $\det \mathbf{A}^{-1} = 1/\det \mathbf{A}$.

Property 5.11 multiply each element of a row by the cofactor of the corresponding elements in another row add and result = 0

DETERMINANTS OF PARTITIONED MATRICES

A *block matrix* is one whose elements are themselves matrices. Property 5.2 can be extended to partitioned matrices in block upper (or lower) triangular form. If

$$\mathbf{A} = \begin{bmatrix} \mathbf{A}_{11} & \mathbf{A}_{12} & \cdots & \mathbf{A}_{1r} \\ \mathbf{0} & \mathbf{A}_{22} & \cdots & \mathbf{A}_{2r} \\ \mathbf{0} & \mathbf{0} & \cdots & \mathbf{A}_{3r} \\ \mathbf{0} & \mathbf{0} & \cdots & \mathbf{A}_{rr} \end{bmatrix}$$

where each of the submatrices $\mathbf{A}_{11}, \mathbf{A}_{22}, \ldots, \mathbf{A}_{rr}$ is square, then

$$\det A = \det \mathbf{A}_{11} \det \mathbf{A}_{22} \det \mathbf{A}_{33} \cdots \det \mathbf{A}_{rr} \tag{5.2}$$

(See Problem 5.8.)

PIVOTAL CONDENSATION

Properties 5.3 through 5.5 describe the effects of elementary row and column operations on a determinant. Combined with Property 5.2, they form the basis for the *pivotal condensation algorithm* for calculating the determinant of a matrix **A**, as follows:

STEP 5.1: Initialize $D = 1$. D is a scalar that will record changes in $\det \mathbf{A}$ as a result of elementary row operations.

STEP 5.2: Use elementary row operations to reduce **A** to row-echelon form. Each time two rows are interchanged, multiply D by -1; each time a row is multiplied by k, multiply D by $1/k$. Do not change D when an elementary row operation of the third kind is used.

STEP 5.3: Calculate $\det \mathbf{A}$ as the product of D and all the diagonal elements of the row-echelon matrix obtained in Step 5.2.

(See Problems 5.6 and 5.7.) This algorithm is easy to program for computer implementation; it becomes increasingly more efficient than expansion by cofactors as the order of **A** becomes larger. If rounding is to be used, then the pivoting strategies given in Chapter 2 are recommended.

INVERSION BY DETERMINANTS

The *cofactor matrix* $\mathbf{A}^c$ associated with a square matrix $\mathbf{A}$ is obtained by replacing each element of $\mathbf{A}$ with its cofactor. If $\det \mathbf{A} \neq 0$, then

$$\mathbf{A}^{-1} = \frac{1}{|\mathbf{A}|}(\mathbf{A}^c)^T \tag{5.3}$$

If $\det \mathbf{A}$ is zero, then $\mathbf{A}$ does not have an inverse. (See Problems 5.9 through 5.11 and Problems 5.18 through 5.20.) The method given in Chapter 4 for inversion is almost always quicker than using *(5.3)*, with 2×2 and 3×3 matrices being exceptions.

Solved Problems

5.1 Calculate the determinants of

$$\mathbf{A} = \begin{bmatrix} 1 & 2 \\ 3 & 4 \end{bmatrix} \quad \text{and} \quad \mathbf{B} = \begin{bmatrix} 2 & -3 \\ 4 & 5 \end{bmatrix}$$

$$\det \mathbf{A} = 1(4) - 2(3) = -2$$
$$\det \mathbf{B} = 2(5) - (-3)(4) = 22$$

5.2 Calculate the determinant of

$$\mathbf{A} = \begin{bmatrix} 2 & 3 & 4 \\ -5 & 5 & 6 \\ 7 & 8 & 9 \end{bmatrix}$$

expanding along (*a*) the first row, (*b*) the first column, and (*c*) the second column.

(*a*) Expanding along the first row, we have

$$\begin{aligned} \det \mathbf{A} &= a_{11}A_{11} + a_{12}A_{12} + a_{13}A_{13} \\ &= 2(-1)^{1+1}\begin{vmatrix} 5 & 6 \\ 8 & 9 \end{vmatrix} + 3(-1)^{1+2}\begin{vmatrix} -5 & 6 \\ 7 & 9 \end{vmatrix} + 4(-1)^{1+3}\begin{vmatrix} -5 & 5 \\ 7 & 8 \end{vmatrix} \\ &= 2(-1)^2\{5(9) - 6(8)\} + 3(-1)^3\{(-5)(9) - 6(7)\} + 4(-1)^4\{(-5)(8) - 5(7)\} \\ &= 2(1)(-3) + 3(-1)(-87) + 4(1)(-75) = -45 \end{aligned}$$

(*b*) Along the first column,

$$\begin{aligned} \det \mathbf{A} &= a_{11}A_{11} + a_{21}A_{21} + a_{31}A_{31} \\ &= 2(-1)^{1+1}\begin{vmatrix} 5 & 6 \\ 8 & 9 \end{vmatrix} + (-5)(-1)^{2+1}\begin{vmatrix} 3 & 4 \\ 8 & 9 \end{vmatrix} + 7(-1)^{3+1}\begin{vmatrix} 3 & 4 \\ 5 & 6 \end{vmatrix} \\ &= 2(-1)^2\{5(9) - 6(8)\} + (-5)(-1)^3\{3(9) - 4(8)\} + 7(-1)^4\{3(6) - 4(5)\} \\ &= 2(1)(-3) + (-5)(-1)(-5) + 7(1)(-2) = -45 \end{aligned}$$

(*c*) Expanding along the second column gives us

$$\begin{aligned}\det \mathbf{A} &= a_{12}A_{12} + a_{22}A_{22} + a_{32}A_{32} \\ &= 3(-1)^{1+2}\begin{vmatrix} -5 & 6 \\ 7 & 9 \end{vmatrix} + 5(-1)^{2+2}\begin{vmatrix} 2 & 4 \\ 7 & 9 \end{vmatrix} + 8(-1)^{3+2}\begin{vmatrix} 2 & 4 \\ -5 & 6 \end{vmatrix} \\ &= 3(-1)^3\{(-5)(9) - 6(7)\} + 5(-1)^4\{2(9) - 4(7)\} + 8(-1)^5\{2(6) - 4(-5)\} \\ &= 3(-1)(-87) + 5(1)(-10) + 8(-1)(32) = -45\end{aligned}$$

5.3 Calculate the determinant of

$$\mathbf{B} = \begin{bmatrix} -3 & 4 & 0 \\ -2 & 7 & 6 \\ 5 & -8 & 0 \end{bmatrix}$$

by expanding along (*a*) the second row and (*b*) the third column.

(*a*) Expanding along the second row gives

$$\begin{aligned}\det \mathbf{B} &= (-2)(-1)^{2+1}\begin{vmatrix} 4 & 0 \\ -8 & 0 \end{vmatrix} + 7(-1)^{2+2}\begin{vmatrix} -3 & 0 \\ 5 & 0 \end{vmatrix} + 6(-1)^{2+3}\begin{vmatrix} -3 & 4 \\ 5 & -8 \end{vmatrix} \\ &= -2(-1)^3\{4(0) - 0(-8)\} + 7(-1)^4\{(-3)(0) - 0(5)\} + 6(-1)^5\{(-3)(-8) - 4(5)\} \\ &= -2(-1)(0) + 7(1)(0) + 6(-1)(4) = -24\end{aligned}$$

(*b*) Along the third column,

$$\begin{aligned}\det \mathbf{B} &= 0B_{13} + 6B_{23} + 0B_{33} = 6B_{23} \\ &= 6(-1)^{2+3}\begin{vmatrix} -3 & 4 \\ 5 & -8 \end{vmatrix} = 6(-1)^5\{(-3)(-8) - 4(5)\} = -24\end{aligned}$$

Part *b* involves less computation because we expanded along a column that has mostly zeros.

5.4 Calculate the determinant of

$$\mathbf{A} = \begin{bmatrix} 1 & -4 & 2 & -2 \\ 4 & 7 & -3 & 5 \\ 3 & 0 & 8 & 0 \\ -5 & -1 & 6 & 9 \end{bmatrix}$$

We expand along the third row, because it is the row or column containing the most zeros:

$$\det \mathbf{A} = 3A_{31} + 0A_{32} + 8A_{33} + 0A_{34} = 3(-1)^{3+1}M_{31} + 8(-1)^{3+3}M_{33}$$

Now we may write

$$\begin{aligned}M_{31} &= \begin{vmatrix} -4 & 2 & -2 \\ 7 & -3 & 5 \\ -1 & 6 & 9 \end{vmatrix} = -4(-1)^2\begin{vmatrix} -3 & 5 \\ 6 & 9 \end{vmatrix} + 2(-1)^3\begin{vmatrix} 7 & 5 \\ -1 & 9 \end{vmatrix} + (-2)(-1)^4\begin{vmatrix} 7 & -3 \\ -1 & 6 \end{vmatrix} \\ &= -4(1)(-57) + 2(-1)(68) + (-2)(1)(39) = 14\end{aligned}$$

and

$$\begin{aligned}M_{33} &= \begin{vmatrix} 1 & -4 & -2 \\ 4 & 7 & 5 \\ -5 & -1 & 9 \end{vmatrix} = 1(-1)^2\begin{vmatrix} 7 & 5 \\ -1 & 9 \end{vmatrix} + (-4)(-1)^3\begin{vmatrix} 4 & 5 \\ -5 & 9 \end{vmatrix} + (-2)(-1)^4\begin{vmatrix} 4 & 7 \\ -5 & -1 \end{vmatrix} \\ &= 1(1)(68) + (-4)(-1)(61) + (-2)(1)(31) = 250\end{aligned}$$

Thus, $\det \mathbf{A} = 3(1)(14) + 8(1)(250) = 2042$

5.5 Verify that det **AB** = det **A** det **B** (Property 5.1) for the matrices given in Problems 5.2 and 5.3.

From the results of those problems, we know that det **A** det **B** = (−45)(−24) = 1080. Now

$$\mathbf{AB} = \begin{bmatrix} 2 & 3 & 4 \\ -5 & 5 & 6 \\ 7 & 8 & 9 \end{bmatrix}\begin{bmatrix} -3 & 4 & 0 \\ -2 & 7 & 6 \\ 5 & -8 & 0 \end{bmatrix} = \begin{bmatrix} 8 & -3 & 18 \\ 35 & -33 & 30 \\ 8 & 12 & 48 \end{bmatrix}$$

To calculate det **AB**, we expand along the first row, finding that

$$\det \mathbf{AB} = 8(-1)^{1+1}\begin{vmatrix} -33 & 30 \\ 12 & 48 \end{vmatrix} + (-3)(-1)^{1+2}\begin{vmatrix} 35 & 30 \\ 8 & 48 \end{vmatrix} + 18(-1)^{1+3}\begin{vmatrix} 35 & -33 \\ 8 & 12 \end{vmatrix}$$

$$= 8(1)(-1944) + (-3)(-1)(1440) + 18(1)(684) = 1080$$

5.6 Use pivotal condensation to evaluate the determinant of

$$\mathbf{A} = \begin{bmatrix} 0 & 2 & 2 \\ 1 & 0 & 3 \\ 2 & 1 & 1 \end{bmatrix}$$

We initialize $D = 1$ and use elementary row operations to reduce **A** to row-echelon form:

$$\begin{matrix} \to \\ \to \\ \end{matrix}\begin{bmatrix} 1 & 0 & 3 \\ 0 & 2 & 2 \\ 2 & 1 & 1 \end{bmatrix}$$ Interchanging the first and second rows: $D \leftarrow D(-1) = 1(-1) = -1$

$$\begin{matrix} \\ \\ \to \end{matrix}\begin{bmatrix} 1 & 0 & 3 \\ 0 & 2 & 2 \\ 0 & 1 & -5 \end{bmatrix}$$ Adding −2 times the first row to the third row: D remains −1

$$\begin{matrix} \\ \to \\ \end{matrix}\begin{bmatrix} 1 & 0 & 3 \\ 0 & 1 & 1 \\ 0 & 1 & -5 \end{bmatrix}$$ Multiplying the second row by 1/2: $D \leftarrow D(2) = -1(2) = -2$

$$\begin{matrix} \\ \\ \to \end{matrix}\begin{bmatrix} 1 & 0 & 3 \\ 0 & 1 & 1 \\ 0 & 0 & -6 \end{bmatrix}$$ Adding −1 times the second row to the third row: D remains −2

$$\begin{matrix} \\ \\ \to \end{matrix}\begin{bmatrix} 1 & 0 & 3 \\ 0 & 1 & 1 \\ 0 & 0 & 1 \end{bmatrix}$$ Multiplying the third row by −1/6: $D \leftarrow D(-6) = (-2)(-6) = 12$

The diagonal elements of this last matrix are all ones, so det **A** = $D(1)(1)(1) = 12$.

5.7 Use pivotal condensation to evaluate the determinant of

$$\mathbf{A} = \begin{bmatrix} 1 & 2 & -3 & 4 \\ 2 & -2 & 5 & -6 \\ -1 & 3 & -4 & 6 \\ 6 & 5 & -3 & 6 \end{bmatrix}$$

We initialize $D = 1$ and reduce **A** to row-echelon form:

$$\begin{matrix} \\ \to \\ \\ \end{matrix}\begin{bmatrix} 1 & 2 & -3 & 4 \\ 0 & -6 & 11 & -14 \\ -1 & 3 & -4 & 6 \\ 6 & 5 & -3 & 6 \end{bmatrix}$$ Adding −2 times the first row to the second row: D remains 1

$$\begin{matrix} \\ \\ \rightarrow \\ \\ \end{matrix}\begin{bmatrix} 1 & 2 & -3 & 4 \\ 0 & -6 & 11 & -14 \\ 0 & 5 & -7 & 10 \\ 6 & 5 & -3 & 6 \end{bmatrix}$$

Adding 1 times the first row to the third row: D remains 1

$$\begin{matrix} \\ \\ \\ \rightarrow \end{matrix}\begin{bmatrix} 1 & 2 & -3 & 4 \\ 0 & -6 & 11 & -14 \\ 0 & 5 & -7 & 10 \\ 0 & -7 & 15 & -18 \end{bmatrix}$$

Adding -6 times the first row to the fourth row: D remains 1

$$\begin{matrix} \\ \rightarrow \\ \\ \\ \end{matrix}\begin{bmatrix} 1 & 2 & -3 & 4 \\ 0 & 1 & -11/6 & 14/6 \\ 0 & 5 & -7 & 10 \\ 0 & -7 & 15 & -18 \end{bmatrix}$$

Multiplying the second row by $-1/6$: $D \leftarrow D(-6) = 1(-6) = -6$

$$\begin{matrix} \\ \\ \rightarrow \\ \\ \end{matrix}\begin{bmatrix} 1 & 2 & -3 & 4 \\ 0 & 1 & -11/6 & 14/6 \\ 0 & 0 & 13/6 & -5/3 \\ 0 & -7 & 15 & -18 \end{bmatrix}$$

Adding -5 times the second row to the third row: D remains -6

$$\begin{matrix} \\ \\ \\ \rightarrow \end{matrix}\begin{bmatrix} 1 & 2 & -3 & 4 \\ 0 & 1 & -11/6 & 14/6 \\ 0 & 0 & 13/6 & -5/3 \\ 0 & 0 & 13/6 & -5/3 \end{bmatrix}$$

Adding 7 times the second row to the fourth row: D remains -6

$$\begin{matrix} \\ \\ \rightarrow \\ \\ \end{matrix}\begin{bmatrix} 1 & 2 & -3 & 4 \\ 0 & 1 & -11/6 & 14/6 \\ 0 & 0 & 1 & -10/13 \\ 0 & 0 & 13/6 & -5/3 \end{bmatrix}$$

Multiplying the third row by $6/13$: $D \leftarrow D(13/6) = -6(13/6) = -13$

$$\begin{matrix} \\ \\ \\ \rightarrow \end{matrix}\begin{bmatrix} 1 & 2 & -3 & 4 \\ 0 & 1 & -11/6 & 14/6 \\ 0 & 0 & 1 & -10/13 \\ 0 & 0 & 0 & 0 \end{bmatrix}$$

Adding $-13/6$ times the third row to the fourth row: D remains -13

The matrix is now in row-echelon form with diagonal elements 1, 1, 1, and 0. Thus, $\det \mathbf{A} = -13(1)(1)(1)(0) = 0$.

5.8 Calculate the determinant of

$$\mathbf{A} = \begin{bmatrix} 5 & 3 & 0 & 2 & 5 \\ 0 & 2 & 1 & 1 & -6 \\ 0 & 1 & 2 & 1 & 1 \\ 0 & 0 & 0 & 3 & 6 \\ 0 & 0 & 0 & 1 & -1 \end{bmatrix}$$

This matrix can be partitioned into block upper triangular form with square matrices on its main diagonal. We introduce the partitioning

$$\mathbf{A} = \left[\begin{array}{c|cc|cc} 5 & 3 & 0 & 2 & 5 \\ \hline 0 & 2 & 1 & 1 & -6 \\ 0 & 1 & 2 & 1 & 1 \\ \hline 0 & 0 & 0 & 3 & 6 \\ 0 & 0 & 0 & 1 & -1 \end{array}\right]$$

and it follows from Eq. (*5.2*) that

$$\det \mathbf{A} = |5| \begin{vmatrix} 2 & 1 \\ 1 & 2 \end{vmatrix} \begin{vmatrix} 3 & 6 \\ 1 & -1 \end{vmatrix} = 5(3)(-9) = -135$$

5.9 Calculate the inverse of

$$\mathbf{A} = \begin{bmatrix} 3 & -1 \\ 5 & 4 \end{bmatrix}$$

We shall use Eq. (5.3). Since the determinant of a 1×1 matrix is the element itself, we have

$$\begin{aligned} A_{11} &= (-1)^{1+1} \det[4] = (1)(4) = 4 \\ A_{12} &= (-1)^{1+2} \det[5] = (-1)(5) = -5 \\ A_{21} &= (-1)^{2+1} \det[-1] = (-1)(-1) = 1 \\ A_{22} &= (-1)^{2+2} \det[3] = (1)(3) = 3 \end{aligned}$$

The determinant of $\mathbf{A}$ is $3(4) - (-1)(5) = 17$, so

$$\mathbf{A}^c = \begin{bmatrix} 4 & -5 \\ 1 & 3 \end{bmatrix} \quad \text{and} \quad \mathbf{A}^{-1} = \frac{1}{17}\begin{bmatrix} 4 & 1 \\ -5 & 3 \end{bmatrix}$$

5.10 Calculate the inverse of

$$\mathbf{A} = \begin{bmatrix} 2 & 3 & 4 \\ -5 & 5 & 6 \\ 7 & 8 & 9 \end{bmatrix}$$

In Problem 5.2, we calculated a number of cofactors for this matrix. In particular,

$$\begin{array}{lll} A_{11} = -3 & A_{12} = 87 & A_{13} = -75 \\ A_{21} = 5 & A_{22} = -10 & \\ A_{31} = -2 & A_{32} = -32 & \end{array}$$

and $\det \mathbf{A} = -45$. In addition,

$$A_{23} = (-1)^{2+3}\begin{vmatrix} 2 & 3 \\ 7 & 8 \end{vmatrix} = (-1)(-5) = 5$$

and

$$A_{33} = (-1)^{3+3}\begin{vmatrix} 2 & 3 \\ -5 & 5 \end{vmatrix} = (1)(25) = 25$$

Thus

$$\mathbf{A}^c = \begin{bmatrix} -3 & 87 & -75 \\ 5 & -10 & 5 \\ -2 & -32 & 25 \end{bmatrix} \quad \text{and} \quad \mathbf{A}^{-1} = -\frac{1}{45}\begin{bmatrix} -3 & 5 & -2 \\ 87 & -10 & -32 \\ -75 & 5 & 25 \end{bmatrix}$$

5.11 Find the inverse of the matrix given in Problem 5.7.

Since the determinant of that matrix was found to be zero, the matrix does not have an inverse.

5.12 Verify Property 5.9 for

$$\mathbf{A} = \begin{bmatrix} 3 & 2 & 1 & -4 & 1 \\ 2 & 3 & 0 & -1 & -1 \\ 1 & -6 & 3 & -8 & 7 \end{bmatrix}$$

The rank of $\mathbf{A}$ was determined in Problem 1.17 to be 2, so there should be at least one 2×2 submatrix of $\mathbf{A}$ having a nonzero determinant. There are many, including the one in the upper left corner:

$$\begin{vmatrix} 3 & 2 \\ 2 & 3 \end{vmatrix} = 5$$

All 3×3 submatrices, obtained by deleting any two columns of $\mathbf{A}$, have zero determinants.

5.13 Prove that the determinant of an elementary matrix of the first kind is -1.

An elementary matrix $\mathbf{E}$ of the first kind is an identity matrix with two rows interchanged. The proof is inductive on the order of $\mathbf{E}$. If $\mathbf{E}$ is 2×2, then

$$\mathbf{E} = \begin{bmatrix} 0 & 1 \\ 1 & 0 \end{bmatrix}$$

and $\det \mathbf{E} = -1$. Now assume the proposition is true for all elementary matrices of the first kind with order $(k-1) \times (k-1)$, and consider an elementary matrix $\mathbf{E}$ of order $k \times k$. Find the first row of $\mathbf{E}$ that was not interchanged, and denote it as row m. Expanding by cofactors along row m yields

$$\det \mathbf{E} = a_{m1}A_{m1} + a_{m2}A_{m2} + \cdots + a_{mm}A_{mm} + \cdots + a_{mk}A_{mk} = A_{mm}$$

because $a_{mj} = 0$ for all $j \neq m$, and $a_{mm} = 1$. Now

$$A_{mm} = (-1)^{m+m}M_{mm} = M_{mm}$$

But M_{mm} is the determinant of an elementary matrix of the first kind having order $(k-1) \times (k-1)$, so by induction it is equal to -1. Thus, $\det \mathbf{E} = A_{mm} = M_{mm} = -1$.

5.14 Prove Property 5.3.

If $\mathbf{B}$ is obtained from $\mathbf{A}$ by interchanging two rows of $\mathbf{A}$, then $\mathbf{B} = \mathbf{EA}$, where $\mathbf{E}$ is an elementary matrix of the first kind. Using Property 5.1 and the result of Problem 5.13, we obtain

$$\det \mathbf{B} = \det \mathbf{EA} = \det \mathbf{E} \det \mathbf{A} = (-1) \det \mathbf{A}$$

from which Property 5.3 immediately follows.

5.15 Prove Property 5.4.

Assume that $\mathbf{B}$ is obtained from an $n \times n$ matrix $\mathbf{A}$ by multiplying the ith row of $\mathbf{A}$ by the scalar k. Evaluating the determinant of $\mathbf{B}$ by expansion of cofactors along the ith row, we obtain

$$\begin{aligned} \det \mathbf{B} &= ka_{i1}A_{i1} + ka_{i2}A_{i2} + \cdots + ka_{in}A_{in} \\ &= k(a_{i1}A_{i1} + a_{i2}A_{i2} + \cdots + a_{in}A_{in}) = k \det \mathbf{A} \end{aligned}$$

from which Property 5.4 follows.

5.16 Prove that the determinant of an elementary matrix of the third kind is 1.

An elementary matrix $\mathbf{E}$ of the third kind is an identity matrix that has been altered by adding a constant times one row of $\mathbf{I}$ to another row of $\mathbf{I}$. The proof is inductive on the order of $\mathbf{E}$. If $\mathbf{E}$ is 2×2, then

$$\mathbf{E} = \begin{bmatrix} 1 & 0 \\ k & 1 \end{bmatrix} \quad \text{or} \quad \mathbf{E} = \begin{bmatrix} 1 & k \\ 0 & 1 \end{bmatrix}$$

In either case, $\det \mathbf{E} = 1$. Now assume the proposition is true for all elementary matrices of the third kind with order $(k-1) \times (k-1)$, and consider an elementary matrix $\mathbf{E}$ of order $k \times k$. Find the first row of $\mathbf{E}$ that was not altered from the $k \times k$ identity matrix, and denote this row as row m. The proof now follows that in Problem 5.13 except that here $M_{mm} = 1$ by induction.

5.17 Prove Property 5.5.

If $\mathbf{B}$ is obtained from square matrix $\mathbf{A}$ by adding to one row of $\mathbf{A}$ a constant times another row of $\mathbf{A}$, then $\mathbf{B} = \mathbf{EA}$, where $\mathbf{E}$ is an elementary matrix of the third kind. Using Property 5.1 and the results of Problem 5.16, we obtain

$$\det \mathbf{B} = \det \mathbf{EA} = \det \mathbf{E} \det \mathbf{A} = 1 \det \mathbf{A} = \det \mathbf{A}$$

5.18 Prove that if the determinant of a matrix **A** is zero, then the matrix does not have an inverse.

Assume that **A** does have an inverse. Then

$$1 = \det \mathbf{I} = \det(\mathbf{A}^{-1}\mathbf{A}) = \det \mathbf{A}^{-1} \det \mathbf{A} = \det \mathbf{A}^{-1}(0) = 0$$

which is absurd. Thus, **A** cannot have an inverse.

5.19 Prove that if each element of the ith row of an $n \times n$ matrix is multiplied by the cofactor of the corresponding element of the kth row $(i, k = 1, 2, \ldots, n;\ i \neq k)$, then the sum of these n products is zero.

For any $n \times n$ matrix **A**, construct a new matrix **B** by replacing the kth row of **A** with its ith row $(i, k = 1, 2, \ldots, n;\ i \neq k)$. The ith and kth rows of **B** are identical, for both are the ith row of **A**; it follows from Property 5.8 that $\det \mathbf{B} = 0$. Thus, evaluating $\det \mathbf{B}$ via expansion by cofactors along its ith row, we may write

$$0 = \det \mathbf{B} = \sum_{j=1}^{n} b_{ij}B_{ij} = \sum_{j=1}^{n} a_{ij}B_{ij} \tag{1}$$

where B_{ij} is the cofactor of b_{ij}.

For each element b_{ij} $(j = 1, 2, \ldots, n)$ in the ith row of **B**, compare the submatrix $\mathbf{B}_{ij}$ obtained from **B** by deleting its ith row and jth column to the submatrix $\mathbf{A}_{kj}$ obtained from **A** by deleting its kth row and jth column. They are the same except for the ordering of their rows; each submatrix contains all the rows of **A** except for the kth and all the columns of **A** except for the jth. Exactly $|i-k|-1$ row reorderings are required to make $\mathbf{B}_{ij}$ equal to $\mathbf{A}_{kj}$, so it follows from Property 5.3 that

$$\det \mathbf{B}_{ij} = (-1)^{|i-k|-1} \det \mathbf{A}_{kj} \tag{2}$$

These determinants are minors of **B** and of **A**, respectively, so *(2)* may be written in cofactor notation as

$$(-1)^{-i-j}B_{ij} = (-1)^{|i-k|-1}(-1)^{-k-j}A_{kj}$$

$$B_{ij} = -A_{kj} \tag{3}$$

Combining *(1)* and *(3)*, we have

$$0 = \sum_{j=1}^{n} a_{ij}B_{ij} = \sum_{j=1}^{n} a_{ij}(-A_{kj}) = -\sum_{j=1}^{n} a_{ij}A_{kj}$$

which, when multiplied by -1, gives the desired result.

5.20 Prove that $\mathbf{A}(\mathbf{A}^c)^T = |\mathbf{A}|\mathbf{I}$.

Consider the (i, k) element of the product $\mathbf{A}(\mathbf{A}^c)^T$; it is

$$\sum_{j=1}^{n} (a_{ij})\{(j, k) \text{ element of } (\mathbf{A}^c)^T\} = \sum_{j=1}^{n} (a_{ij})\{(k, j) \text{ element of } \mathbf{A}^c\} = \sum_{j=1}^{n} a_{ij}A_{kj}$$

It follows from Problem 5.19 that this sum is zero when $i \neq k$. When $i = k$, the sum is $\det \mathbf{A}$ because it is an expansion by cofactors along the kth row of **A**. Therefore, we may write

$$\mathbf{A}(\mathbf{A}^c)^T = \begin{bmatrix} |\mathbf{A}| & 0 & \cdots & 0 \\ 0 & |\mathbf{A}| & \cdots & 0 \\ \cdots & \cdots & \cdots & \cdots \\ 0 & 0 & \cdots & |\mathbf{A}| \end{bmatrix} = |\mathbf{A}|\mathbf{I}$$

Note that if $|\mathbf{A}| \neq 0$, then $\mathbf{A}(\mathbf{A}^c)^T/|\mathbf{A}| = \mathbf{I}$, from which *(5.3)* follows.

Supplementary Problems

In Problems 5.21 through 5.26, let

$$\mathbf{A}=\begin{bmatrix}3 & 6\\ 4 & 7\end{bmatrix} \qquad \mathbf{B}=\begin{bmatrix}8 & -3\\ 5 & -6\end{bmatrix} \qquad \mathbf{C}=\begin{bmatrix}3 & 0 & 0\\ -1 & 2 & 0\\ 1 & 1 & 0\end{bmatrix}$$

$$\mathbf{D}=\begin{bmatrix}2 & 1 & 3\\ 4 & 2 & -1\\ 2 & -1 & 1\end{bmatrix} \qquad \mathbf{E}=\begin{bmatrix}1 & -2 & 3\\ 3 & 5 & 1\\ 6 & 4 & 2\end{bmatrix} \qquad \mathbf{F}=\begin{bmatrix}3 & 4 & 1 & 1\\ 6 & 9 & 2 & 5\\ -1 & 1 & 2 & -5\end{bmatrix}$$

5.21 Find (*a*) det **A** and (*b*) det **B**, and (*c*) show that det **AB** = det **A** det **B**.

5.22 Find (*a*) det **C** and (*b*) det **D**, and (*c*) show that det **CD** = det **C** det **D**.

5.23 Find (*a*) det **E** and (*b*) det **F**.

5.24 Use determinants to find (*a*) $\mathbf{A}^{-1}$ and (*b*) $\mathbf{B}^{-1}$.

5.25 Use determinants to find $\mathbf{D}^{-1}$.

5.26 Use determinants to find $\mathbf{E}^{-1}$.

In Problems 5.27 and 5.28, find the determinant of the given matrix.

5.27 $\begin{bmatrix}2 & 0 & 1 & 3\\ -1 & 1 & 0 & 2\\ 1 & 1 & 2 & 1\\ -4 & 2 & 0 & -3\end{bmatrix}$ **5.28** $\begin{bmatrix}2 & 2 & 3 & 3\\ 2 & 3 & 3 & 2\\ 5 & 3 & 7 & 9\\ 3 & 2 & 4 & 7\end{bmatrix}$

5.29 Use determinants to find the inverse of the matrix given in Problem 5.28.

5.30 Prove Property 5.8.

5.31 Prove that if **A** has order $n \times n$, then det $k\mathbf{A} = k^n$ det **A**.

5.32 Prove that if **A** has order $n \times n$, then $|\mathbf{A}^2| = |\mathbf{A}|^2$.

5.33 Prove that if **A** and **B** are square matrices of the same order, then det **AB** = det **BA**.

5.34 Prove that if **A** is invertible, then det $\mathbf{A}^{-1} = 1/\det \mathbf{A}$.

5.35 Let **A** = **LU** be an **LU** decomposition of **A** (see Chapter 3). Show that det **A** is equal to the product of the diagonal elements of **L**.

Chapter 6

Vectors

DIMENSION

A *vector* is a matrix having either one row or one column. The number of elements in a row vector or a column vector is its *dimension*, and the elements are called *components*. The transpose of a row vector is a column vector, and vice versa.

LINEAR DEPENDENCE AND INDEPENDENCE

A set of m-dimensional vectors $\{\mathbf{V}_1, \mathbf{V}_2, \ldots, \mathbf{V}_n\}$ of the same type (row or column) is *linearly dependent* if there exist constants $c_1, c_2, \ldots, c_n$ not all zero such that

$$c_1\mathbf{V}_1 + c_2\mathbf{V}_2 + \cdots + c_n\mathbf{V}_n = \mathbf{0} \tag{6.1}$$

Example 6.1 The set of five-dimensional vectors

$$\{[1, 0, -2, 0, 0]^T, [2, 0, 3, 0, 0]^T, [0, 2, 0, 0, 1]^T, \text{ and } [5, 0, 4, 0, 0]^T\}$$

is linearly dependent because

$$1\begin{bmatrix}1\\0\\-2\\0\\0\end{bmatrix} + 2\begin{bmatrix}2\\0\\3\\0\\0\end{bmatrix} + 0\begin{bmatrix}0\\2\\0\\0\\1\end{bmatrix} + (-1)\begin{bmatrix}5\\0\\4\\0\\0\end{bmatrix} = \begin{bmatrix}0\\0\\0\\0\\0\end{bmatrix}$$

A set of m-dimensional vectors $\{\mathbf{V}_1, \mathbf{V}_2, \ldots, \mathbf{V}_n\}$ of the same type is *linearly independent* if the only constants for which Eq. (*6.1*) holds are $c_1 = c_2 = \cdots = c_n = 0$.

The following algorithm may be used to determine whether a set of row vectors is linearly independent or dependent. The algorithm is applicable to column vectors too, if their transposes are considered instead. (See Problems 6.2 and 6.3.)

STEP 6.1: Construct a matrix $\mathbf{V}$ whose rows are the row vectors under consideration. That is, the first row of $\mathbf{V}$ is $\mathbf{V}_1$, the second row of $\mathbf{V}$ is $\mathbf{V}_2$, and so on.

STEP 6.2: Determine the rank of $\mathbf{V}$.

STEP 6.3: If the rank of $\mathbf{V}$ is smaller than the number of vectors in the set under consideration (i.e., the number of rows of $\mathbf{V}$), then the vectors are linearly dependent; otherwise, they are linearly independent.

LINEAR COMBINATIONS

A vector $\mathbf{B}$ is a *linear combination* of vectors $\mathbf{V}_1, \mathbf{V}_2, \ldots, \mathbf{V}_n$ if there exist constants $d_1, d_2, \ldots, d_n$ such that

$$\mathbf{B} = d_1\mathbf{V}_1 + d_2\mathbf{V}_2 + \cdots + d_n\mathbf{V}_n \tag{6.2}$$

For the matrix addition and equality of (*6.2*) to be defined, the vectors must all be of the same type (row or column) and have the same dimension.

Example 6.2 The vector $[-3, 4, -1, 0, 2]^T$ is a linear combination of the vectors of Example 6.1 because

$$\begin{bmatrix} -3 \\ 4 \\ -1 \\ 0 \\ 2 \end{bmatrix} = 0 \begin{bmatrix} 1 \\ 0 \\ -2 \\ 0 \\ 0 \end{bmatrix} + 1 \begin{bmatrix} 2 \\ 0 \\ 3 \\ 0 \\ 0 \end{bmatrix} + 2 \begin{bmatrix} 0 \\ 2 \\ 0 \\ 0 \\ 1 \end{bmatrix} + (-1) \begin{bmatrix} 5 \\ 0 \\ 4 \\ 0 \\ 0 \end{bmatrix}$$

Equation *(6.2)* represents a set of simultaneous linear equations in the unknowns $d_1, d_2, \ldots, d_n$. The algorithms given in Chapter 2 may be used to determine whether or not the d_i $(i = 1, 2, \ldots, n)$ exist and what they are. (See Problems 6.4 and 6.5.)

PROPERTIES OF LINEARLY DEPENDENT VECTORS

Property 6.1: Every set of $m + 1$ or more m-dimensional vectors of the same type (either row or column) is linearly dependent.

Property 6.2: An ordered set of nonzero vectors is linearly dependent if and only if one vector can be written as a linear combination of the vectors that precede it.

Property 6.3: If a set of vectors is linearly independent, then any subset of those vectors is also linearly independent.

Property 6.4: If a set of vectors is linearly dependent, then any larger set containing this set is also linearly dependent.

Property 6.5: Any set of vectors of the same dimension that contains the zero vector is linearly dependent.

Property 6.6: The set consisting of a single vector is linearly dependent if and only if that vector is the zero vector.

ROW RANK AND COLUMN RANK

Consider each row of a matrix **A** to be a row vector. The *row rank* of **A** is the maximum number of linearly independent vectors that can be formed from these row vectors; it is the rank of **A** (see Problem 6.11). Similarly, the *column rank* of **A** is the maximum number of linearly independent vectors that can be formed from the columns of **A**. It may be obtained by calculating the rank of $\mathbf{A}^T$, because the rows of $\mathbf{A}^T$ are the columns of **A**. The row rank of a matrix equals its column rank (see Problem 6.10); so the column rank of **A** is also the rank of **A**.

Solved Problems

6.1 Determine whether the set {[1, 1, 3], [2, −1, 3], [0, 1, 1], [4, 4, 3]} is linearly independent.

Since the set contains more vectors (four) than the dimension of its member vectors (three), the vectors are linearly dependent by Property 6.1. They are thus *not* linearly independent.

6.2 Determine whether the set {[1, 2, −1, 6], [3, 8, 9, 10], [2, −1, 2, −2]} is linearly independent.

Using Steps 6.1 through 6.3, we first construct

$$\mathbf{V} = \begin{bmatrix} 1 & 2 & -1 & 6 \\ 3 & 8 & 9 & 10 \\ 2 & -1 & 2 & -2 \end{bmatrix}$$

Matrix **V** was transformed in Problem 1.13 into the row-echelon form:

$$\begin{bmatrix} 1 & 2 & -1 & 6 \\ 0 & 1 & 6 & -4 \\ 0 & 0 & 1 & -1 \end{bmatrix}$$

By inspection, the rank of **V** is 3, which equals the number of vectors in the given set; hence, the given set of vectors is linearly independent.

6.3 Determine whether the set $\{[3, 2, 1, -4, 1]^T, [2, 3, 0, -1, -1]^T, [1, -6, 3, -8, 7]^T\}$ is linearly independent.

Using the algorithm of this chapter, we construct

$$\mathbf{V} = \begin{bmatrix} 3 & 2 & 1 & -4 & 1 \\ 2 & 3 & 0 & -1 & -1 \\ 1 & -6 & 3 & -8 & 7 \end{bmatrix}$$

which we transformed into row-echelon form in Problem 1.15:

$$\begin{bmatrix} 1 & 2/3 & 1/3 & -4/3 & 1/3 \\ 0 & 1 & -2/5 & 1 & -1 \\ 0 & 0 & 0 & 0 & 0 \end{bmatrix}$$

Since the rank of **V** is 2, which is less than the number of vectors in the given set, that set is linearly dependent.

6.4 Determine whether $[6, 10, 2]^T$ is a linear combination of $[1, 3, 2]^T$, $[2, 8, -1]^T$, and $[-1, 9, 2]^T$.

It is a linear combination if and only if there exist constants d_1, d_2, and d_3 such that

$$\begin{bmatrix} 6 \\ 10 \\ -2 \end{bmatrix} = d_1 \begin{bmatrix} 1 \\ 3 \\ 2 \end{bmatrix} + d_2 \begin{bmatrix} 2 \\ 8 \\ -1 \end{bmatrix} + d_3 \begin{bmatrix} -1 \\ 9 \\ 2 \end{bmatrix}$$

Solving this system is equivalent to solving the systems of Problem 2.8 with each x replaced by a d. In that problem we found that this system is consistent; hence $[6, 10, -2]^T$ is a linear combination of the other three vectors—in particular, for $d_1 = 1$, $d_2 = 2$, and $d_3 = -1$.

6.5 Determine whether [5, 1, 8] is a linear combination of [2, 3, 5], [1, 6, 7], and [0, 1, 1].

It is a linear combination if and only if there exist constants d_1, d_2, and d_3 such that

$$[5,1,8] = d_1[2,3,5] + d_2[1,6,7] + d_3[0,1,1]$$
$$= [2d_1 + d_2, 3d_1 + 6d_2 + d_3, 5d_1 + 7d_2 + d_3]$$

which is equivalent to the system

$$\begin{aligned} 2d_1 + d_2 \quad\quad &= 5 \\ 3d_1 + 6d_2 + d_3 &= 1 \\ 5d_1 + 7d_2 + d_3 &= 8 \end{aligned}$$

This system was shown in Problem 2.5 to be inconsistent, so [5, 1, 8] is not a linear combination of the other three vectors.

6.6 Prove that every set of $m+1$ or more m-dimensional vectors of the same type (either row or column) is linearly dependent.

Consider a set of n-such vectors, with $n > m$. Equation (*6.1*) generates m-homogeneous equations (one for each component of the vectors under consideration) in the n-unknowns $c_1, c_2, \ldots, c_n$. If we were to solve those equations by Gaussian elimination (see Chapter 2), we would find that the solution set has at least $n - m$ arbitrary unknowns. Since these arbitrary unknowns may be chosen to be nonzero, there exists a solution set for (*6.1*) which is not all zero; thus the n vectors are linearly dependent.

6.7 Prove that an elementary row operation of the first kind does not alter the row rank of a matrix.

Let **B** be obtained from matrix **A** by interchanging two rows. Clearly the rows of **A** form the same set of row vectors as the rows of **B**; so **A** and **B** must have the same row rank.

6.8 Prove that if $\mathbf{AX} = \mathbf{0}$ and $\mathbf{BX} = \mathbf{0}$ have the same solution set, then the $n \times n$ matrices **A** and **B** have the same column rank.

The system $\mathbf{AX} = \mathbf{0}$ can be written as

$$x_1\mathbf{A}_1 + x_2\mathbf{A}_2 + \cdots + x_n\mathbf{A}_n = \mathbf{0} \qquad (1)$$

where $\mathbf{A}_1$ is the first column of **A**, $\mathbf{A}_2$ is the second column of **A**, and so on, and $\mathbf{X} = [x_1, x_2, \ldots, x_n]^T$. Similarly, the system $\mathbf{BX} = \mathbf{0}$ can be written as

$$x_1\mathbf{B}_1 + x_2\mathbf{B}_2 + \cdots + x_n\mathbf{B}_n = \mathbf{0} \qquad (2)$$

Denote the column rank of **A** as a, and the column rank of **B** as b. Assume that the column rank of **A** is greater than the column rank of **B**, so that $a > b$. Now there must exist a columns of **A** which are linearly independent. Without loss of generality, we can assume that these are the first a columns of **A**. (If not, rearrange **A** so that they are; this column rearrangement does not change the column rank of **A**, by reasoning analogous to that used in Problem 6.8.) However, the first a columns of **B** are linearly dependent, because b is assumed to be smaller than a. Thus, there exist constants $d_1, d_2, \ldots, d_a$, not all zero, such that

$$d_1\mathbf{B}_1 + d_2\mathbf{B}_2 + \cdots + d_a\mathbf{B}_a = \mathbf{0}$$

From this, it follows that

$$d_1\mathbf{B}_1 + d_2\mathbf{B}_2 + \cdots + d_a\mathbf{B}_a + 0\mathbf{B}_{a+1} + \cdots + 0\mathbf{B}_n = \mathbf{0}$$

and that

$$x_1 = d_1 \qquad x_2 = d_2 \qquad \cdots \qquad x_a = d_a \qquad x_{a+1} = x_{a+2} = \cdots = x_n = 0$$

is a solution of system (*2*). Since these same values are given to be a solution of system (*1*), it follows that

$$d_1\mathbf{A}_1 + d_2\mathbf{A}_2 + \cdots + d_a\mathbf{A}_a = \mathbf{0}$$

where, as noted, the constants $d_1, d_2, \ldots, d_a$ are not all zero. But this implies that $\mathbf{A}_1, \mathbf{A}_2, \ldots, \mathbf{A}_a$ are linearly dependent, which is a contraction. Thus the column rank of **A** cannot be greater than the column rank of **B**.

A similar argument, with the roles of **A** and **B** reversed, shows that the column rank of **B** cannot be greater than the column rank of **A**, so the two column ranks must be equal.

6.9 Prove that an elementary row operation of any kind does not alter the column rank of a matrix.

Denote the original matrix as **A**, and the matrix obtained by applying an elementary row operation to **A** as **B**. The two homogeneous systems of equations $\mathbf{AX} = \mathbf{0}$ and $\mathbf{BX} = \mathbf{0}$ have the same set of solutions (see Chapter 2). Thus, as a result of Problem 6.8, **A** and **B** have the same column rank.

6.10 Prove that the row rank and column rank of any matrix are identical.

Assume that the row rank of an $m \times n$ matrix **A** is r, and its column rank is c. We wish to show that $r = c$. Rearrange the rows of **A** so that the first r rows are linearly independent and the remaining $m - r$ rows are the linear combinations of the first r rows. It follows from Problems 6.7 and 6.9 that the column rank and row rank of **A** remain unaltered. Denote the rows of **A** as $\mathbf{A}_1, \mathbf{A}_2, \ldots, \mathbf{A}_m$, in order, and define

$$\mathbf{B} = \begin{bmatrix} \mathbf{A}_1 \\ \mathbf{A}_2 \\ \cdots \\ \mathbf{A}_r \end{bmatrix} \quad \text{and} \quad \mathbf{C} = \begin{bmatrix} \mathbf{A}_{r+1} \\ \mathbf{A}_{r+2} \\ \cdots \\ \mathbf{A}_m \end{bmatrix}$$

Then **A** is the partitioned matrix $\begin{bmatrix} \mathbf{B} \\ \mathbf{C} \end{bmatrix}$. Furthermore, since every row of **C** is a linear combination of rows of **B**, there exists a matrix **T** such that $\mathbf{C} = \mathbf{TB}$. In particular, if

$$\mathbf{A}_{r+1} = d_1\mathbf{A}_1 + d_2\mathbf{A}_2 + \cdots + d_r\mathbf{A}_r$$

then $[d_1, d_2, \ldots, d_r]$ is the first row of **T**. Now for any n-dimensional vector **X**,

$$\mathbf{AX} = \begin{bmatrix} \mathbf{BX} \\ \mathbf{CX} \end{bmatrix} = \begin{bmatrix} \mathbf{BX} \\ \mathbf{TBX} \end{bmatrix}$$

Hence, $\mathbf{AX} = \mathbf{0}$ if and only if $\mathbf{BX} = \mathbf{0}$, and it follows from Problem 6.8 that **A** and **B** have the same column rank c. But the columns of **B** are r-dimensional vectors, so the column rank of **B** cannot be greater than r. That is,

$$c \le r \tag{1}$$

By repeating this reasoning on $\mathbf{A}^T$, we conclude that the column rank of $\mathbf{A}^T$ cannot be greater than the row rank of $\mathbf{A}^T$. But since the columns of $\mathbf{A}^T$ are the rows of **A** and vice versa, this means that the row rank of **A** cannot be greater than the column rank of **A**; that is,

$$r \le c \tag{2}$$

We conclude from (*1*) and (*2*) that $r = c$.

6.11 Prove that both the row rank and the column rank of a matrix equals its rank.

Let **U** be a matrix in row-echelon form obtained from **A** by elementary row operations. Then it follows from Problem 6.9, that **A** and **U** have the same column rank. Now denote the rank of **A** as r. From the definition of rank, r is the number of nonzero rows in **U**. Since the first nonzero element in each of the first r rows of **U** appears in a different column, it is simple to show that the first r rows of **U** are linearly independent and, therefore, that the row rank of **U** is r. The result of Problem 6.10 tells us that the column rank of **U** is also r. And, since **U** and **A** have the same column rank, the column and row-ranks of **A** are equal to its rank r.

6.12 Problem 6.8 suggests the following algorithm for choosing a maximal subset of linearly independent vectors from any given set: Construct a matrix **A** whose columns are the given set of vectors, and transform the matrix into row-echelon form **U** using elementary row operations. Then $\mathbf{AX} = \mathbf{0}$ has the same solution set as $\mathbf{UX} = \mathbf{0}$, which implies that any subset of the columns of **A** are linearly independent vectors if and only if the same subset of columns of **U** are linearly independent. Now the columns of **U** containing the first nonzero element in each of the nonzero rows of **U** are a maximal set of linearly independent column vectors for **U**, so those same columns in **A** are a maximal set of linearly independent column vectors for **A**.

Use this algorithm to choose a maximal set of linearly independent vectors from $[3, 2, 1]$, $[2, 3, -6]$, $[1, 0, 3]$, $[-4, -1, -8]$, and $[1, -1, 7]$.

We form the matrix

$$\mathbf{A} = \begin{bmatrix} 3 & 2 & 1 & -4 & 1 \\ 2 & 3 & 0 & -1 & -1 \\ 1 & -6 & 3 & -8 & 7 \end{bmatrix}$$

which, as shown in Problem 1.15, has the row-echelon form

$$\mathbf{U} = \begin{bmatrix} 1 & 2/3 & 1/3 & -4/3 & 1/3 \\ 0 & 1 & -2/5 & 1 & -1 \\ 0 & 0 & 0 & 0 & 0 \end{bmatrix}$$

The first and second columns of **U** contain the first nonzero element in each nonzero row of **U**. Therefore, the first and second columns of **A** constitute a maximal set of linearly independent vectors for the columns of **A**. That is, $[3, 2, 1]$ and $[2, 3, -6]$ are linearly independent, and all the other vectors in the original set are linear combinations of those two. In particular,

$$[1, 0, 3] = \tfrac{3}{5}[3, 2, 1] - \tfrac{2}{5}[2, 3, -6]$$
$$[-4, -1, -8] = (-2)[3, 2, 1] + (1)[2, 3, -6]$$
$$[1, -1, 7] = (1)[3, 2, 1] + (-1)[2, 3, -6]$$

6.13 Choose a maximal subset of linearly independent vectors from the following set:

$$\begin{bmatrix} 1 \\ 2 \\ 3 \end{bmatrix} \quad \begin{bmatrix} 2 \\ 4 \\ 6 \end{bmatrix} \quad \begin{bmatrix} 1 \\ 0 \\ -1 \end{bmatrix} \quad \begin{bmatrix} 4 \\ 6 \\ 8 \end{bmatrix} \quad \begin{bmatrix} 1 \\ 1 \\ 1 \end{bmatrix} \quad \begin{bmatrix} 1 \\ 1 \\ 2 \end{bmatrix}$$

We form the matrix

$$\mathbf{A} = \begin{bmatrix} 1 & 2 & 1 & 4 & 1 & 1 \\ 2 & 4 & 0 & 6 & 1 & 1 \\ 3 & 6 & -1 & 8 & 1 & 2 \end{bmatrix}$$

which has the row-echelon form

$$\mathbf{U} = \begin{bmatrix} 1 & 2 & 1 & 4 & 1 & 1 \\ 0 & 0 & 1 & 1 & 1/2 & 1/2 \\ 0 & 0 & 0 & 0 & 0 & 1 \end{bmatrix}$$

The first, third, and sixth columns of **U** contain the first nonzero element in each of its nonzero rows. Therefore, the first, third, and sixth columns of **A** constitute a maximal set of linearly independent vectors for the original set of vectors.

6.14 Prove that an ordered set of nonzero vectors $\{\mathbf{V}_1, \mathbf{V}_2, \ldots, \mathbf{V}_n\}$ is linearly dependent if and only if one of its vectors can be written as a linear combination of the vectors that precede it.

Suppose the set is linearly dependent, and let i be the first integer between 2 and n for which $\{\mathbf{V}_1, \mathbf{V}_2, \ldots, \mathbf{V}_i\}$ forms a linearly dependent set. Such an integer must exist, and at the very worst $i = n$. Then there exists a set of constants $d_1, d_2, \ldots, d_i$, not all zero, such that

$$d_1\mathbf{V}_1 + d_2\mathbf{V}_2 + \cdots + d_{i-1}\mathbf{V}_{i-1} + d_i\mathbf{V}_i = \mathbf{0}$$

Furthermore, $d_i \neq 0$, for otherwise the set $\{\mathbf{V}_1, \mathbf{V}_2, \ldots, \mathbf{V}_{i-1}\}$ would be linearly dependent, contradicting the defining property of i. Hence,

$$\mathbf{V}_i = -\frac{d_1}{d_i}\mathbf{V}_1 - \frac{d_2}{d_i}\mathbf{V}_2 - \cdots - \frac{d_{i-1}}{d_i}\mathbf{V}_{i-1}$$

That is, $\mathbf{V}_i$ can be written as a linear combination of $\mathbf{V}_1, \ldots, \mathbf{V}_{i-1}$.

On the other hand, suppose that for some i $(i = 2, 3, \ldots, n)$

$$\mathbf{V}_i = d_1\mathbf{V}_1 + d_2\mathbf{V}_2 + \cdots + d_{i-1}\mathbf{V}_{i-1}$$

Then

$$d_1\mathbf{V}_1 + d_2\mathbf{V}_2 + \cdots + d_{i-1}\mathbf{V}_{i-1} + (-1)\mathbf{V}_i + 0\mathbf{V}_{i+1} + \cdots + 0\mathbf{V}_n = \mathbf{0}$$

This is (*6.1*) with $c_i = -1 \neq 0$, $c_k = d_k$ $(k = 1, \ldots, i-1)$, and $c_k = 0$ $(k = i+1, i+2, \ldots, n)$. So the set of vectors is linearly dependent.

Supplementary Problems

In Problems 6.15 through 6.20, determine whether the given set of vectors is linearly independent.

6.15 $\{[1, 2]^T, [2, 4]^T\}$

6.16 $\{[1, 1, 2], [2, 2, 2], [2, 2, 1]\}$

6.17 $\{[1, 0, 1], [1, 2, 0], [0, 1, 2]\}$

6.18 $\{[1, 0, 2, 0]^T, [2, 2, 0, 1]^T, [1, -2, 6, -1]^T\}$

6.19 $\{[2, 0, 1, 1], [0, 1, 2, -1], [1, -1, -1, 1], [0, 0, 1, 2]\}$

6.20 $\{[1, 2, 1]^T, [1, 1, 0]^T, [0, 1, -1]^T, [2, 1, 3]^T\}$

6.21 Is $[1, 3]^T$ a linear combination of the vectors given in Problem 6.15?

6.22 (*a*) Determine whether $[0, 0, 1]$ can be written as a linear combination of the vectors given in Problem 6.16. (*b*) Repeat part *a* for the vector $[1, 2, 0]$.

6.23 (*a*) Determine whether $[2, 1, 2, 1]$ can be written as a linear combination of the vectors given in Problem 6.19. (*b*) Repeat part *a* for the vector $[0, 0, 0, 1]$.

6.24 Show that any 3-dimensional row vector can be expressed as a linear combination of the vectors given in Problem 6.17.

6.25 Choose a maximal subset of linearly independent vectors from those given in Problem 6.15.

6.26 Choose a maximal subset of linearly independent vectors from those given in Problem 6.16.

6.27 Choose a maximal set of linearly independent vectors from the following: $[1,2,1,-1]$, $[1,0,-1,2]$, $[2,2,0,1]$, $[3,2,-1,3]$, $[0,1,1,0]$, $[3,3,0,3]$.

6.28 An m-dimensional vector $\mathbf{V}$ is a *convex combination* of the m-dimensional vectors $\mathbf{V}_1, \mathbf{V}_2, \ldots, \mathbf{V}_n$ of the same type (row or column) if there exist nonnegative constants $d_1, d_2, \ldots, d_n$ whose sum is 1, such that $\mathbf{V} = d_1\mathbf{V}_1 + d_2\mathbf{V}_2 + \cdots + d_n\mathbf{V}_n$. Show that $[5/3, 5/6]$ is a convex combination of the vectors $[1,1]$, $[3,0]$, and $[1,2]$.

6.29 Determine whether $[0,7]^T$ can be written as a convex combination of the vectors

$$\begin{bmatrix} 3 \\ 6 \end{bmatrix} \quad \begin{bmatrix} -6 \\ 9 \end{bmatrix} \quad \begin{bmatrix} 2 \\ 1 \end{bmatrix} \quad \begin{bmatrix} -1 \\ 1 \end{bmatrix}$$

6.30 Prove that if $\{\mathbf{V}_1, \mathbf{V}_2, \ldots, \mathbf{V}_r\}$ is linearly independent and $\mathbf{V}$ cannot be written as a linear combination of this set, then $\{\mathbf{V}_1, \mathbf{V}_2, \ldots, \mathbf{V}_r, \mathbf{V}\}$ is also linearly independent.

6.31 Prove Property 6.5.

6.32 The *null space* of a matrix $\mathbf{A}$ is the set of all vectors which are solutions of $\mathbf{AX} = \mathbf{0}$. Determine the null space of

$$\mathbf{A} = \begin{bmatrix} 1 & 2 \\ 3 & 4 \end{bmatrix}$$

6.33 Determine the null space of the matrix

$$\mathbf{A} = \begin{bmatrix} 1 & 3 \\ 2 & 6 \end{bmatrix}$$

6.34 Determine the null space of the matrix

$$\mathbf{A} = \begin{bmatrix} 1 & 1 & 0 & 2 \\ 2 & 2 & 0 & 4 \\ 0 & 1 & 1 & 0 \\ 2 & 3 & 1 & 4 \end{bmatrix}$$

Chapter 7

Eigenvalues and Eigenvectors

CHARACTERISTIC EQUATION

A nonzero column vector $\mathbf{X}$ is an *eigenvector* (or *right eigenvector* or *right characteristic vector*) of a square matrix $\mathbf{A}$ if there exists a scalar λ such that

$$\mathbf{AX} = \lambda\mathbf{X} \tag{7.1}$$

Then λ is an *eigenvalue* (or *characteristic value*) of $\mathbf{A}$. Eigenvalues may be zero; an eigenvector may not be the zero vector.

Example 7.1 $[1, -1]^T$ is an eigenvector corresponding to the eigenvalue $\lambda = -2$ for the matrix

$$\mathbf{A} = \begin{bmatrix} 3 & 5 \\ -2 & -4 \end{bmatrix}$$

because

$$\begin{bmatrix} 3 & 5 \\ -2 & -4 \end{bmatrix}\begin{bmatrix} 1 \\ -1 \end{bmatrix} = \begin{bmatrix} -2 \\ 2 \end{bmatrix} = -2\begin{bmatrix} -1 \\ 1 \end{bmatrix}$$

The *characteristic equation* of an $n \times n$ matrix $\mathbf{A}$ is the nth-degree polynomial equation

$$\det(\mathbf{A} - \lambda\mathbf{I}) = 0 \tag{7.2}$$

Solving the characteristic equation for λ gives the eigenvalues of $\mathbf{A}$, which may be real, complex, or multiples of each other. Once an eigenvalue is determined, it may be substituted into (*7.1*), and then that equation may be solved for the corresponding eigenvectors. (See Problems 7.1 through 7.3.) The polynomial $\det(\mathbf{A} - \lambda\mathbf{I})$ is called the *characteristic polynomial* of $\mathbf{A}$.

PROPERTIES OF EIGENVALUES AND EIGENVECTORS

Property 7.1: The sum of the eigenvalues of a matrix is equal to its *trace*, which is the sum of the elements on its main diagonal.

Property 7.2: Eigenvectors corresponding to different eigenvalues are linearly independent.

Property 7.3: A matrix is singular if and only if it has a zero eigenvalue.

Property 7.4: If $\mathbf{X}$ is an eigenvector of $\mathbf{A}$ corresponding to the eigenvalue λ and $\mathbf{A}$ is invertible, then $\mathbf{X}$ is an eigenvector of $\mathbf{A}^{-1}$ corresponding to its eigenvalue $1/\lambda$.

Property 7.5: If $\mathbf{X}$ is an eigenvector of a matrix, then so too is $k\mathbf{X}$ for any nonzero constant k, and both $\mathbf{X}$ and $k\mathbf{X}$ correspond to the same eigenvalue.

Property 7.6: A matrix and its transpose have the same eigenvalues.

Property 7.7: The eigenvalues of an upper or lower triangular matrix are the elements on its main diagonal.

Property 7.8: The product of the eigenvalues (counting multiplicities) of a matrix equals the determinant of the matrix.

Property 7.9: If $\mathbf{X}$ is an eigenvector of $\mathbf{A}$ corresponding to eigenvalue λ, then $\mathbf{X}$ is an an eigenvector of $\mathbf{A} - c\mathbf{I}$ corresponding to the eigenvalue $\lambda - c$ for any scalar c.

LINEARLY INDEPENDENT EIGENVECTORS

The eigenvectors corresponding to a particular eigenvalue contain one or more arbitrary scalars. (See Problems 7.1 through 7.3.) The number of arbitrary scalars is the number of linearly independent eigenvectors associated with that eigenvalue. To obtain a maximal set of linearly independent eigenvectors corresponding to an eigenvalue, sequentially set each of these arbitrary scalars equal to a convenient nonzero number (usually chosen to avoid fractions) with all other arbitrary scalars set equal to zero. It follows from Property 7.2 that when the sets corresponding to all the eigenvalues are combined, the result is a maximal set of linearly independent eigenvectors for the matrix. (See Problems 7.4 through 7.6.)

COMPUTATIONAL CONSIDERATIONS

There are no theoretical difficulties in determining eigenvalues, but there are practical ones. First, evaluating the determinant in *(7.2)* for an $n \times n$ matrix requires approximately $n!$ multiplications, which for large n is a prohibitive number. Second, obtaining the roots of a general characteristic polynomial poses an intractable algebraic problem. Consequently, numerical algorithms are employed for determining the eigenvalues of large matrices (see Chapters 19 and 20).

THE CAYLEY-HAMILTON THEOREM

Theorem 7.1: Every square matrix satisfies its own characteristic equation. That is, if

$$\det(\mathbf{A} - \lambda\mathbf{I}) = b_n\lambda^n + b_{n-1}\lambda^{n-1} + \cdots + b_2\lambda^2 + b_1\lambda + b_0$$

then

$$b_n\mathbf{A}^n + b_{n-1}\mathbf{A}^{n-1} + \cdots + b_2\mathbf{A}^2 + b_1\mathbf{A} + b_0\mathbf{I} = \mathbf{0}$$

(See Problems 7.15 through 7.17.)

Solved Problems

7.1 Determine the eigenvalues and eigenvectors of

$$\mathbf{A} = \begin{bmatrix} 3 & 5 \\ -2 & -4 \end{bmatrix}$$

For this matrix,

$$\mathbf{A} - \lambda\mathbf{I} = \begin{bmatrix} 3 & 5 \\ -2 & -4 \end{bmatrix} - \lambda\begin{bmatrix} 1 & 0 \\ 0 & 1 \end{bmatrix} = \begin{bmatrix} 3-\lambda & 5 \\ -2 & -4-\lambda \end{bmatrix}$$

hence

$$\det(\mathbf{A} - \lambda\mathbf{I}) = (3-\lambda)(-4-\lambda) - 5(-2) = \lambda^2 + \lambda - 2$$

The characteristic equation of $\mathbf{A}$ is $\lambda^2 + \lambda - 2 = 0$; when solved for λ, it gives the two eigenvalues $\lambda = 1$ and $\lambda = -2$. As a check, we utilize Property 7.1: The trace of $\mathbf{A}$ is $3 + (-4) = -1$, which is also the sum of the eigenvalues.

The eigenvectors corresponding to $\lambda = 1$ are obtained by solving Eq. (7.1) for $\mathbf{X} = [x_1, x_2]^T$ with this value of λ. After substituting and rearranging, we have

$$\left(\begin{bmatrix} 3 & 5 \\ -2 & -4 \end{bmatrix} - 1\begin{bmatrix} 1 & 0 \\ 0 & 1 \end{bmatrix}\right)\begin{bmatrix} x_1 \\ x_2 \end{bmatrix} = \begin{bmatrix} 0 \\ 0 \end{bmatrix}$$

or

$$\begin{bmatrix} 2 & 5 \\ -2 & -5 \end{bmatrix}\begin{bmatrix} x_1 \\ x_2 \end{bmatrix} = \begin{bmatrix} 0 \\ 0 \end{bmatrix}$$

which is equivalent to the set of linear equations

$$2x_1 + 5x_2 = 0$$
$$-2x_1 - 5x_2 = 0$$

The solution to this system is $x_1 = -\frac{5}{2}x_2$ with x_2 arbitrary, so the eigenvectors corresponding to $\lambda = 1$ are

$$\mathbf{X} = \begin{bmatrix} x_1 \\ x_2 \end{bmatrix} = \begin{bmatrix} -\frac{5}{2}x_2 \\ x_2 \end{bmatrix} = x_2\begin{bmatrix} -5/2 \\ 1 \end{bmatrix}$$

with x_2 arbitrary.

When $\lambda = -2$, (*7.1*) may be written

$$\left\{\begin{bmatrix} 3 & 5 \\ -2 & -4 \end{bmatrix} - (-2)\begin{bmatrix} 1 & 0 \\ 0 & 1 \end{bmatrix}\right\}\begin{bmatrix} x_1 \\ x_2 \end{bmatrix} = \begin{bmatrix} 0 \\ 0 \end{bmatrix}$$

or

$$\begin{bmatrix} 5 & 5 \\ -2 & -2 \end{bmatrix}\begin{bmatrix} x_1 \\ x_2 \end{bmatrix} = \begin{bmatrix} 0 \\ 0 \end{bmatrix}$$

which is equivalent to the set of linear equations

$$5x_1 + 5x_2 = 0$$
$$-2x_1 - 2x_2 = 0$$

The solution to this system is $x_1 = -x_2$ with x_2 arbitrary, so the eigenvectors corresponding to $\lambda = -2$ are

$$\mathbf{X} = \begin{bmatrix} x_1 \\ x_2 \end{bmatrix} = \begin{bmatrix} -x_2 \\ x_2 \end{bmatrix} = x_2\begin{bmatrix} -1 \\ 1 \end{bmatrix}$$

with x_2 arbitrary.

7.2 Determine the eigenvalues and eigenvectors of

$$\mathbf{A} = \begin{bmatrix} 5 & 2 & 2 \\ 3 & 6 & 3 \\ 6 & 6 & 9 \end{bmatrix}$$

For this matrix,

$$\mathbf{A} - \lambda\mathbf{I} = \begin{bmatrix} 5 & 2 & 2 \\ 3 & 6 & 3 \\ 6 & 6 & 9 \end{bmatrix} - \lambda\begin{bmatrix} 1 & 0 & 0 \\ 0 & 1 & 0 \\ 0 & 0 & 1 \end{bmatrix} = \begin{bmatrix} 5-\lambda & 2 & 2 \\ 3 & 6-\lambda & 3 \\ 6 & 6 & 9-\lambda \end{bmatrix}$$

The determinant of this last matrix may be obtained by expansion by cofactors (see Chapter 5); it is

$$-\lambda^3 + 20\lambda^2 - 93\lambda + 126 = -(\lambda-3)^2(\lambda-14)$$

The characteristic equation of **A** is $-(\lambda-3)^2(\lambda-14) = 0$, which has as its solution the eigenvalue $\lambda = 3$ of multiplicity two and the eigenvalue $\lambda = 14$ of multiplicity one. As a check, we utilize Property 7.1: The trace of **A** is $5 + 6 + 9 = 20$, which equals the sum of the three eigenvalues.

The eigenvectors corresponding to $\lambda = 3$ are obtained by solving (*7.1*) for $\mathbf{X} = [x_1, x_2, x_3]^T$ with this value of λ. Thus, we may write

$$\left(\begin{bmatrix} 5 & 2 & 2 \\ 3 & 6 & 3 \\ 6 & 6 & 9 \end{bmatrix} - 3\begin{bmatrix} 1 & 0 & 0 \\ 0 & 1 & 0 \\ 0 & 0 & 1 \end{bmatrix}\right)\begin{bmatrix} x_1 \\ x_2 \\ x_3 \end{bmatrix} = \begin{bmatrix} 0 \\ 0 \\ 0 \end{bmatrix}$$

or

$$\begin{bmatrix} 2 & 2 & 2 \\ 3 & 3 & 3 \\ 6 & 6 & 6 \end{bmatrix}\begin{bmatrix} x_1 \\ x_2 \\ x_3 \end{bmatrix} = \begin{bmatrix} 0 \\ 0 \\ 0 \end{bmatrix}$$

which is equivalent to the set of linear equations

$$\begin{aligned} 2x_1 + 2x_2 + 2x_3 &= 0 \\ 3x_1 + 3x_2 + 3x_3 &= 0 \\ 6x_1 + 6x_2 + 6x_3 &= 0 \end{aligned}$$

The solution to this system is $x_1 = -x_2 - x_3$ with x_2 and x_3 arbitrary; the eigenvectors corresponding to $\lambda = 3$ are thus

$$\mathbf{X} = \begin{bmatrix} x_1 \\ x_2 \\ x_3 \end{bmatrix} = \begin{bmatrix} -x_2 - x_3 \\ x_2 \\ x_3 \end{bmatrix} = x_2 \begin{bmatrix} -1 \\ 1 \\ 0 \end{bmatrix} + x_3 \begin{bmatrix} -1 \\ 0 \\ 1 \end{bmatrix}$$

with x_2 and x_3 arbitrary.

When $\lambda = 14$, (*7.1*) becomes

$$\left(\begin{bmatrix} 5 & 2 & 2 \\ 3 & 6 & 3 \\ 6 & 6 & 9 \end{bmatrix} - 14 \begin{bmatrix} 1 & 0 & 0 \\ 0 & 1 & 0 \\ 0 & 0 & 1 \end{bmatrix} \right) \begin{bmatrix} x_1 \\ x_2 \\ x_3 \end{bmatrix} = \begin{bmatrix} 0 \\ 0 \\ 0 \end{bmatrix}$$

or

$$\begin{bmatrix} -9 & 2 & 2 \\ 3 & -8 & 3 \\ 6 & 6 & -5 \end{bmatrix} \begin{bmatrix} x_1 \\ x_2 \\ x_3 \end{bmatrix} = \begin{bmatrix} 0 \\ 0 \\ 0 \end{bmatrix}$$

which is equivalent to the set of linear equations

$$\begin{aligned} -9x_1 + 2x_2 + 2x_3 &= 0 \\ 3x_1 - 8x_2 + 3x_3 &= 0 \\ 6x_1 + 6x_2 - 5x_3 &= 0 \end{aligned}$$

The solution to this system is $x_1 = \frac{1}{3}x_3$ and $x_2 = \frac{1}{2}x_3$ with x_3 arbitrary; the eigenvectors corresponding to $\lambda = 14$ are thus

$$\mathbf{X} = \begin{bmatrix} x_1 \\ x_2 \\ x_3 \end{bmatrix} = \begin{bmatrix} \frac{1}{3}x_3 \\ \frac{1}{2}x_3 \\ x_3 \end{bmatrix} = x_3 \begin{bmatrix} 1/3 \\ 1/2 \\ 1 \end{bmatrix}$$

with x_3 arbitrary.

7.3 Determine the eigenvalues and eigenvectors of

$$\mathbf{A} = \begin{bmatrix} 3 & 4 \\ -5 & -5 \end{bmatrix}$$

For this matrix,

$$\mathbf{A} - \lambda\mathbf{I} = \begin{bmatrix} 3 & 4 \\ -5 & -5 \end{bmatrix} - \lambda \begin{bmatrix} 1 & 0 \\ 0 & 1 \end{bmatrix} = \begin{bmatrix} 3-\lambda & 4 \\ -5 & -5-\lambda \end{bmatrix}$$

hence

$$\det(\mathbf{A} - \lambda\mathbf{I}) = (3-\lambda)(-5-\lambda) - 4(-5) = \lambda^2 + 2\lambda + 5$$

The characteristic equation of $\mathbf{A}$ is $\lambda^2 + 2\lambda + 5 = 0$; when solved for λ, it gives the two complex eigenvalues $\lambda = -1 + i2$ and $\lambda = -1 - i2$. As a check, we note that the trace of $\mathbf{A}$ is -2, which is the sum of these eigenvalues.

The eigenvectors corresponding to $\lambda = 1$ are obtained by solving Eq. (*7.1*) for $\mathbf{X} = [x_1, x_2]^T$ with this value of λ. After substituting and rearranging, we have

$$\left\{ \begin{bmatrix} 3 & 4 \\ -5 & -5 \end{bmatrix} - (-1 + i2) \begin{bmatrix} 1 & 0 \\ 0 & 1 \end{bmatrix} \right\} \begin{bmatrix} x_1 \\ x_2 \end{bmatrix} = \begin{bmatrix} 0 \\ 0 \end{bmatrix}$$

or

$$\begin{bmatrix} 4 - i2 & 4 \\ -5 & -4 - i2 \end{bmatrix} \begin{bmatrix} x_1 \\ x_2 \end{bmatrix} = \begin{bmatrix} 0 \\ 0 \end{bmatrix}$$

which is equivalent to the set of linear equations

$$\begin{aligned}(4-i2)x_1 + \qquad 4x_2 &= 0\\ -5x_1 + (-4-i2)\,x_2 &= 0\end{aligned}$$

The solution to this system is $x_1 = (-4/5 - i2/5)x_2$ with x_2 arbitrary; the eigenvectors corresponding to $\lambda = -1 + i2$ are thus

$$\mathbf{X} = \begin{bmatrix} x_1 \\ x_2 \end{bmatrix} = \begin{bmatrix} (-4/5 - i2/5)x_2 \\ x_2 \end{bmatrix} = x_2\begin{bmatrix} -4/5 - i2/5 \\ 1 \end{bmatrix}$$

with x_2 arbitrary.

With $\lambda = -1 - i2$, the corresponding eigenvectors are found in a similar manner to be

$$\mathbf{X} = \begin{bmatrix} x_1 \\ x_2 \end{bmatrix} = \begin{bmatrix} (-4/5 + i2/5)x_2 \\ x_2 \end{bmatrix} = x_2\begin{bmatrix} -4/5 + i2/5 \\ 1 \end{bmatrix}$$

with x_2 arbitrary.

7.4 Choose a maximal set of linearly independent eigenvectors for the matrix given in Problem 7.2.

The eigenvectors associated with $\lambda = 3$ were found in Problem 7.2 to be

$$x_2\begin{bmatrix} -1 \\ 1 \\ 0 \end{bmatrix} + x_3\begin{bmatrix} -1 \\ 0 \\ 1 \end{bmatrix} \qquad x_2, x_3 \text{ arbitrary}$$

There are two linearly independent eigenvectors associated with $\lambda = 3$, one for each arbitrary scalar. One of them may be obtained by setting $x_2 = 1$, $x_3 = 0$; the other, by setting $x_2 = 0$, $x_3 = 1$.

The eigenvectors associated with $\lambda = 14$ are

$$x_2\begin{bmatrix} 1/3 \\ 1/2 \\ 1 \end{bmatrix} \qquad x_2 \text{ arbitrary}$$

Since there is only one arbitrary constant here, there is only one linearly independent eigenvector associated with $\lambda = 14$. It may be obtained by choosing x_2 to be any nonzero scalar. A convenient choice, to avoid fractions, is $x_2 = 6$. Combining the linearly independent eigenvectors corresponding to the two eigenvalues, we obtain

$$\begin{bmatrix} -1 \\ 1 \\ 0 \end{bmatrix} \quad \begin{bmatrix} -1 \\ 0 \\ 1 \end{bmatrix} \quad \begin{bmatrix} 2 \\ 3 \\ 6 \end{bmatrix}$$

as a maximal set of linearly independent eigenvectors for the matrix.

7.5 Choose a maximal set of linearly independent eigenvectors for the matrix given in Problem 7.1.

The eigenvectors corresponding to $\lambda = 1$ were found in Problem 7.1 to be

$$x_2\begin{bmatrix} -5/2 \\ 1 \end{bmatrix} \qquad x_2 \text{ arbitrary}$$

Since there is only one arbitrary scalar, there is only one linearly independent eigenvector associated with this eigenvalue. It may be obtained by choosing x_2 to be any nonzero scalar. A convenient choice, to avoid fractions, is $x_2 = 2$.

The eigenvectors corresponding to $\lambda = -2$ are

$$x_2\begin{bmatrix} -1 \\ 1 \end{bmatrix} \qquad x_2 \text{ arbitrary}$$

There is one linearly independent eigenvector associated with this eigenvalue, and it may be obtained by choosing x_2 to be any nonzero scalar. A convenient choice here is $x_2 = 1$. Collecting the linearly independent eigenvectors for the two eigenvalues, we have

$$\begin{bmatrix} -5 \\ 2 \end{bmatrix} \quad \begin{bmatrix} -1 \\ 1 \end{bmatrix}$$

as a maximal set of linearly independent eigenvectors for the matrix.

7.6 Choose a maximal set of linearly independent eigenvectors for the matrix

$$\mathbf{A} = \begin{bmatrix} 2 & 1 & 0 & 0 & 0 \\ 0 & 2 & 0 & 0 & 0 \\ 0 & 0 & 2 & 0 & 0 \\ 0 & 0 & 0 & 2 & 1 \\ 0 & 0 & 0 & 0 & 2 \end{bmatrix}$$

Since this matrix is upper triangular its eigenvalues are the elements on its main diagonal. Thus, $\lambda = 2$ is an eigenvalue of multiplicity five. The eigenvectors associated with this eigenvalue are

$$\begin{bmatrix} x_1 \\ 0 \\ x_3 \\ x_4 \\ 0 \end{bmatrix} = x_1 \begin{bmatrix} 1 \\ 0 \\ 0 \\ 0 \\ 0 \end{bmatrix} + x_3 \begin{bmatrix} 0 \\ 0 \\ 1 \\ 0 \\ 0 \end{bmatrix} + x_4 \begin{bmatrix} 0 \\ 0 \\ 0 \\ 1 \\ 0 \end{bmatrix}$$

with x_1, x_3, and x_4 arbitrary. Because there are three arbitrary scalars, there are three linearly independent eigenvectors associated with **A**. One may be obtained by setting $x_1 = 1$, $x_3 = x_4 = 0$; another by setting $x_3 = 1$, $x_1 = x_4 = 0$; and the third by setting $x_4 = 1$ and $x_1 = x_3 = 0$. Note that this matrix has only three linearly independent eigenvectors, even though it has order 5×5.

7.7 Show if λ is an eigenvalue of a matrix **A**, then it is a solution to (*7.2*).

If λ is an eigenvalue of **A**, there must exist a nonzero vector **X** such that $\mathbf{AX} = \lambda\mathbf{X}$. Thus, $\mathbf{AX} - \lambda\mathbf{X} = \mathbf{0}$, and $(\mathbf{A} - \lambda\mathbf{I})\mathbf{X} = \mathbf{0}$. This implies that $\mathbf{A} - \lambda\mathbf{I}$ is singular, for otherwise $\mathbf{X} = (\mathbf{A} - \lambda\mathbf{I})^{-1}\mathbf{0} = \mathbf{0}$, which is not the case. But if $\mathbf{A} - \lambda\mathbf{I}$ is singular, then $\det(\mathbf{A} - \lambda\mathbf{I}) = 0$ (see Chapter 5).

7.8 Show that eigenvectors corresponding to different eigenvalues are linearly independent.

Let $\lambda_1, \lambda_2, \ldots, \lambda_m$ be different eigenvalues of a matrix **A**, and let $\mathbf{X}_1, \mathbf{X}_2, \ldots, \mathbf{X}_m$ be associated eigenvectors. We must show that the only solution to

$$c_1\mathbf{X}_1 + c_2\mathbf{X}_2 + \cdots + c_m\mathbf{X}_m = \mathbf{0} \tag{1}$$

is $c_1 = c_2 = \cdots = c_m = 0$. Multiplying (*1*) on the left by **A**, we obtain

$$c_1\mathbf{AX}_1 + c_2\mathbf{AX}_2 + \cdots + c_m\mathbf{AX}_m = \mathbf{A0} = \mathbf{0}$$

Since each vector here is an eigenvector, we use (*7.1*) to write

$$c_1\lambda_1\mathbf{X}_1 + c_2\lambda_2\mathbf{X}_2 + \cdots + c_m\lambda_m\mathbf{X}_m = \mathbf{0} \tag{2}$$

Multiplying (*2*) on the left by **A** and again using (*7.1*), we obtain

$$c_2\lambda_1^2\mathbf{X}_1 + c_2\lambda_2^2\mathbf{X}_2 + \cdots + c_m\lambda_m^2\mathbf{X}_m = \mathbf{0} \tag{3}$$

Equations (*1*) through (*3*) are the first three equations of the set

$$\begin{array}{llll}
c_1\mathbf{X}_1 & + c_2\mathbf{X}_2 & + \cdots + c_m\mathbf{X}_m & = \mathbf{0} \\
c_1\lambda_1\mathbf{X}_1 & + c_2\lambda_2\mathbf{X}_2 & + \cdots + c_m\lambda_m\mathbf{X}_m & = \mathbf{0} \\
c_1\lambda_1^2\mathbf{X}_1 & + c_2\lambda_2^2\mathbf{X}_2 & + \cdots + c_m\lambda_m^2\mathbf{X}_m & = \mathbf{0} \\
c_1\lambda_1^3\mathbf{X} & + c_2\lambda_2^3\mathbf{X}_2 & + \cdots + c_m\lambda_m^3\mathbf{X}_m & = \mathbf{0} \\
\cdots & \cdots & \cdots & \cdots \\
c_1\lambda_1^{m-1}\mathbf{X}_1 & + c_2\lambda_2^{m-1}\mathbf{X}_2 & + \cdots + c_m\lambda_m^{m-1}\mathbf{X}_m & = \mathbf{0}
\end{array}$$

generated by sequentially multiplying each equation on the left by $\mathbf{A}$. This system can be written in the matrix form

$$\begin{bmatrix} 1 & 1 & 1\cdots & 1 \\ \lambda_1 & \lambda_2 & \lambda_3\cdots & \lambda_m \\ \lambda_1^2 & \lambda_2^2 & \lambda_3^2\cdots & \lambda_m^2 \\ \cdots & \cdots & \cdots & \cdots \\ \lambda_1^{m-1} & \lambda_2^{m-1} & \lambda_3^{m-1}\cdots & \lambda_m^{m-1} \end{bmatrix} \begin{bmatrix} c_1\mathbf{X}_1 \\ c_2\mathbf{X}_2 \\ c_3\mathbf{X}_3 \\ \cdots \\ c_m\mathbf{X}_m \end{bmatrix} = \begin{bmatrix} \mathbf{0} \\ \mathbf{0} \\ \mathbf{0} \\ \cdots \\ \mathbf{0} \end{bmatrix} \tag{4}$$

The first matrix on the left is an $m \times m$ matrix which we shall denote as $\mathbf{Q}$. Its determinant is called the *Vandermonde determinant* and is

$$(\lambda_2 - \lambda_1)(\lambda_3 - \lambda_2)(\lambda_3 - \lambda_1)(\lambda_4 - \lambda_3)(\lambda_4 - \lambda_2)(\lambda_4 - \lambda_1)\cdots(\lambda_m - \lambda_1)$$

which is not zero in this situation because all the eigenvalues are different. As a result $\mathbf{Q}$ is nonsingular, and the system (*4*) can be written as

$$\begin{bmatrix} c_1\mathbf{X}_1 \\ c_2\mathbf{X}_2 \\ c_3\mathbf{X}_3 \\ \cdots \\ c_m\mathbf{X}_m \end{bmatrix} = \mathbf{Q}^{-1} \begin{bmatrix} \mathbf{0} \\ \mathbf{0} \\ \mathbf{0} \\ \cdots \\ \mathbf{0} \end{bmatrix} = \begin{bmatrix} \mathbf{0} \\ \mathbf{0} \\ \mathbf{0} \\ \cdots \\ \mathbf{0} \end{bmatrix}$$

It follows that $c_i\mathbf{X}_i = \mathbf{0}$ $(i = 1, 2, \ldots, m)$. But since each $\mathbf{X}_i$ is an eigenvector, it is not zero; so $c_i = 0$ for each i.

7.9 Prove that a matrix is singular if and only if it has a zero eigenvalue.

A matrix $\mathbf{A}$ has a zero eigenvalue if and only if $\det(\mathbf{A} - 0\mathbf{I}) = 0$, which is true if and only if $\det \mathbf{A} = 0$, which in turn is true if and only if $\mathbf{A}$ is singular (see Chapter 5).

7.10 Prove that if $\mathbf{X}$ is an eigenvector corresponding to the eigenvalue λ of an invertible matrix $\mathbf{A}$, then $\mathbf{X}$ is an eigenvector of $\mathbf{A}^{-1}$ corresponding to its eigenvalue $1/\lambda$.

It follows from Problem 7.9 that $\lambda \neq 0$. We are given $\mathbf{AX} = \lambda\mathbf{X}$, so $\mathbf{A}^{-1}(\mathbf{AX}) = \mathbf{A}^{-1}(\lambda\mathbf{X})$ and $\mathbf{X} = \lambda(\mathbf{A}^{-1}\mathbf{X})$. Dividing by λ, we obtain $\mathbf{A}^{-1}\mathbf{X} = (1/\lambda)\mathbf{X}$, which implies the desired result.

7.11 Prove that a matrix and its transpose have the same eigenvalues.

If λ is an eigenvalue of $\mathbf{A}$, then

$$0 = \det(\mathbf{A} - \lambda\mathbf{I}) = \det\{(\mathbf{A}^T)^T - \lambda\mathbf{I}^T\} = \det(\mathbf{A}^T - \lambda\mathbf{I})^T = \det(\mathbf{A}^T - \lambda\mathbf{I})$$

by Property 5.7. Thus, λ is also an eigenvalue of $\mathbf{A}^T$.

7.12 Prove that if $\mathbf{X}_1, \mathbf{X}_2, \ldots, \mathbf{X}_n$ are all eigenvectors of a matrix $\mathbf{A}$ corresponding to the same eigenvalue λ, then any nonzero linear combination of these vectors is also an eigenvector of $\mathbf{A}$ corresponding to λ.

Set $\mathbf{X} = d_1\mathbf{X}_1 + d_2\mathbf{X}_2 + \cdots + d_n\mathbf{X}_n$. Then

$$\begin{aligned}\mathbf{AX} &= \mathbf{A}(d_1\mathbf{X}_1 + d_2\mathbf{X}_2 + \cdots + d_n\mathbf{X}_n)\\ &= d_1\mathbf{AX}_1 + d_2\mathbf{AX}_2 + \cdots + d_n\mathbf{AX}_n\\ &= d_1\lambda\mathbf{X}_1 + d_2\lambda\mathbf{X}_2 + \cdots + d_n\lambda\mathbf{X}_n\\ &= \lambda(d_1\mathbf{X}_1 + d_2\mathbf{X}_2 + \cdots + d_n\mathbf{X}_n) = \lambda\mathbf{X}\end{aligned}$$

Thus, **X** is an eigenvector of **A**. Note that a nonzero constant times an eigenvector is also an eigenvector corresponding to the same eigenvalue.

7.13 A *left eigenvector* of a matrix **A** is a nonzero row vector **X** having the property that $\mathbf{XA} = \lambda\mathbf{X}$ or, equivalently, that

$$\mathbf{X}(\mathbf{A} - \lambda\mathbf{I}) = \mathbf{0} \tag{1}$$

for some scalar λ. Again λ is an eigenvalue for **A**, and it is found as before. Once λ is determined, it is substituted into (*1*) and then that equation is solved for **X**. Find the eigenvalues and left eigenvectors for

$$\mathbf{A} = \begin{bmatrix} 3 & 5 \\ -2 & -4 \end{bmatrix}$$

The eigenvalues were found in Problem 7.1 to be $\lambda = 1$ and $\lambda = -2$. Set $\mathbf{X} = [x_1, x_2]$. With $\lambda = 1$, (*1*) becomes

$$[x_1, x_2]\left(\begin{bmatrix} 3 & 5 \\ -2 & -4 \end{bmatrix} - 1\begin{bmatrix} 1 & 0 \\ 0 & 1 \end{bmatrix}\right) = [0, 0]$$

or

$$[x_1, x_2]\begin{bmatrix} 2 & 5 \\ -2 & -5 \end{bmatrix} = [0, 0]$$

or

$$[2x_1 - 2x_2, 5x_1 - 5x_2] = [0, 0]$$

which is equivalent to the set of equations

$$\begin{aligned}2x_1 - 2x_2 &= 0\\ 5x_1 - 5x_2 &= 0\end{aligned}$$

The solution to this system is $x_1 = x_2$, with x_2 arbitrary. The left eigenvectors corresponding to $\lambda = 1$ are thus $[x_1, x_2] = [x_2, x_2] = x_2[1, 1]$ with x_2 arbitrary.

For $\lambda = -2$, (*1*) reduces to

$$[x_1, x_2]\begin{bmatrix} 5 & 5 \\ -2 & -2 \end{bmatrix} = [0, 0]$$

or

$$[5x_1 - 2x_2, 5x_1 - 2x_2] = [0, 0]$$

which is equivalent to the set of equations

$$\begin{aligned}5x_1 - 2x_2 &= 0\\ 5x_1 - 2x_2 &= 0\end{aligned}$$

The solution to this system is $x_1 = \frac{2}{5}x_2$, with x_2 arbitrary. The left eigenvectors corresponding to $\lambda = -2$ are $[x_1, x_2] = [\frac{2}{5}x_2, x_2] = x_2[2/5, 1]$ with x_2 arbitrary.

7.14 Prove that the transpose of a right eigenvector of **A** is a left eigenvector of $\mathbf{A}^T$ corresponding to the same eigenvalue.

If **X** is a right eigenvector of **A** corresponding to the eigenvalue λ, then $\mathbf{AX} = \lambda\mathbf{X}$. Taking the transpose of both sides of this equation, we obtain $\mathbf{X}^T\mathbf{A}^T = \lambda\mathbf{X}^T$.

7.15 Verify the Cayley-Hamilton theorem for

$$\mathbf{A} = \begin{bmatrix} 3 & 5 \\ -2 & -4 \end{bmatrix}$$

The characteristic equation for **A** was determined in Problem 7.1 to be $\lambda^2 + \lambda - 2 = 0$. Substituting **A** for λ, we obtain

$$\mathbf{A}^2 + \mathbf{A} - 2\mathbf{I} = \begin{bmatrix} -1 & -5 \\ 2 & 6 \end{bmatrix} + \begin{bmatrix} 3 & 5 \\ -2 & -4 \end{bmatrix} - 2\begin{bmatrix} 1 & 0 \\ 0 & 1 \end{bmatrix} = \begin{bmatrix} 0 & 0 \\ 0 & 0 \end{bmatrix}$$

7.16 Verify the Cayley-Hamilton theorem for

$$\mathbf{A} = \begin{bmatrix} 5 & 2 & 2 \\ 3 & 6 & 3 \\ 6 & 6 & 9 \end{bmatrix}$$

The characteristic equation for **A** was found in Problem 7.2 to be $-\lambda^3 + 20\lambda^2 - 93\lambda + 126 = 0$. Therefore, we evaluate

$$-\mathbf{A}^3 + 20\mathbf{A}^2 - 93\mathbf{A} + 126\mathbf{I} = -\begin{bmatrix} 521 & 494 & 494 \\ 741 & 768 & 741 \\ 1{,}482 & 1{,}482 & 1{,}509 \end{bmatrix} + 20\begin{bmatrix} 43 & 34 & 34 \\ 51 & 60 & 51 \\ 102 & 102 & 111 \end{bmatrix} - 93\begin{bmatrix} 5 & 2 & 2 \\ 3 & 6 & 3 \\ 6 & 6 & 9 \end{bmatrix}$$

$$+ 126\begin{bmatrix} 1 & 0 & 1 \\ 0 & 1 & 0 \\ 0 & 0 & 1 \end{bmatrix} = \begin{bmatrix} 0 & 0 & 0 \\ 0 & 0 & 0 \\ 0 & 0 & 0 \end{bmatrix}$$

7.17 Prove the Cayley-Hamilton theorem.

We denote the characteristic polynomial of an $n \times n$ matrix **A** as

$$d(\lambda) = b_n\lambda^n + b_{n-1}\lambda^{n-1} + \cdots + b_2\lambda^2 + b_1\lambda + b_0 \quad (1)$$

and set

$$\mathbf{C} = \mathbf{A} - \lambda\mathbf{I} \quad (2)$$

Then

$$d(\lambda) = \det(\mathbf{A} - \lambda\mathbf{I}) = \det \mathbf{C} \quad (3)$$

Since **C** is an $n \times n$ matrix having first-degree polynomials in λ for its diagonal elements and scalars elsewhere, it follows that the cofactor matrix $\mathbf{C}^c$ associated with **C** (see Chapter 5) will have elements that are polynomials of degree $n-1$ or $n-2$ in λ. Elements on the main diagonal of $\mathbf{C}^c$ will be polynomials of degree $n-1$; all other elements will be polynomials of degree $n-2$. The same will be true of the transpose of this cofactor matrix; hence $(\mathbf{C}^c)^T$ may be written as the sum of products of distinct powers of λ and scalar matrices $\mathbf{M}_i$:

$$(\mathbf{C}^c)^T = \mathbf{M}_{n-1}\lambda^{n-1} + \mathbf{M}_{n-2}\lambda^{n-2} + \cdots + \mathbf{M}_1\lambda + \mathbf{M}_0 \quad (4)$$

where $\mathbf{M}_0, \mathbf{M}_1, \ldots, \mathbf{M}_{n-1}$ are all $n \times n$ scalar matrices.

It follows from Problem 5.20 and (*3*) that

$$\mathbf{C}(\mathbf{C}^c)^T = (\det \mathbf{C})\mathbf{I} = d(\lambda)\mathbf{I} \quad (5)$$

Using (*2*), we obtain

$$\mathbf{C}(\mathbf{C}^c)^T = (\mathbf{A} - \lambda\mathbf{I})(\mathbf{C}^c)^T = \mathbf{A}(\mathbf{C}^c)^T - \lambda(\mathbf{C}^c)^T$$

With (*5*), this yields

$$d(\lambda)\mathbf{I} = \mathbf{A}(\mathbf{C}^c)^T - \lambda(\mathbf{C}^c)^T \quad (6)$$

Substituting (*1*) and (*4*) into (*6*), we obtain

$$b_n\lambda^n\mathbf{I} + b_{n-1}\lambda^{n-1}\mathbf{I} + \cdots + b_1\lambda\mathbf{I} + b_0\mathbf{I} = \mathbf{AM}_{n-1}\lambda^{n-1} + \mathbf{AM}_{n-2}\lambda^{n-2} + \cdots + \mathbf{AM}_1\lambda + \mathbf{AM}_0$$

$$- \mathbf{M}_{n-1}\lambda^n - \mathbf{M}_{n-2}\lambda^{n-1} - \cdots - \mathbf{M}_1\lambda^2 - \mathbf{M}_0\lambda$$

Both sides of this matrix equation are polynomials in λ. Since two matrix polynomials are equal if and only if their corresponding coefficients are equal, it follows that

$$b_n\mathbf{I} = -\mathbf{M}_{n-1}$$
$$b_{n-1}\mathbf{I} = \mathbf{A}\mathbf{M}_{n-1} - \mathbf{M}_{n-2}$$
$$b_{n-2}\mathbf{I} = \mathbf{A}\mathbf{M}_{n-2} - \mathbf{M}_{n-3}$$
$$\cdots\cdots\cdots\cdots\cdots\cdots$$
$$b_1\mathbf{I} = \mathbf{A}\mathbf{M}_1 - \mathbf{M}_0$$
$$b_0\mathbf{I} = \mathbf{A}\mathbf{M}_0$$

Multiplying the first or these equations by $\mathbf{A}^n$, the second by $\mathbf{A}^{n-1}$, the third by $\mathbf{A}^{n-2}$, and so on (the last equation will be multiplied by $\mathbf{A}^0 = \mathbf{I}$) and then adding, we find that terms on the right side cancel, leaving

$$b_n\mathbf{A}^n + b_{n-1}\mathbf{A}^{n-1} + b_{n-2}\mathbf{A}^{n-2} + \cdots + b_1\mathbf{A} + b_0\mathbf{I} = \mathbf{0}$$

which is the Cayley-Hamilton theorem for **A** with characteristic polynomial given by (*1*).

Supplementary Problems

In Problems 7.18 through 7.26, find the eigenvalues and corresponding eigenvectors for the given matrix

7.18 $\begin{bmatrix} 1 & 3 \\ 1 & 3 \end{bmatrix}$ **7.19** $\begin{bmatrix} 5 & -4 \\ 4 & -3 \end{bmatrix}$ **7.20** $\begin{bmatrix} 5 & 8 \\ 1 & 7 \end{bmatrix}$

7.21 $\begin{bmatrix} -8 & 7 \\ -4 & 8 \end{bmatrix}$ **7.22** $\begin{bmatrix} 3 & 2 \\ -2 & -3 \end{bmatrix}$ **7.23** $\begin{bmatrix} 3 & 5 \\ -1 & 1 \end{bmatrix}$

7.24 $\begin{bmatrix} 1 & 1 & -1 \\ 0 & 2 & 1 \\ 0 & 0 & 3 \end{bmatrix}$ **7.25** $\begin{bmatrix} 0 & 1 & -1 \\ 0 & -1 & 1 \\ 0 & 0 & 0 \end{bmatrix}$ **7.26** $\begin{bmatrix} -2 & 2 & 3 \\ 2 & 1 & 6 \\ 3 & 6 & 6 \end{bmatrix}$

In Problems 7.27 through 7.34, find the eigenvalues and a maximal set of linearly independent eigenvectors for the given matrix

7.27 $\begin{bmatrix} 5 & 1 & 0 \\ 0 & 5 & 1 \\ 0 & 0 & 5 \end{bmatrix}$ **7.28** $\begin{bmatrix} 5 & 1 & 0 \\ 0 & 5 & 0 \\ 0 & 0 & 5 \end{bmatrix}$ **7.29** $\begin{bmatrix} 5 & 0 & 0 \\ 0 & 5 & 0 \\ 0 & 0 & 5 \end{bmatrix}$

7.30 $\begin{bmatrix} 3 & 1 & 0 & 0 \\ 0 & 3 & 0 & 0 \\ 0 & 0 & 3 & 1 \\ 0 & 0 & 0 & 3 \end{bmatrix}$ **7.31** $\begin{bmatrix} 3 & 1 & 0 & 0 \\ 0 & 3 & 1 & 0 \\ 0 & 0 & 3 & 1 \\ 0 & 0 & 0 & 3 \end{bmatrix}$ **7.32** $\begin{bmatrix} 4 & 3 & 0 & 0 \\ -2 & -1 & 0 & 0 \\ 0 & 0 & 4 & 3 \\ 0 & 0 & -3 & -2 \end{bmatrix}$

7.33 $\begin{bmatrix} 3 & 1 & 1 \\ 1 & 5 & 1 \\ 1 & 1 & 3 \end{bmatrix}$ **7.34** $\begin{bmatrix} 6 & 2 & -2 \\ 2 & 6 & -2 \\ -2 & -2 & 10 \end{bmatrix}$

In Problems 7.35 through 7.40, find the eigenvalues and a maximal set of linearly independent left eigenvectors for the given matrix

7.35 The matrix in Problem 7.18.

7.36 The matrix in Problem 7.19.

7.37 The matrix in Problem 7.20.

7.38 $\begin{bmatrix} 3 & 2 & -1 \\ 2 & 3 & -1 \\ -1 & -1 & 4 \end{bmatrix}$ **7.39** $\begin{bmatrix} 1 & 1 & -1 \\ 1 & 1 & -1 \\ -1 & -1 & 1 \end{bmatrix}$ **7.40** $\begin{bmatrix} 3 & -1 & 1 \\ -1 & 3 & -1 \\ 1 & -1 & 3 \end{bmatrix}$

7.41 Verify the Cayley-Hamilton theorem for the matrix in (*a*) Problem 7.18; (*b*) Problem 7.24; and (*c*) Problem 7.30.

7.42 Show that if λ is an eigenvalue of $\mathbf{A}$ with corresponding eigenvector $\mathbf{X}$, then $\mathbf{X}$ is also an eigenvector of $\mathbf{A}^2$ corresponding to λ^2.

7.43 Show that if λ is an eigenvalue of $\mathbf{A}$ with corresponding eigenvector $\mathbf{X}$, then for any scalar c, $\mathbf{X}$ is an eigenvector of $\mathbf{A} - c\mathbf{I}$ corresponding to the eigenvalue $\lambda - c$.

7.44 Prove that if $\mathbf{A}$ has order $n \times n$, then

$$\det(\mathbf{A} - \lambda\mathbf{I}) = (-1)^n\{\lambda^n - (\text{trace }\mathbf{A})\lambda^{n-1} + O(\lambda^{n-2})\}$$

where $O(\lambda^{n-2})$ denotes a polynomial in λ of degree $n-2$ or less.

7.45 Prove that the trace of a square matrix is equal to the sum of the eigenvalues of that matrix.

7.46 Prove that trace $(\mathbf{A} + \mathbf{B}) =$ trace $\mathbf{A}$ + trace $\mathbf{B}$ if $\mathbf{A}$ and $\mathbf{B}$ are square matrices of the same order.

7.47 Prove that trace $\mathbf{AB}$ = trace $\mathbf{BA}$ if $\mathbf{A}$ and $\mathbf{B}$ are square matrices of the same order.

7.48 Show that if $\mathbf{S}$ is an invertible matrix of the same order as $\mathbf{A}$ then trace$(\mathbf{S}^{-1}\mathbf{AS}) =$ trace $\mathbf{A}$.

7.49 Prove that the determinant of a square matrix equals the product of all the eigenvalues of that matrix.

7.50 Show that the $n \times n$ matrix

$$\mathbf{C} = \begin{bmatrix} 0 & 1 & 0 & \cdots & 0 & 0 \\ 0 & 0 & 1 & \cdots & 0 & 0 \\ 0 & 0 & 0 & \cdots & 0 & 0 \\ \cdots & \cdots & \cdots & \cdots & \cdots & \cdots \\ 0 & 0 & 0 & \cdots & 0 & 1 \\ -a_0 & -a_1 & -a_2 & \cdots & -a_{n-2} & -a_{n-1} \end{bmatrix}$$

has as its characteristic equation

$$(-1)^n(\lambda^n + a_{n-1}\lambda^{n-1} + a_{n-2}\lambda^{n-2} + \cdots + a_2\lambda^2 + a_1\lambda + a_0) = 0$$

The matrix $\mathbf{C}$ is called the *companion matrix* for this characteristic equation.

Chapter 8

Functions of Matrices

SEQUENCES AND SERIES OF MATRICES

A sequence $\{\mathbf{B}_k\}$ of matrices $\mathbf{B}_k = [b_{ij}^{(k)}]$, all of the same order, converges to a matrix $\mathbf{B} = [b_{ij}]$ if the elements $b_{ij}^{(k)}$ converge to b_{ij} for every i and j. The infinite series $\Sigma_{n=0}^{\infty} \mathbf{B}_n$ converges to $\mathbf{B}$ if the sequence of partial sums $\{\mathbf{S}_k = \Sigma_{n=0}^{k} \mathbf{B}_n\}$ converges to $\mathbf{B}$. (See Problem 8.1.)

WELL-DEFINED FUNCTIONS

If a function $f(z)$ of a complex variable z has a Maclaurin series expansion

$$f(z) = \sum_{n=0}^{\infty} a_n z^n$$

which converges for $|z| < R$, then the matrix series $\Sigma_{n=0}^{\infty} a_n \mathbf{A}^n$ converges, provided $\mathbf{A}$ is square and each of its eigenvalues has absolute value less than R. In such a case, $f(\mathbf{A})$ is defined as

$$f(\mathbf{A}) = \sum_{n=0}^{\infty} a_n \mathbf{A}^n$$

and is called a *well-defined function*. By convention, $\mathbf{A}^0 = \mathbf{I}$. (See Problems 8.2 and 8.3.)

Example 8.1

$$e^z = 1 + \frac{1}{1!} z + \frac{1}{2!} z^2 + \cdots = \sum_{j=0}^{\infty} \frac{1}{j!} z^j$$

converges for all values of z (that is, $R = \infty$). Since every eigenvalue λ of any square matrix satisfies the condition that $|\lambda| < \infty$,,

$$e^{\mathbf{A}} = \mathbf{I} + \frac{1}{1!}\mathbf{A} + \frac{1}{2!}\mathbf{A}^2 + \cdots = \sum_{j=1}^{\infty} \frac{1}{j!} \mathbf{A}^j$$

is well defined for all square matrices $\mathbf{A}$.

COMPUTING FUNCTIONS OF MATRICES

An infinite series expansion for $f(\mathbf{A})$ is not generally useful for computing the elements of $f(\mathbf{A})$. It follows (with some effort) from the Cayley-Hamilton theorem that every well-defined function of an $n \times n$ matrix $\mathbf{A}$ can be expressed as a polynomial of degree $n-1$ in $\mathbf{A}$. Thus,

$$f(\mathbf{A}) = a_{n-1}\mathbf{A}^{n-1} + a_{n-2}\mathbf{A}^{n-2} + \cdots + a_2\mathbf{A}^2 + a_1\mathbf{A} + a_0\mathbf{I} \tag{8.1}$$

where the scalars $a_{n-1}, a_{n-2}, \ldots, a_2, a_1, a_0$ are determined as follows:

STEP 8.1: Let

$$r(\lambda) = a_{n-1}\lambda^{n-1} + a_{n-2}\lambda^{n-2} + \cdots + a_2\lambda^2 + a_1\lambda + a_0$$

which is the right side of (*8.1*) with $\mathbf{A}^j$ replaced by λ^j $(j = 0, 1, \ldots, n-1)$.

STEP 8.2: For each distinct eigenvalue λ_i of $\mathbf{A}$, formulate the equation

$$f(\lambda_i) = r(\lambda_i) \tag{8.2}$$

STEP 8.3: If λ_i is an eigenvalue of multiplicity k, for $k > 1$, then formulate also the following equations, involving derivatives of $f(\lambda)$ and $r(\lambda)$ with respect to λ:

$$\begin{gathered} f'(\lambda)\big|_{\lambda=\lambda_i} = r'(\lambda)\big|_{\lambda=\lambda_i} \\ f''(\lambda)\big|_{\lambda=\lambda_i} = r''(\lambda)\big|_{\lambda=\lambda_i} \\ \cdots\cdots\cdots\cdots\cdots\cdots\cdots \\ f^{(k-1)}(\lambda)\big|_{\lambda=\lambda_i} = r^{(k-1)}(\lambda)\big|_{\lambda=\lambda_i} \end{gathered} \tag{8.3}$$

STEP 8.4: Solve the set of all equations obtained in Steps 8.2 and 8.3 for the unknown scalars $a_0, a_1, \ldots, a_{n-1}$.

Once the scalars determined in Step 8.4 are substituted into (*8.1*), $f(\mathbf{A})$ may be calculated. (See Problems 8.4 through 8.6.)

THE FUNCTION $e^{\mathbf{A}t}$

For any constant square matrix $\mathbf{A}$ and real variable t, the matrix function $e^{\mathbf{A}t}$ is computed by setting $\mathbf{B} = \mathbf{A}t$ and then calculating $e^{\mathbf{B}}$ as described in the preceding section. (See Problems 8.7 through 8.10.)

The eigenvalues of $\mathbf{B} = \mathbf{A}t$ are the eigenvalues of $\mathbf{A}$ multiplied by t (see Property 7.5). Note that (*8.3*) involves derivatives with respect to λ and not t; the correct sequence of steps is to first take the necessary derivatives of $f(\lambda)$ and $r(\lambda)$ with respect to λ and then substitute $\lambda = \lambda_i$. The reverse procedure—first substituting $\lambda = \lambda_i$ (a function of t) into (*8.2*) and then taking derivatives with respect to t—can give erroneous results.

DIFFERENTIATION AND INTEGRATION OF MATRICES

The derivative of $\mathbf{A} = [a_{ij}]$ is the matrix obtained by differentiating each element of $\mathbf{A}$; that is, $d\mathbf{A}/dt = [da_{ij}/dt]$. Similarly, the integral of $\mathbf{A}$, either definite or indefinite, is obtained by integrating each element of $\mathbf{A}$. Thus,

$$\int_a^b \mathbf{A}\,dt = \left[\int_a^b a_{ij}dt\right] \qquad \text{and} \qquad \int \mathbf{A}\,dt = \left[\int a_{ij}\,dt\right]$$

(See Problems 8.11 and 8.12.)

DIFFERENTIAL EQUATIONS

The initial-value matrix differential equation

$$\dot{\mathbf{X}}(t) = \mathbf{A}\mathbf{X}(t) + \mathbf{F}(t) \qquad \mathbf{X}(t_0) = \mathbf{C}$$

has the solution

$$\mathbf{X}(t) = e^{\mathbf{A}(t-t_0)}\mathbf{C} + e^{\mathbf{A}t}\int_{t_0}^{t} e^{-\mathbf{A}s}\mathbf{F}(s)\,ds \tag{8.4}$$

or, equivalently,

$$\mathbf{X}(t) = e^{\mathbf{A}(t-t_0)}\mathbf{C} + \int_{t_0}^{t} e^{\mathbf{A}(t-s)}\mathbf{F}(s)\,ds \tag{8.5}$$

If the differential equation is homogeneous [i.e., $\mathbf{F}(t) = \mathbf{0}$], then (*8.4*) and (*8.5*) reduce to $\mathbf{X}(t) = e^{\mathbf{A}(t-t_0)}\mathbf{C}$.

In (*8.4*) and (*8.5*), the matrices $e^{\mathbf{A}(t-t_0)}$, $e^{-\mathbf{A}s}$, and $e^{\mathbf{A}(t-s)}$ are easily computed from $e^{\mathbf{A}t}$ by replacing the variable t with $t-t_0$, $-s$, and $t-s$, respectively. Usually, $\mathbf{X}(t)$ is obtained more easily from (*8.5*) than from (*8.4*), because the former involves one fewer matrix multiplication. However, the integrals arising in (*8.5*) are generally more difficult to evaluate that those in (*8.4*). (See Problems 8.13 and 8.14.)

THE MATRIX EQUATION AX + XB = C

The equation $\mathbf{AX}+\mathbf{XB}=\mathbf{C}$, where $\mathbf{A}$, $\mathbf{B}$, and $\mathbf{C}$ denote constant square matrices of the same order, has a unique solution if and only if $\mathbf{A}$ and $\mathbf{B}$ have no eigenvalues in common. This unique solution is given by

$$\mathbf{X}=-\int_0^{\infty} e^{\mathbf{A}t}\mathbf{C}e^{\mathbf{B}t}\,dt \tag{8.6}$$

provided the integral exists (see Problem 8.15).

Example 8.2 For $\mathbf{A}=\mathbf{I}$ and $\mathbf{B}=\mathbf{0}$ the matrix equation has the unique solution $\mathbf{X}=\mathbf{C}$, but the integral (*8.6*) diverges.

Solved Problems

8.1 Determine $\lim_{k\to\infty}\mathbf{B}_k$ when

$$\mathbf{B}_k=\begin{bmatrix}\dfrac{1}{k^2} & \dfrac{2+k}{3+2k}\\[2ex] 5 & \left(1+\dfrac{2}{k}\right)^k\end{bmatrix}$$

Since $\quad \lim_{k\to\infty}\dfrac{1}{k^2}=0 \quad \lim_{k\to\infty}\dfrac{2+k}{3+2k}=\dfrac{1}{2} \quad \lim_{k\to\infty}5=5 \quad$ and $\quad \lim_{k\to\infty}\left(1+\dfrac{2}{k}\right)^k=e^2$

we have
$$\lim_{k\to\infty}\mathbf{B}_k=\begin{bmatrix}0 & 1/2\\ 5 & e^2\end{bmatrix}$$

8.2 For which matrices $\mathbf{A}$ is the function $\cos\mathbf{A}$ well defined?

The Maclaurin series for $\cos z$ is

$$\cos z = 1-\frac{z^2}{2!}+\frac{z^4}{4!}-\frac{z^6}{6!}\cdots=\sum_{n=0}^{\infty}\frac{(-1)^n z^{2n}}{(2n)!}$$

which converges for all values of z (that is, $R=\infty$). Every eigenvalue of any square matrix satisfies the condition that $|\lambda|<\infty$, so

$$\cos\mathbf{A}=\mathbf{I}-\frac{\mathbf{A}^2}{2!}+\frac{\mathbf{A}^4}{4!}-\frac{\mathbf{A}^6}{6!}\cdots=\sum_{n=0}^{\infty}\frac{(-1)^n\mathbf{A}^{2n}}{(2n)!}$$

is well defined for every square matrix $\mathbf{A}$.

8.3 Determine whether arctan **A** is well defined for

$$\mathbf{A} = \begin{bmatrix} 2 & 4 \\ 1 & 2 \end{bmatrix}$$

The Maclaurin series for arctan z is

$$\arctan z = z - \frac{z^3}{3} + \frac{z^5}{5} - \frac{z^7}{7} \cdots = \sum_{n=0}^{\infty} \frac{(-1)^n z^{2n+1}}{2n+1}$$

which converges for all values of z having absolute value less than 1. Therefore,

$$\arctan \mathbf{A} = \mathbf{A} - \frac{\mathbf{A}^3}{3} + \frac{\mathbf{A}^5}{5} - \frac{\mathbf{A}^7}{7} \cdots = \sum_{n=0}^{\infty} \frac{(-1)^n \mathbf{A}^{2n+1}}{2n+1}$$

is well defined for any square matrix whose eigenvalues are all less than 1 in absolute value. The given matrix **A** has eigenvalues $\lambda_1 = 0$ and $\lambda_2 = 4$. Since the second of these eigenvalues has absolute value greater than 1, arctan **A** is not defined for this matrix.

8.4 Find cos **A** for the matrix given in Problem 8.3.

From Problem 8.2 we know that cos **A** is well defined for all matrices. For this particular 2×2 matrix **A**, (*8.1*) becomes

$$\cos \mathbf{A} = a_1\mathbf{A} + a_0\mathbf{I} = \begin{bmatrix} 2a_1 + a_0 & 4a_1 \\ a_1 & 2a_1 + a_0 \end{bmatrix} \tag{1}$$

Now $f(\lambda) = \cos \lambda$, $r(\lambda) = a_1\lambda + a_0$, and the distinct eigenvalues of **A** are $\lambda_1 = 0$ and $\lambda_2 = 4$. Substituting these quantities into (*8.2*) once for each distinct eigenvalue, we formulate the two equations

$$\cos 0 = a_1(0) + a_0$$
$$\cos 4 = a_1(4) + a_0$$

Solving these equations for a_1 and a_0, we obtain $a_0 = \cos 0 = 1$ and $a_1 = (\cos 4 - 1)/4 = -0.413411$. Substituting these values into (*1*) and simplifying give us

$$\cos \mathbf{A} = \begin{bmatrix} 0.173178 & -1.653644 \\ -0.413411 & 0.173178 \end{bmatrix}$$

8.5 Find $e^{\mathbf{A}}$ for the matrix given in Problem 8.3.

It follows from Example 8.1 that $e^{\mathbf{A}}$ is defined for all matrices. For this particular 2×2 matrix **A**, (*8.1*) becomes

$$e^{\mathbf{A}} = a_1\mathbf{A} + a_0\mathbf{I} = \begin{bmatrix} 2a_1 + a_0 & 4a_1 \\ a_1 & 2a_1 + a_0 \end{bmatrix} \tag{1}$$

Now $f(\lambda) = e^{\lambda}$, $r(\lambda) = a_1\lambda + a_0$, and the distinct eigenvalues of **A** are $\lambda_1 = 0$ and $\lambda_2 = 4$. Substituting these quantities into (*8.2*) once for each eigenvalue, we formulate the two equations

$$e^0 = a_1(0) + a_0$$
$$e^4 = a_1(4) + a_0$$

Thus $a_0 = e^0 = 1$ and $a_1 = (e^4 - 1)/4 = 13.3995$. Substituting these values into (*1*) and simplifying give us

$$\cos \mathbf{A} = \begin{bmatrix} 27.7991 & 53.5982 \\ 13.3995 & 27.7991 \end{bmatrix}$$

8.6 Find sin **A** for

$$\mathbf{A} = \begin{bmatrix} -2 & 2 & 0 \\ 0 & -2 & 1 \\ 0 & 0 & -2 \end{bmatrix}$$

The Maclaurin series for sin z converges for all finite values of z, so sin **A** is well defined for all matrices. For the given matrix 3×3 (*8.1*) becomes

$$\sin \mathbf{A} = a_2\mathbf{A}^2 + a_1\mathbf{A} + a_0\mathbf{I} = a_2\begin{bmatrix} 4 & -8 & 2 \\ 0 & 4 & -4 \\ 0 & 0 & 4 \end{bmatrix} + a_1\begin{bmatrix} -2 & 2 & 0 \\ 0 & -2 & 1 \\ 0 & 0 & -2 \end{bmatrix} + a_0\begin{bmatrix} 1 & 0 & 0 \\ 0 & 1 & 0 \\ 0 & 0 & 1 \end{bmatrix}$$

$$= \begin{bmatrix} 4a_2 - 2a_1 + a_0 & -8a_2 + 2a_1 & 2a_2 \\ 0 & 4a_2 - 2a_1 + a_0 & -4a_2 + a_1 \\ 0 & 0 & 4a_2 - 2a_1 + a_0 \end{bmatrix} \qquad (1)$$

Matrix **A** has eigenvalue $\lambda = -2$ with multiplicity three, so we will have to use Step 8.3. We determine

$$f(\lambda) = \sin\lambda \qquad r(\lambda) = a_2\lambda^2 + a_1\lambda + a_0$$
$$f'(\lambda) = \cos\lambda \qquad r'(\lambda) = 2a_2\lambda + a_1$$
$$f''(\lambda) = -\sin\lambda \qquad r''(\lambda) = 2a_2$$

and write (*8.2*) and (*8.3*) as, respectively,

$$\sin(-2) = a_2(-2)^2 + a_1(-2) + a_0$$
$$\cos(-2) = 2a_2(-2) + a_1$$
$$-\sin(-2) = 2a_2$$

We thus obtain $a_2 = -\frac{1}{2}\sin(-2) = 0.454649$; $a_1 = \cos(-2) - 2\sin(-2) = 1.40245$; and $a_0 = 2\cos(-2) - \sin(-2) = 0.0770038$. Substituting these values into (*1*) and simplifying give us

$$\sin \mathbf{A} = \begin{bmatrix} -0.909297 & -0.832294 & 0.909297 \\ 0 & -0.909297 & -0.416147 \\ 0 & 0 & -0.909297 \end{bmatrix}$$

8.7 Find $e^{\mathbf{A}t}$ for

$$\mathbf{A} = \begin{bmatrix} 0 & 1 \\ -1 & 0 \end{bmatrix}$$

We set

$$\mathbf{B} = \mathbf{A}t = \begin{bmatrix} 0 & t \\ -t & 0 \end{bmatrix}$$

and compute $e^{\mathbf{B}}$. Since **B** is of order 2×2, (*8.1*) becomes

$$e^{\mathbf{B}} = a_1\mathbf{B} + a_0\mathbf{I} = \begin{bmatrix} a_0 & a_1 t \\ -a_1 t & a_0 \end{bmatrix} \qquad (1)$$

Here $f(\lambda) = e^{\lambda}$, $r(\lambda) = a_1\lambda + a_0$, and the distinct eigenvalues of **B** are $\lambda_1 = it$ and $\lambda_2 = -it$. Substituting these quantities into (*8.2*) separately for each eigenvalue, we obtain the two equations

$$e^{it} = a_1 it + a_0$$
$$e^{-it} = a_1(-it) + a_0$$

Solving these equations for a_1 and a_0, we obtain

$$a_1 = \frac{1}{2it}(e^{it} - e^{-it}) = \frac{\sin t}{t} \qquad \text{and} \qquad a_0 = \frac{1}{2}(e^{it} + e^{-it}) = \cos t$$

Substituting these values into (*1*), we determine

$$e^{\mathbf{A}t} = e^{\mathbf{B}} = \begin{bmatrix} \cos t & \sin t \\ -\sin t & \cos t \end{bmatrix}$$

8.8 Find $e^{\mathbf{A}t}$ for

$$\mathbf{A} = \begin{bmatrix} 0 & 1 \\ 8 & -2 \end{bmatrix}$$

We set

$$\mathbf{B} = \mathbf{A}t = \begin{bmatrix} 0 & t \\ 8t & -2t \end{bmatrix}$$

and compute $e^{\mathbf{B}}$. Since $\mathbf{B}$ is of order 2×2, (*8.1*) becomes

$$e^{\mathbf{B}} = a_1\mathbf{B} + a_0\mathbf{I} = \begin{bmatrix} a_0 & a_1 t \\ 8a_1 t & -2a_1 t + a_0 \end{bmatrix} \qquad (1)$$

Here $f(\lambda) = e^{\lambda}$, $r(\lambda) = a_1\lambda + a_0$, and the distinct eigenvalues of $\mathbf{B}$ are $\lambda_1 = 2t$ and $\lambda_2 = -4t$. Substituting these quantities into (*8.2*) once for each eigenvalue, we obtain the two equations

$$e^{2t} = a_1(2t) + a_0$$
$$e^{-4t} = a_1(-4t) + a_0$$

Solving these equations for a_1 and a_0, we obtain $a_1 = (e^{2t} - e^{-4t})/6t$ and $a_0 = (2e^{2t} + e^{-4t})/3$. Substituting these values into (*1*), we get

$$e^{\mathbf{A}t} = \frac{1}{6}\begin{bmatrix} 4e^{2t} + 2e^{-4t} & e^{2t} - e^{-4t} \\ 8e^{2t} - 8e^{-4t} & 2e^{2t} + 4e^{-4t} \end{bmatrix}$$

8.9 Find $e^{\mathbf{A}t}$ for

$$\mathbf{A} = \begin{bmatrix} 0 & 0 & 0 \\ 1 & 0 & 0 \\ 1 & 0 & 1 \end{bmatrix}$$

We set $\mathbf{B} = \mathbf{A}t$ and compute $e^{\mathbf{B}}$. Since $\mathbf{B}$ is a 3×3 matrix, (*8.1*) becomes

$$e^{\mathbf{B}} = a_2\mathbf{B}^2 + a_1\mathbf{B} + a_0\mathbf{I} = \begin{bmatrix} a_0 & 0 & 0 \\ a_1 t & a_0 & 0 \\ a_2t^2 + a_1 t & 0 & a_2t^2 + a_1t + a_0 \end{bmatrix} \qquad (1)$$

Now $f(\lambda) = e^{\lambda}$, $r(\lambda) = a_2\lambda^2 + a_1\lambda + a_0$, and the distinct eigenvalues of $\mathbf{B}$ are $\lambda_1 = 0$ with multiplicity two and $\lambda_2 = t$ with multiplicity one. Substituting these quantities, along with $f'(\lambda) = e^{\lambda}$ and $r'(\lambda) = 2a_2\lambda + a_1$, into (*8.2*) and (*8.3*), we formulate the three equations

$$e^0 = a_2(0)^2 + a_1(0) + a_0$$
$$e^0 = 2a_2(0) + a_1$$
$$e^t = a_2t^2 + a_1t + a_0$$

Thus, $a_2 = (e^t - t - 1)/t^2$, $a_1 = 1$, and $a_0 = 1$. Substituting these values into (*1*) and simplifying, we obtain

$$e^{\mathbf{A}t} = e^{\mathbf{B}} = \begin{bmatrix} 1 & 0 & 0 \\ t & 1 & 0 \\ e^t - 1 & 0 & e^t \end{bmatrix}$$

8.10 Establish the equations that are needed to find $e^{\mathbf{A}t}$ if

$$\mathbf{A} = \begin{bmatrix} 1 & 2 & 3 & 4 & 5 & 6 \\ 0 & 1 & 2 & 3 & 4 & 5 \\ 0 & 0 & 2 & 3 & 4 & 5 \\ 0 & 0 & 0 & 2 & 3 & 4 \\ 0 & 0 & 0 & 0 & 0 & 0 \\ 0 & 0 & 0 & 0 & 0 & 1 \end{bmatrix}$$

We set $\mathbf{B} = \mathbf{A}t$ and compute $e^{\mathbf{B}}$. Since $\mathbf{B}$ is a 6×6 matrix (*8.1*) becomes

$$e^{\mathbf{B}} = a_5\mathbf{B}^5 + a_4\mathbf{B}^4 + a_3\mathbf{B}^3 + a_2\mathbf{B}^2 + a_1\mathbf{B} + a_0\mathbf{I} \tag{1}$$

The distinct eigenvalues of $\mathbf{B}$ are $\lambda_1 = t$ with multiplicity three, $\lambda_2 = 2t$ with multiplicity two, and $\lambda_3 = 0$ with multiplicity one. We determine

$$\begin{aligned} f(\lambda) &= e^{\lambda} & r(\lambda) &= a_5\lambda^5 + a_4\lambda^4 + a_3\lambda^3 + a_2\lambda^2 + a_1\lambda + a_0 \\ f'(\lambda) &= e^{\lambda} & r'(\lambda) &= 5a_5\lambda^4 + 4a_4{}^3 + 3a_3\lambda^2 + 2a_2\lambda + a_1 \\ f''(\lambda) &= e^{\lambda} & r''(\lambda) &= 20a_5\lambda^3 + 12a_4\lambda^2 + 6a_3\lambda + 2a_2 \end{aligned}$$

and (*8.2*) and (*8.3*) become

$$\begin{aligned} e^t &= a_5t^5 + a_4t^4 + a_3t^3 + a_2t^2 + a_1t + a_0 \\ e^t &= 5a_5t^4 + 4a_4t^3 + 3a_3t^2 + 2a_2t + a_1 \\ e^t &= 20a_5t^3 + 12a_4t^2 + 6a_3t + 2a_2 \\ e^{2t} &= a_5(2t)^5 + a_4(2t)^4 + a_3(2t)^3 + a_2(2t)^2 + a_1(2t) + a_0 \\ e^{2t} &= 5a_5(2t)^4 + 4a_4(2t)^3 + 3a_3(2t)^2 + 2a_2(2t) + a_1 \\ e^0 &= a_5(0)^5 + a_4(0)^4 + a_3(0)^3 + a_2(0)^2 + a_1(0) + a_0 \end{aligned}$$

which should be simplified before they are solved.

8.11 Find $d\mathbf{A}/dt$ if

$$\mathbf{A} = \begin{bmatrix} t^2+1 & e^{2t} \\ \sin t & 45 \end{bmatrix}$$

$$\frac{d\mathbf{A}}{dt} = \begin{bmatrix} \dfrac{d}{dt}(t^2+1) & \dfrac{d}{dt}(e^{2t}) \\ \dfrac{d}{dt}(\sin t) & \dfrac{d}{dt}(45) \end{bmatrix} = \begin{bmatrix} 2t & 2e^{2t} \\ \cos t & 0 \end{bmatrix}$$

8.12 Find $\int \mathbf{A}\,dt$ for $\mathbf{A}$ as given in Problem 8.11.

$$\int \mathbf{A}\,dt = \begin{bmatrix} \displaystyle\int (t^2+1)\,dt & \displaystyle\int e^{2t}\,dt \\ \displaystyle\int \sin t\,dt & \displaystyle\int 45\,dt \end{bmatrix} = \begin{bmatrix} \frac{1}{3}t^3 + t + c_1 & \frac{1}{2}e^{2t} + c_2 \\ -\cos t + c_3 & 45t + c_4 \end{bmatrix}$$

8.13 Solve $\dot{\mathbf{X}}(t) = \mathbf{A}\mathbf{X}(t) + \mathbf{F}(t)$ with initial value $\mathbf{X}(0) = \mathbf{C}$ when

$$\mathbf{A} = \begin{bmatrix} 0 & 1 \\ 8 & -2 \end{bmatrix} \qquad \mathbf{F}(t) = \begin{bmatrix} 0 \\ e^t \end{bmatrix} \qquad \mathbf{C} = \begin{bmatrix} 1 \\ -4 \end{bmatrix}$$

The solution is given by either (*8.4*) or (*8.5*). We shall use (*8.4*) here, and (*8.5*) in Problem 8.14. For $\mathbf{A}$ as given, $e^{\mathbf{A}t}$ has already been calculated in Problem 8.8. Therefore, we can compute

$$e^{\mathbf{A}(t-t_0)}\mathbf{C} = e^{\mathbf{A}t}\mathbf{C} = \frac{1}{6}\begin{bmatrix} 4e^{2t}+2e^{-4t} & e^{2t}-e^{-4t} \\ 8e^{2t}-8e^{-4t} & 2e^{2t}+4e^{-4t} \end{bmatrix}\begin{bmatrix} 1 \\ -4 \end{bmatrix} = \begin{bmatrix} e^{-4t} \\ -4e^{-4t} \end{bmatrix}$$

$$e^{-\mathbf{A}s}\mathbf{F}(s) = \frac{1}{6}\begin{bmatrix} 4e^{-2s}+2e^{4s} & e^{-2s}-e^{4s} \\ 8e^{-2s}-8e^{4s} & 2e^{-2s}+4e^{4s} \end{bmatrix}\begin{bmatrix} 0 \\ e^{s} \end{bmatrix} = \begin{bmatrix} \frac{1}{6}e^{-s}-\frac{1}{6}e^{5s} \\ \frac{2}{6}e^{-s}+\frac{4}{6}e^{5s} \end{bmatrix}$$

$$\int_{t_0}^{t} e^{-\mathbf{A}s}\mathbf{F}(s)\,ds = \begin{bmatrix} \int_0^t (\frac{1}{6}e^{-s}-\frac{1}{6}e^{5s})\,ds \\ \int_0^t (\frac{1}{3}e^{-s}+\frac{2}{3}e^{5s})\,ds \end{bmatrix} = \frac{1}{30}\begin{bmatrix} -5e^{-t}-e^{5t}+6 \\ -10e^{-t}+4e^{5t}+6 \end{bmatrix}$$

$$e^{\mathbf{A}t}\int_{t_0}^{t} e^{-\mathbf{A}s}\mathbf{F}(s)\,ds = \frac{1}{6}\cdot\frac{1}{30}\begin{bmatrix} 4e^{2t}+2e^{-4t} & e^{2t}-e^{-4t} \\ 8e^{2t}-8e^{-4t} & 2e^{2t}+4e^{-4t} \end{bmatrix}\begin{bmatrix} -5e^{-t}-e^{5t}+6 \\ -10e^{-t}+4e^{5t}+6 \end{bmatrix}$$

$$= \frac{1}{180}\begin{bmatrix} (4e^{2t}+2e^{-4t})(-5e^{-t}-e^{5t}+6)+(e^{2t}-e^{-4t})(-10e^{-t}+4e^{5t}+6) \\ (8e^{2t}-8e^{-4t})(-5e^{-t}-e^{5t}+6)+(2e^{2t}+4e^{-4t})(-10e^{-t}+4e^{5t}+6) \end{bmatrix}$$

$$= \frac{1}{30}\begin{bmatrix} -6e^{t}+5e^{2t}+e^{-4t} \\ -6e^{t}+10e^{2t}-4e^{-4t} \end{bmatrix}$$

Thus

$$\mathbf{X}(t) = e^{\mathbf{A}(t-t_0)}\mathbf{C} + e^{\mathbf{A}t}\int_{t_0}^{t} e^{-\mathbf{A}s}\mathbf{F}(s)\,ds$$

$$= \begin{bmatrix} e^{-4t} \\ -4e^{-4t} \end{bmatrix} + \frac{1}{30}\begin{bmatrix} -6e^{t}+5e^{2t}+e^{-4t} \\ -6e^{t}+10e^{2t}-4e^{-4t} \end{bmatrix} = \begin{bmatrix} \frac{31}{30}e^{-4t}+\frac{1}{6}e^{2t}-\frac{1}{5}e^{t} \\ -\frac{62}{15}e^{-4t}+\frac{1}{3}e^{2t}-\frac{1}{5}e^{t} \end{bmatrix}$$

8.14 Use (*8.5*) to solve Problem 8.13.

The vector $e^{\mathbf{A}(t-t_0)}\mathbf{C}$ was found in Problem 8.13. Furthermore,

$$e^{\mathbf{A}(t-s)}\mathbf{F}(s) = \frac{1}{6}\begin{bmatrix} 4e^{2(t-s)}+2e^{-4(t-s)} & e^{2(t-s)}-e^{-4(t-s)} \\ 8e^{2(t-s)}-8e^{-4(t-s)} & 2e^{2(t-s)}+4e^{-4(t-s)} \end{bmatrix}\begin{bmatrix} 0 \\ e^{s} \end{bmatrix} = \frac{1}{6}\begin{bmatrix} e^{(2t-s)}-e^{(-4t+5s)} \\ 2e^{(2t-s)}+4e^{(-4t+5s)} \end{bmatrix}.$$

$$\int_{t_0}^{t} e^{\mathbf{A}(t-s)}\mathbf{F}(s)\,ds = \frac{1}{6}\begin{bmatrix} \int_0^t (e^{(2t-s)}-e^{(-4t+5s)})\,ds \\ \int_0^t (2e^{(2t-s)}+4e^{(-4t+5s)})\,ds \end{bmatrix}$$

$$= \frac{1}{6}\begin{bmatrix} (-e^{(2t-s)}-\frac{1}{5}e^{(-4t+5s)})_{s=0}^{s=t} \\ -2e^{(2t-s)}+\frac{4}{5}e^{(-4t+5s)})_{s=0}^{s=t} \end{bmatrix} = \frac{1}{6}\begin{bmatrix} -\frac{6}{5}e^{t}+e^{2t}+\frac{1}{5}e^{-4t} \\ -\frac{6}{5}e^{t}+2e^{2t}-\frac{4}{5}e^{-4t} \end{bmatrix}$$

Thus,

$$\mathbf{X}(t) = e^{\mathbf{A}(t-t_0)}\mathbf{C} + \int_{t_0}^{t} e^{\mathbf{A}(t-s)}\mathbf{F}(s)\,ds$$

$$= \begin{bmatrix} e^{-4t} \\ -4e^{-4t} \end{bmatrix} + \frac{1}{6}\begin{bmatrix} -\frac{6}{5}e^{t}+e^{2t}+\frac{1}{5}e^{-4t} \\ -\frac{6}{5}e^{t}+2e^{2t}-\frac{4}{5}e^{-4t} \end{bmatrix} = \begin{bmatrix} \frac{31}{30}e^{-4t}+\frac{1}{6}e^{2t}-\frac{1}{5}e^{t} \\ -\frac{62}{15}e^{-4t}+\frac{1}{3}e^{2t}-\frac{1}{5}e^{t} \end{bmatrix}$$

as before.

8.15 Solve the matrix equation $\mathbf{AX}+\mathbf{XB}=\mathbf{C}$ for $\mathbf{X}$ when

$$\mathbf{A}=\begin{bmatrix}-6 & 0\\ 0 & -8\end{bmatrix}\qquad \mathbf{B}=\begin{bmatrix}3 & 2\\ 4 & 1\end{bmatrix}\qquad \mathbf{C}=\begin{bmatrix}2 & 0\\ -1 & 0\end{bmatrix}$$

In preparation for the use of (*8.6*), we calculate

$$e^{\mathbf{A}t}=\begin{bmatrix}e^{-6t} & 0\\ 0 & e^{-8t}\end{bmatrix}\qquad\text{and}\qquad e^{\mathbf{B}t}=\begin{bmatrix}\frac{2}{3}e^{5t}+\frac{1}{3}e^{-t} & \frac{1}{3}e^{5t}-\frac{1}{3}e^{-t}\\ \frac{2}{3}e^{5t}-\frac{2}{3}e^{-t} & \frac{1}{3}e^{5t}+\frac{2}{3}e^{-t}\end{bmatrix}$$

Then

$$e^{\mathbf{A}t}\mathbf{C}e^{\mathbf{B}t}=\begin{bmatrix}\frac{4}{3}e^{-t}+\frac{2}{3}e^{-7t} & \frac{2}{3}e^{-t}-\frac{2}{3}e^{-7t}\\ -\frac{2}{3}e^{-3t}-\frac{1}{3}e^{-9t} & -\frac{1}{3}e^{-3t}+\frac{1}{3}e^{-9t}\end{bmatrix}$$

and

$$\mathbf{X}=-\int_0^\infty e^{\mathbf{A}t}\mathbf{C}e^{\mathbf{B}t}dt=-\begin{bmatrix}-\frac{4}{3}e^{-t}-\frac{2}{21}e^{-7t}\Big|_0^\infty & -\frac{2}{3}e^{-t}+\frac{2}{21}e^{-7t}\Big|_0^\infty\\ \frac{2}{9}e^{-3t}+\frac{1}{27}e^{-9t}\Big|_0^\infty & \frac{1}{9}e^{-3t}-\frac{1}{27}e^{-9t}\Big|_0^\infty\end{bmatrix}=\begin{bmatrix}-10/7 & -4/7\\ 7/27 & 2/27\end{bmatrix}$$

8.16 Prove that $e^{\mathbf{A}t}e^{\mathbf{B}t}=e^{(\mathbf{A}+\mathbf{B})t}$ if and only if the matrices **A** and **B** commute (that is, if and only if the commutative property for multiplication holds for **A** and **B**).

If $\mathbf{AB}=\mathbf{BA}$, and only then, we have

$$(\mathbf{A}+\mathbf{B})^2=(\mathbf{A}+\mathbf{B})(\mathbf{A}+\mathbf{B})=\mathbf{A}^2+\mathbf{AB}+\mathbf{BA}+\mathbf{B}^2=\mathbf{A}^2+2\mathbf{AB}+\mathbf{B}^2$$

$$=\sum_{k=0}^{2}\binom{2}{k}\mathbf{A}^{n-k}\mathbf{B}^k$$

and, in general,

$$(\mathbf{A}+\mathbf{B})^n=\sum_{k=0}^{n}\binom{n}{k}\mathbf{A}^{n-k}\mathbf{B}^k \tag{1}$$

where

$$\binom{n}{k}=\frac{n!}{k!(n-k)!}$$

is the binomial coefficient ("n things taken k at a time").

Now according to the defining equation, we have for any **A** and **B**:

$$e^{\mathbf{A}t}e^{\mathbf{B}t}=\left(\sum_{n=0}^{\infty}\frac{1}{n!}\mathbf{A}^nt^n\right)\left(\sum_{n=0}^{\infty}\frac{1}{n!}\mathbf{B}^nt^n\right)=\sum_{n=0}^{\infty}\sum_{k=0}^{n}\frac{\mathbf{A}^{n-k}t^{n-k}}{(n-k)!}\frac{\mathbf{B}^kt^k}{k!}$$

$$=\sum_{n=0}^{\infty}\left[\sum_{k=0}^{n}\frac{\mathbf{A}^{n-k}\mathbf{B}^k}{(n-k)!k!}\right]t^n=\sum_{n=0}^{\infty}\left[\sum_{k=0}^{n}\binom{n}{k}\mathbf{A}^{n-k}\mathbf{B}^k\right]\frac{t^n}{n!} \tag{2}$$

and

$$e^{(\mathbf{A}+\mathbf{B})t}=\sum_{n=0}^{\infty}\frac{1}{n!}(\mathbf{A}+\mathbf{B})^nt^n=\sum_{n=0}^{\infty}(\mathbf{A}+\mathbf{B})^n\frac{t^n}{n!} \tag{3}$$

The last series in (*3*) is equal to the last series in (*2*) if and only if (*1*) holds; that is, if and only if **A** and **B** commute.

8.17 Prove that $e^{\mathbf{A}t}e^{-\mathbf{A}s}=e^{\mathbf{A}(t-s)}$.

Setting $t=1$ in Problem 8.16, we conclude that $e^{\mathbf{A}}e^{\mathbf{B}}=e^{(\mathbf{A}+\mathbf{B})}$ if **A** and **B** commute. But the matrices $\mathbf{A}t$ and $-\mathbf{A}s$ commute, since

$$(\mathbf{A}t)(-\mathbf{A}s)=(\mathbf{AA})(-ts)=(\mathbf{AA})(-st)=(-\mathbf{A}s)(\mathbf{A}t)$$

Consequently, $e^{\mathbf{A}t}e^{-\mathbf{A}s}=s^{(\mathbf{A}t-\mathbf{A}s)}=e^{\mathbf{A}(t-s)}$.

8.18 Prove that $e^0 = I$.

From the definition of matrix multiplication, $\mathbf{0}^n = \mathbf{0}$ for $n \geq 1$. Hence,

$$e^{\mathbf{0}} = e^{\mathbf{0}t} = \sum_{n=0}^{\infty} \frac{1}{n!} \mathbf{0}^n t^n = \mathbf{I} + \sum_{n=1}^{\infty} \frac{1}{n!} \mathbf{0}^n t^n = \mathbf{I} + \mathbf{0} = \mathbf{I}$$

Supplementary Problems

8.19 Determine the limit of each of the following sequences of matrices as k goes to ∞:

$$\mathbf{A}_k = \begin{bmatrix} 1 - \dfrac{1}{k} & \dfrac{2k-1}{3k+1} \\ \dfrac{2}{k+1} & \dfrac{k-1}{k+1} \end{bmatrix} \qquad \mathbf{B}_k = \frac{1}{k^2}\begin{bmatrix} k^2+1 & k+1 & 1 \\ 2k^2 & k^2+k & 2k \end{bmatrix}$$

$$\mathbf{C}_k = \left[5, \frac{5}{k}, \frac{k-k^2}{k+1}\right]$$

8.20 The *Bessel function of the first kind of order zero* is defined as

$$J_0(x) = \sum_{k=1}^{\infty} \frac{(-1)^k x^{2k}}{2^{2k}(k!)^2}$$

For which matrices $\mathbf{A}$ is $J_0(\mathbf{A})$ well defined?

8.21 Determine the conditions on matrix $\mathbf{A}$ that will make the following function well defined:

$$f(\mathbf{A}) = \sum_{k=0}^{\infty} \frac{1}{4^k(k+1)} \mathbf{A}^k$$

8.22 Find (*a*) $\sin \mathbf{A}$ and (*b*) $e^{\mathbf{A}}$ for $\mathbf{A} = \begin{bmatrix} 3 & 2 \\ 4 & 1 \end{bmatrix}$.

8.23 Find (*a*) $\cos \mathbf{A}$ and (*b*) $3\mathbf{A}^{57} + 2\mathbf{A}^{18}$ for $\mathbf{A} = \begin{bmatrix} 3 & -2 \\ 4 & -3 \end{bmatrix}$.

8.24 Find (*a*) $\sin \mathbf{A}$ and (*b*) $\cos \mathbf{A}$ for the 3×3 zero matrix

In Problems *8.25* through *8.31*, find $e^{\mathbf{A}t}$

8.25 $\begin{bmatrix} -2 & 0 \\ 0 & -3 \end{bmatrix}$ **8.26** $\begin{bmatrix} 5 & 6 \\ -4 & -5 \end{bmatrix}$ **8.27** $\begin{bmatrix} 2 & 1 \\ 0 & 2 \end{bmatrix}$

8.28 $\begin{bmatrix} 4 & 5 \\ -4 & -4 \end{bmatrix}$ **8.29** $\begin{bmatrix} 2 & 1 & 0 \\ 0 & 2 & 1 \\ 0 & 0 & 2 \end{bmatrix}$ **8.30** $\begin{bmatrix} 2 & 0 & 0 \\ 0 & 2 & 1 \\ 0 & 0 & 2 \end{bmatrix}$

8.31 $\begin{bmatrix} -1 & 1 & 0 \\ 0 & 2 & 1 \\ 0 & 0 & 2 \end{bmatrix}$

8.32 Find sin $\mathbf{A}t$ for

$$\mathbf{A} = \begin{bmatrix} 2 & 0 & 0 & 0 \\ 0 & 2 & 1 & 0 \\ 0 & 0 & 2 & 0 \\ 0 & 0 & 0 & 3 \end{bmatrix}$$

8.33 Solve $\dot{\mathbf{X}}(t) = \mathbf{A}\mathbf{X}(t) + \mathbf{F}(t)$; $\mathbf{X}(0) = \mathbf{C}$ when

$$\mathbf{A} = \begin{bmatrix} 0 & 1 \\ 8 & -2 \end{bmatrix} \qquad \mathbf{F}(t) = \mathbf{0} \qquad \mathbf{C} = \begin{bmatrix} 2 \\ 3 \end{bmatrix}$$

8.34 Solve $\dot{\mathbf{X}}(t) = \mathbf{A}\mathbf{X}(t) + \mathbf{F}(t)$ when

$$\mathbf{A} = \begin{bmatrix} 0 & 1 \\ -9 & 6 \end{bmatrix} \qquad \mathbf{F}(t) = \begin{bmatrix} 0 \\ t \end{bmatrix}$$

8.35 Solve $\dot{\mathbf{X}}(t) = \mathbf{A}\mathbf{X}(t) + \mathbf{F}(t)$; $\mathbf{X}(0) = \mathbf{C}$ when

$$\mathbf{A} = \begin{bmatrix} 0 & 1 \\ 8 & -2 \end{bmatrix} \qquad \mathbf{F}(t) = \begin{bmatrix} 0 \\ 4 \end{bmatrix} \qquad \mathbf{C} = \mathbf{0}$$

8.36 Solve Problem 8.35 with $\mathbf{C} = \begin{bmatrix} 1 \\ 2 \end{bmatrix}$.

8.37 Solve $\mathbf{AX} + \mathbf{XB} = \mathbf{C}$ when

$$\mathbf{A} = \begin{bmatrix} -2 & 0 \\ 0 & -3 \end{bmatrix} \qquad \mathbf{B} = \begin{bmatrix} 5 & 6 \\ -4 & -5 \end{bmatrix} \qquad \mathbf{C} = \begin{bmatrix} 1 & 1 \\ 1 & 1 \end{bmatrix}$$

Chapter 9

Canonical Bases

GENERALIZED EIGENVECTORS

A vector $\mathbf{X}_m$ is a *generalized* (*right*) *eigenvector of rank m* for the square matrix $\mathbf{A}$ and associated eigenvector λ if

$$(\mathbf{A}-\lambda\mathbf{I})^m\mathbf{X}_m=\mathbf{0} \qquad \text{but} \qquad (\mathbf{A}-\lambda\mathbf{I})^{m-1}\mathbf{X}_m\neq\mathbf{0}$$

(See Problem 9.1 through 9.4.) Right eigenvectors, as defined in Chapter 7, are generalized eigenvectors of rank 1.

CHAINS

A *chain* generated by a generalized eigenvector $\mathbf{X}_m$ of rank m associated with the eigenvalue λ is a set of vectors $\{\mathbf{X}_m, \mathbf{X}_{m-1}, \mathbf{X}_{m-2}, \ldots, X_1\}$ defined recursively as

$$\mathbf{X}_j=(\mathbf{A}-\lambda\mathbf{I})\mathbf{X}_{j+1} \qquad (j=m-1, m-2, \ldots, 1) \tag{9.1}$$

(See Problems 9.5 and 9.6.) A chain is a linearly independent set of generalized eigenvectors of descending rank. The number of vectors in the set is called the *length* of the chain.

CANONICAL BASIS

A *canonical basis* for an $n \times n$ matrix $\mathbf{A}$ is a set of n linearly independent generalized eigenvectors composed entirely of chains. That is, if a generalized eigenvector of rank m appears in the basis, so too does the complete chain generated by that vector.

The simplest canonical bases, when they exist, are those consisting solely of chains of length one (i.e., of linearly independent eigenvectors). Such bases always exist when the eigenvalues of a matrix are distinct. (See Problem 9.9.) The chains associated with an eigenvalue of multiplicity greater than one are determined with the following algorithm, which first establishes the number of generalized eigenvectors of each rank that will appear in a canonical basis and then provides a means for obtaining them:

STEP 9.1: Denote the multiplicity of λ as m, and determine the smallest positive integer p for which the rank of $(\mathbf{A}-\lambda\mathbf{I})^p$ equals $n-m$, where n denotes the number of rows (and columns) in $\mathbf{A}$.

STEP 9.2: For each integer k between 1 and p, inclusive, compute the *eigenvalue rank number* N_k as

$$N_k=\operatorname{rank}(\mathbf{A}-\lambda\mathbf{I})^{k-1}-\operatorname{rank}(\mathbf{A}-\lambda\mathbf{I})^k \tag{9.2}$$

Each N_k is the number of generalized eigenvectors of rank k that will appear in the canonical basis.

STEP 9.3: Determine a generalized eigenvector of rank p, and construct the chain generated by this vector. Each of these vectors is part of the canonical basis.

STEP 9.4: Reduce each positive N_k $(k=1, 2, \ldots, p)$ by 1. If all N_k are zero, stop; the procedure is completed. If not, continue to Step. 9.5.

STEP 9.5: Find the highest value of k for which N_k is not zero, and determine a generalized eigenvector of that rank which is linearly independent of all previously determined generalized eigenvectors associated with λ. Form the chain generated by this vector, and include it in the basis. Return to Step 9.4.

(See Problems 9.10 through 9.13.)

THE MINIMUM POLYNOMIAL

The *minimum polynomial* $m(\lambda)$ for an $n \times n$ matrix $\mathbf{A}$ is the monic polynomial of least degree for which $m(\mathbf{A}) = \mathbf{0}$. Designate the distinct eigenvalues of $\mathbf{A}$ as $\lambda_1, \lambda_2, \ldots, \lambda_s$ $(1 \le s \le n)$, and for each λ_i determine a p_i as in Step 9.1 above. The minimum polynomial for $\mathbf{A}$ is then

$$m(\lambda) = (\lambda - \lambda_1)^{p_1}(\lambda - \lambda_2)^{p_2} \cdots (\lambda - \lambda_s)^{p_s} \tag{9.3}$$

(See Problems 9.14 and 9.15.)

Solved Problems

9.1 Show that $\mathbf{X} = [1, 0, 0]^T$ is a generalized eigenvector of rank 2 corresponding to the eigenvalue $\lambda = 3$ for the matrix

$$\mathbf{A} = \begin{bmatrix} -7 & -25 & 1 \\ 4 & 13 & 1 \\ 0 & 0 & 2 \end{bmatrix}$$

$$\mathbf{A} - 3\mathbf{I} = \begin{bmatrix} -10 & -25 & 1 \\ 4 & 10 & 1 \\ 0 & 0 & -1 \end{bmatrix} \quad \text{and} \quad (\mathbf{A} - 3\mathbf{I})^2 = \begin{bmatrix} 0 & 0 & -36 \\ 0 & 0 & 13 \\ 0 & 0 & 1 \end{bmatrix}$$

For $\mathbf{X} = [1, 0, 0]^T$, we have $(\mathbf{A} - 3\mathbf{I})\mathbf{X} = [-10, 4, 0]^T \neq \mathbf{0}$ and $(\mathbf{A} - 3\mathbf{I})^2\mathbf{X} = \mathbf{0}$, which implies that $\mathbf{X}$ is a generalized eigenvector of rank 2.

9.2 Find a generalized eigenvector of rank 3 corresponding to the eigenvalue $\lambda = 7$ for the matrix

$$\mathbf{A} = \begin{bmatrix} 7 & 1 & 2 \\ 0 & 7 & 1 \\ 0 & 0 & 7 \end{bmatrix}$$

We seek a three-dimensional vector $\mathbf{X}_3 = [x_1, x_2, x_3]^T$ such that $(\mathbf{A} - 7\mathbf{I})^3\mathbf{X}_3 = \mathbf{0}$ and $(\mathbf{A} - 7\mathbf{I})^2\mathbf{X}_3 \neq \mathbf{0}$. We have

$$(\mathbf{A} - 7\mathbf{I})^3\mathbf{X}_3 = \begin{bmatrix} 0 & 0 & 0 \\ 0 & 0 & 0 \\ 0 & 0 & 0 \end{bmatrix}\begin{bmatrix} x_1 \\ x_2 \\ x_3 \end{bmatrix} = \begin{bmatrix} 0 \\ 0 \\ 0 \end{bmatrix}$$

and

$$(\mathbf{A} - 7\mathbf{I})^2\mathbf{X}_3 = \begin{bmatrix} 0 & 0 & 1 \\ 0 & 0 & 0 \\ 0 & 0 & 0 \end{bmatrix}\begin{bmatrix} x_1 \\ x_2 \\ x_3 \end{bmatrix} = \begin{bmatrix} x_3 \\ 0 \\ 0 \end{bmatrix}$$

The condition $(\mathbf{A} - 7\mathbf{I})^3\mathbf{X}_3 = \mathbf{0}$ is automatically satisfied; the condition $(\mathbf{A} - 7\mathbf{I})^2\mathbf{X}_3 \neq \mathbf{0}$ is satisfied only if $x_3 \neq 0$. Thus, x_1 and x_2 are arbitrary, whereas x_3 is constrained to be nonzero. A simple choice is $x_1 = x_2 = 0$, $x_3 = 1$, yielding $\mathbf{X}_3 = [0, 0, 1]^T$.

9.3 Find a generalized eigenvector of rank 2 corresponding to the eigenvalue $\lambda = 4$ for the matrix

$$\mathbf{A} = \begin{bmatrix} 4 & 0 & 0 & 0 \\ 1 & 5 & 1 & 0 \\ -1 & -1 & 3 & 0 \\ 0 & 0 & 0 & 3 \end{bmatrix}$$

We seek a four-dimensional vector $\mathbf{X}_2 = [x_1, x_2, x_3, x_4]^T$ such that $(\mathbf{A} - 4\mathbf{I})^2\mathbf{X}_2 = \mathbf{0}$ and $(\mathbf{A} - 4\mathbf{I})\mathbf{X}_2 \neq \mathbf{0}$. We have

$$(\mathbf{A} - 4\mathbf{I})^2\mathbf{X}_2 = \begin{bmatrix} 0 & 0 & 0 & 0 \\ 0 & 0 & 0 & 0 \\ 0 & 0 & 0 & 0 \\ 0 & 0 & 0 & 1 \end{bmatrix}\begin{bmatrix} x_1 \\ x_2 \\ x_3 \\ x_4 \end{bmatrix} = \begin{bmatrix} 0 \\ 0 \\ 0 \\ x_4 \end{bmatrix}$$

and
$$(\mathbf{A} - 4\mathbf{I})\mathbf{X}_2 = \begin{bmatrix} 0 & 0 & 0 & 0 \\ 1 & 1 & 1 & 0 \\ -1 & -1 & -1 & 0 \\ 0 & 0 & 0 & -1 \end{bmatrix}\begin{bmatrix} x_1 \\ x_2 \\ x_3 \\ x_4 \end{bmatrix} = \begin{bmatrix} 0 \\ x_1 + x_2 + x_3 \\ -x_1 - x_2 - x_3 \\ -x_4 \end{bmatrix}$$

To satisfy $(\mathbf{A} - 4\mathbf{I})^2\mathbf{X}_2 = \mathbf{0}$, we must have $x_4 = 0$. Then, to satisfy $(\mathbf{A} - 4\mathbf{I})\mathbf{X}_2 \neq \mathbf{0}$, we must guarantee $x_1 \neq -x_2 - x_3$. A simple choice is $x_1 = 1$, $x_2 = x_3 = x_4 = 0$. This gives us $\mathbf{X}_2 = [1, 0, 0, 0]^T$.

9.4 Show that there is no generalized eigenvector of rank 3 corresponding to the eigenvalue $\lambda = 4$ for the matrix given in Problem 9.3.

For such a vector $\mathbf{X}_3 = [x_1, x_2, x_3, x_4]^T$ to exist, the conditions $(\mathbf{A} - 4\mathbf{I})^3\mathbf{X}_3 = \mathbf{0}$ and $(\mathbf{A} - 4\mathbf{I})^2\mathbf{X}_3 \neq \mathbf{0}$ must be satisfied. For the given matrix **A**,

$$(\mathbf{A} - 4\mathbf{I})^3\mathbf{X}_3 = [0, 0, 0, -x_4]^T \qquad \text{while} \qquad (\mathbf{A} - 4\mathbf{I})^2\mathbf{X}_3 = [0, 0, 0, x_4]^T$$

To satisfy both conditions, x_4 must be zero and nonzero simultaneously, which is impossible. Therefore, **A** has no generalized eigenvector of rank 3 corresponding to $\lambda = 4$.

9.5 Determine the chain that is generated by the generalized eigenvector of rank 3 found in Problem 9.2.

From Problem 9.2, we have $\mathbf{X}_3 = [0, 0, 1]^T$ corresponding to the eigenvalue $\lambda = 7$. Furthermore,

$$\mathbf{A} - 7\mathbf{I} = \begin{bmatrix} 0 & 1 & 2 \\ 0 & 0 & 1 \\ 0 & 0 & 0 \end{bmatrix}$$

It follows from (*9.1*) that

$$\mathbf{X}_2 = (\mathbf{A} - 7\mathbf{I})\mathbf{X}_3 = \begin{bmatrix} 0 & 1 & 2 \\ 0 & 0 & 1 \\ 0 & 0 & 0 \end{bmatrix}\begin{bmatrix} 0 \\ 0 \\ 1 \end{bmatrix} = \begin{bmatrix} 2 \\ 1 \\ 0 \end{bmatrix}$$

and
$$\mathbf{X}_1 = (\mathbf{A} - 7\mathbf{I})\mathbf{X}_2 = \begin{bmatrix} 0 & 1 & 2 \\ 0 & 0 & 1 \\ 0 & 0 & 0 \end{bmatrix}\begin{bmatrix} 2 \\ 1 \\ 0 \end{bmatrix} = \begin{bmatrix} 1 \\ 0 \\ 0 \end{bmatrix}$$

The chain is

$$\{\mathbf{X}_3, \mathbf{X}_2, \mathbf{X}_1\} = \left\{\begin{bmatrix} 0 \\ 0 \\ 1 \end{bmatrix}, \begin{bmatrix} 2 \\ 1 \\ 0 \end{bmatrix}, \begin{bmatrix} 1 \\ 0 \\ 0 \end{bmatrix}\right\}$$

9.6 Determine the chain that is generated by the generalized eigenvector of rank 2 found in Problem 9.3.

From Problem 9.3 we have $\mathbf{X}_2 = [1, 0, 0, 0]^T$, corresponding to $\lambda = 4$. Using (*9.1*), we write

$$\mathbf{X}_1 = (\mathbf{A} - 4\mathbf{I})\mathbf{X}_2 = \begin{bmatrix} 0 & 0 & 0 & 0 \\ 1 & 1 & 1 & 0 \\ -1 & -1 & -1 & 0 \\ 0 & 0 & 0 & -1 \end{bmatrix} \begin{bmatrix} 1 \\ 0 \\ 0 \\ 0 \end{bmatrix} = \begin{bmatrix} 0 \\ 1 \\ -1 \\ 0 \end{bmatrix}$$

The chain is $\{\mathbf{X}_2, \mathbf{X}_1\} = \{[1, 0, 0, 0]^T, [0, 1, -1, 0]^T\}$.

9.7 Show that if $\mathbf{X}_m$ is a generalized eigenvector of rank m for matrix $\mathbf{A}$ and eigenvalue λ, then $\mathbf{X}_j$ as defined by (*9.1*) is a generalized eigenvector of rank j corresponding to the same matrix and eigenvalue.

Since $\mathbf{X}_m$ is a generalized eigenvector of rank m,

$$(\mathbf{A} - \lambda\mathbf{I})^m \mathbf{X}_m = \mathbf{0} \qquad \text{and} \qquad (\mathbf{A} - \lambda\mathbf{I})^{m-1}\mathbf{X}_m \neq \mathbf{0}$$

It follows from Eq. (*9.1*) that

$$\mathbf{X}_j = (\mathbf{A} - \lambda\mathbf{I})\mathbf{X}_{j+1} = (\mathbf{A} - \lambda\mathbf{I})^{m-j}\mathbf{X}_m$$

Therefore
$$(\mathbf{A} - \lambda\mathbf{I})^j\mathbf{X}_j = (\mathbf{A} - \lambda\mathbf{I})^j(\mathbf{A} - \lambda\mathbf{I})^{m-j}\mathbf{X}_m = (\mathbf{A} - \lambda\mathbf{I})^m\mathbf{X}_m = \mathbf{0}$$

and
$$(\mathbf{A} - \lambda\mathbf{I})^{j-1}\mathbf{X}_j = (\mathbf{A} - \lambda\mathbf{I})^{j-1}(\mathbf{A} - \lambda\mathbf{I})^{m-j}\mathbf{X}_m = (\mathbf{A} - \lambda\mathbf{I})^{m-1}\mathbf{X}_m \neq \mathbf{0}$$

which together imply that $\mathbf{X}_j$ is a generalized eigenvector of rank j for $\mathbf{A}$ and λ.

9.8 Show that a chain is a linearly independent set of vectors.

The proof is inductive on the length of the chain. For chains of length one, the generating generalized eigenvector $\mathbf{X}_1$ must be an eigenvector, so $\mathbf{X}_1 \neq \mathbf{0}$. Therefore, the only solution to the vector equation $c_1 X_1 = \mathbf{0}$ is $c_1 = 0$, and the chain is independent.

Assume that all chains containing exactly $k - 1$ vectors are linearly independent, and consider a chain consisting of the k-vector set $\{\mathbf{X}_k, \mathbf{X}_{k-1}, \ldots, \mathbf{X}_1\}$ for matrix $\mathbf{A}$ and eigenvalue λ. We must show that the only solution to the vector equation

$$c_k\mathbf{X}_k + c_{k-1}\mathbf{X}_{k-1} + \cdots + c_1\mathbf{X}_1 = \mathbf{0} \tag{1}$$

is $c_k = c_{k-1} = \cdots = c_1 = 0$. Multiply (*1*) by $(\mathbf{A} - \lambda\mathbf{I})^{k-1}$, and observe that for each $\mathbf{X}_j$ ($j = k-1, k-2, \ldots, 1$) in that equation,

$$(\mathbf{A} - \lambda\mathbf{I})^{k-1}c_j\mathbf{X}_j = c_j(\mathbf{A} - \lambda\mathbf{I})^{k-j-1}(\mathbf{A} - \lambda\mathbf{I})^j\mathbf{X}_j = c_j(\mathbf{A} - \lambda\mathbf{I})^{k-j-1}\mathbf{0} = \mathbf{0}$$

because each $\mathbf{X}_j$ is a generalized eigenvector of rank j (see Problem 9.7). What remains, then, is

$$c_k(\mathbf{A} - \lambda\mathbf{I})^{k-1}\mathbf{X}_k = \mathbf{0} \tag{2}$$

Since $\mathbf{X}_k$ is a generalized eigenvector of rank k, $(\mathbf{A} - \lambda\mathbf{I})^{k-1}\mathbf{X}_k \neq 0$, and it follows from (*2*) that $c_k = 0$. Equation (*1*) thus reduces to

$$c_{k-1}\mathbf{X}_{k-1} + \cdots + c_1\mathbf{X}_1 = \mathbf{0} \tag{3}$$

But $\mathbf{X}_{k-1}, \ldots, \mathbf{X}_1$ is a chain of length $k - 1$, which we assumed to be linearly independent, so the constants $c_{k-1}, \ldots, c_1$ in (*3*) must all be zero. Therefore, the only solution to (*1*) is $c_k = c_{k-1} = \cdots = c_1 = 0$, from which it follows that the chain of length k is linearly independent.

9.9 Determine a canonical basis for

$$\mathbf{A} = \begin{bmatrix} 3 & 5 \\ -2 & -4 \end{bmatrix}$$

The eigenvalues for this matrix were found in Problem 7.1 to be $\lambda = 1$ and $\lambda = -2$. Since they are distinct, a canonical basis for **A** will consist of one eigenvector for each eigenvalue. Eigenvectors corresponding to $\lambda = 1$ were determined in Problem 7.1 as $x_2[-5/2, 1]^T$ with x_2 arbitrary. We set $x_2 = 2$ to avoid fractions, and obtain the single eigenvector $[-5, 2]^T$. The eigenvectors associated with $\lambda = -2$ are $x_2[-1, 1]^T$ with x_2 again arbitrary. Selecting $x_2 = 1$ in this case, we obtain the single eigenvector $[-1, 1]^T$. A canonical basis for **A** is thus $[-5, 2]^T$, $[-1, 1]^T$.

9.10 Determine the number of generalized eigenvectors of each rank corresponding to $\lambda = 4$ that will appear in a canonical basis for

$$\mathbf{A} = \begin{bmatrix} 4 & 2 & 1 & 0 & 0 & 0 \\ 0 & 4 & -1 & 0 & 0 & 0 \\ 0 & 0 & 4 & 0 & 0 & 0 \\ 0 & 0 & 0 & 4 & 2 & 0 \\ 0 & 0 & 0 & 0 & 4 & 0 \\ 0 & 0 & 0 & 0 & 0 & 7 \end{bmatrix}$$

For this 6×6 matrix, the eigenvalue 4 has multiplicity five (while $\lambda = 7$ has multiplicity one), so $n = 6$, $m = 5$, and $n - m = 1$ for Step 9.1. Now

$$\mathbf{A} - 4\mathbf{I} = \begin{bmatrix} 0 & 2 & 1 & 0 & 0 & 0 \\ 0 & 0 & -1 & 0 & 0 & 0 \\ 0 & 0 & 0 & 0 & 0 & 0 \\ 0 & 0 & 0 & 0 & 2 & 0 \\ 0 & 0 & 0 & 0 & 0 & 0 \\ 0 & 0 & 0 & 0 & 0 & 3 \end{bmatrix}$$

has rank 4, while

$$(\mathbf{A} - 4\mathbf{I})^2 = \begin{bmatrix} 0 & 0 & -2 & 0 & 0 & 1 \\ 0 & 0 & 0 & 0 & 0 & -1 \\ 0 & 0 & 0 & 0 & 0 & 0 \\ 0 & 0 & 0 & 0 & 0 & 0 \\ 0 & 0 & 0 & 0 & 0 & 0 \\ 0 & 0 & 0 & 0 & 0 & 0 \\ 0 & 0 & 0 & 0 & 0 & 9 \end{bmatrix}$$

has rank 2, and

$$(\mathbf{A} - 4\mathbf{I})^3 = \begin{bmatrix} 0 & 0 & 0 & 0 & 0 & -2 \\ 0 & 0 & 0 & 0 & 0 & 0 \\ 0 & 0 & 0 & 0 & 0 & 0 \\ 0 & 0 & 0 & 0 & 0 & 0 \\ 0 & 0 & 0 & 0 & 0 & 0 \\ 0 & 0 & 0 & 0 & 0 & 27 \end{bmatrix}$$

has rank $1 = n - m$. Therefore, $p = 3$. Using Step 9.2, we compute

$$N_3 = \text{rank}(\mathbf{A} - 4\mathbf{I})^2 - \text{rank}(\mathbf{A} - 4\mathbf{I})^3 = 2 - 1 = 1$$
$$N_2 = \text{rank}(\mathbf{A} - 4\mathbf{I})^1 - \text{rank}(\mathbf{A} - 4\mathbf{I})^2 = 4 - 2 = 2$$
$$N_1 = \text{rank}(\mathbf{A} - 4\mathbf{I})^0 - \text{rank}(\mathbf{A} - 4\mathbf{I})^1 = \text{rank}(\mathbf{I}) - \text{rank}(\mathbf{A} - 4\mathbf{I}) = 6 - 4 = 2$$

A canonical basis will contain one generalized eigenvector of rank 3, two generalized eigenvectors of rank 2, and two generalized eigenvectors of rank 1, all corresponding to $\lambda = 4$. (It will also contain one generalized eigenvector corresponding to $\lambda = 7$.)

9.11 Find a canonical basis for the matrix given in Problem 9.10.

We first find the vectors in the basis corresponding to $\lambda = 4$, using the information obtained in the solution to Problem 9.10. There is one generalized eigenvector of rank $p = 3$, which we denote as $\mathbf{X}_3 = [x_1, x_2, x_3, x_4, x_5, x_6]^T$. We note that to have $(\mathbf{A} - 4\mathbf{I})^3\mathbf{X}_3 = \mathbf{0}$, we must set $x_6 = 0$; and to have $(\mathbf{A} - 4\mathbf{I})^2\mathbf{X}_3 \neq 0$, we must have $x_3 \neq 0$. A simple choice is $\mathbf{X}_3 = [0, 0, 1, 0, 0, 0]^T$, which generates as the rest of its chain

$$\mathbf{X}_2 = (\mathbf{A} - 4\mathbf{I})\mathbf{X}_3 = \begin{bmatrix} 0 & 2 & 1 & 0 & 0 & 0 \\ 0 & 0 & -1 & 0 & 0 & 0 \\ 0 & 0 & 0 & 0 & 0 & 0 \\ 0 & 0 & 0 & 0 & 2 & 0 \\ 0 & 0 & 0 & 0 & 0 & 0 \\ 0 & 0 & 0 & 0 & 0 & 3 \end{bmatrix} \begin{bmatrix} 0 \\ 0 \\ 1 \\ 0 \\ 0 \\ 0 \end{bmatrix} = \begin{bmatrix} 1 \\ -1 \\ 0 \\ 0 \\ 0 \\ 0 \end{bmatrix}$$

and

$$\mathbf{X}_1 = (\mathbf{A} - 4\mathbf{I})\mathbf{X}_2 = \begin{bmatrix} 0 & 2 & 1 & 0 & 0 & 0 \\ 0 & 0 & -1 & 0 & 0 & 0 \\ 0 & 0 & 0 & 0 & 0 & 0 \\ 0 & 0 & 0 & 0 & 2 & 0 \\ 0 & 0 & 0 & 0 & 0 & 0 \\ 0 & 0 & 0 & 0 & 0 & 3 \end{bmatrix} \begin{bmatrix} 1 \\ -1 \\ 0 \\ 0 \\ 0 \\ 0 \end{bmatrix} = \begin{bmatrix} -2 \\ 0 \\ 0 \\ 0 \\ 0 \\ 0 \end{bmatrix}$$

We next reduce each nonzero N_k by 1, obtaining $N_3 = 0$, $N_2 = 1$, and $N_1 = 1$; thus, one generalized eigenvector of rank 2 and one generalized eigenvector of rank 1 associated with $\lambda = 4$ remain to be found. We first seek another generalized eigenvector of rank 2, which we denote as $\mathbf{Y}_2 = [y_1, y_2, y_3, y_4, y_5, y_6]^T$. If we are to have $(\mathbf{A} - 4\mathbf{I})^2\mathbf{Y}_2 = \mathbf{0}$, then both y_3 and y_6 must be zero; and if $(\mathbf{A} - 4\mathbf{I})\mathbf{Y}_2 \neq \mathbf{0}$, then either y_2 or y_5 must be nonzero. A convenient choice *that is linearly independent of* $\mathbf{X}_1, \mathbf{X}_2$, and $\mathbf{X}_3$ is $\mathbf{Y}_2 = [0, 0, 0, 0, 1, 0]^T$, which generates, as the remaining vector of its chain,

$$\mathbf{Y}_1 = (\mathbf{A} - 4\mathbf{I})\mathbf{Y}_2 = \begin{bmatrix} 0 & 2 & 1 & 0 & 0 & 0 \\ 0 & 0 & -1 & 0 & 0 & 0 \\ 0 & 0 & 0 & 0 & 0 & 0 \\ 0 & 0 & 0 & 0 & 2 & 0 \\ 0 & 0 & 0 & 0 & 0 & 0 \\ 0 & 0 & 0 & 0 & 0 & 3 \end{bmatrix} \begin{bmatrix} 0 \\ 0 \\ 0 \\ 0 \\ 1 \\ 0 \end{bmatrix} = \begin{bmatrix} 0 \\ 0 \\ 0 \\ 2 \\ 0 \\ 0 \end{bmatrix}$$

Reducing each nonzero N_k by 1 again, we obtain $N_3 = 0$, $N_2 = 0$, and $N_1 = 0$; so all the necessary basis vectors corresponding to $\lambda = 4$ have been found.

The eigenvector $\lambda = 7$ has multiplicity one, so its contribution to a canonical basis is any eigenvector associated with it. One such eigenvector is $\mathbf{Z}_1 = [0, 0, 0, 0, 0, 1]^T$. A complete canonical basis for the 6×6 matrix $\mathbf{A}$ is the set of six vectors $\{\mathbf{X}_3, \mathbf{X}_2, \mathbf{X}_1, \mathbf{Y}_2, \mathbf{Y}_1, \mathbf{Z}_1\}$ consisting of one chain of length three, one chain of length two, and one chain of length one.

9.12 Find a canonical basis for

$$\mathbf{A} = \begin{bmatrix} 3 & 2 & 0 & 1 \\ 0 & 3 & 0 & 0 \\ 0 & 0 & 3 & -1 \\ 0 & 0 & 0 & 3 \end{bmatrix}$$

Matrix $\mathbf{A}$ has order 4×4 and eigenvalue $\lambda = 3$ with multiplicity four. Thus, $n = 4$, $m = 4$, and $n - m = 0$. Here

$$\mathbf{A} - 3\mathbf{I} = \begin{bmatrix} 0 & 2 & 0 & 1 \\ 0 & 0 & 0 & 0 \\ 0 & 0 & 0 & -1 \\ 0 & 0 & 0 & 0 \end{bmatrix}$$

has rank 2, while

$$(\mathbf{A}-3\mathbf{I})^2 = \begin{bmatrix} 0 & 0 & 0 & 0 \\ 0 & 0 & 0 & 0 \\ 0 & 0 & 0 & 0 \\ 0 & 0 & 0 & 0 \end{bmatrix}$$

has rank 0, so $p = 2$. Then

$$N_2 = \text{rank}(\mathbf{A}-3\mathbf{I})^1 - \text{rank}(\mathbf{A}-3\mathbf{I})^2 = 2 - 0 = 2$$
$$N_1 = \text{rank}(\mathbf{A}-3\mathbf{I})^0 - \text{rank}(\mathbf{A}-3\mathbf{I})^1 = \text{rank}(\mathbf{I}) - \text{rank}(\mathbf{A}-3\mathbf{I}) = 4 - 2 = 2$$

A canonical basis for **A** will contain two generalized eigenvectors of rank 2. We denote one of these as $\mathbf{X}_2 = [x_1, x_2, x_3, x_4]^T$. The condition $(\mathbf{A}-3\mathbf{I})^2\mathbf{X}_2 = \mathbf{0}$ is satisfied by all four-dimensional vectors, so it places no constraints on $\mathbf{X}_2$. The requirement $(\mathbf{A}-3\mathbf{I})\mathbf{X}_2 \neq \mathbf{0}$ is satisfied if either

$$x_4 \neq 0 \quad \text{or} \quad 2x_2 + x_4 \neq 0 \tag{1}$$

A convenient choice, therefore, is $\mathbf{X}_2 = [0, 0, 0, 1]^T$, which generates the remaining vector of its chain:

$$\mathbf{X}_1 = (\mathbf{A}-3\mathbf{I})\mathbf{X}_2 = \begin{bmatrix} 0 & 2 & 0 & 1 \\ 0 & 0 & 0 & 0 \\ 0 & 0 & 0 & -1 \\ 0 & 0 & 0 & 0 \end{bmatrix} \begin{bmatrix} 0 \\ 0 \\ 0 \\ 1 \end{bmatrix} = \begin{bmatrix} 1 \\ 0 \\ -1 \\ 0 \end{bmatrix}$$

We reduce N_1 and N_2 by 1, obtaining $N_2 = N_1 = 1$. Another generalized eigenvector of rank 2 for $\lambda = 3$, *linearly independent of* $\mathbf{X}_1$ and $\mathbf{X}_2$ but satisfying (*1*), is $\mathbf{Y}_2 = [0, 1, 0, 0]^T$. This vector generates

$$\mathbf{Y}_1 = (\mathbf{A}-3\mathbf{I})\mathbf{Y}_2 = \begin{bmatrix} 0 & 2 & 0 & 1 \\ 0 & 0 & 0 & 0 \\ 0 & 0 & 0 & -1 \\ 0 & 0 & 0 & 0 \end{bmatrix} \begin{bmatrix} 0 \\ 1 \\ 0 \\ 0 \end{bmatrix} = \begin{bmatrix} 2 \\ 0 \\ 0 \\ 0 \end{bmatrix}$$

Now N_1 and N_2 are reduced to zero; a canonical basis for **A** is thus $\{\mathbf{X}_2, \mathbf{X}_1, \mathbf{Y}_2, \mathbf{Y}_1\}$, comprised of two chains, both of length two.

9.13 Determine a canonical basis for

$$\mathbf{A} = \begin{bmatrix} 4 & 1 & 1 & 2 & 2 \\ -1 & 2 & 1 & 3 & 0 \\ 0 & 0 & 3 & 0 & 0 \\ 0 & 0 & 0 & 2 & 1 \\ 0 & 0 & 0 & 1 & 2 \end{bmatrix}$$

The eigenvalues for this matrix are $\lambda = 3$ with multiplicity four and $\lambda = 1$ with multiplicity one. For $\lambda = 3$, $n - m = 5 - 4 = 1$. Also,

$$\mathbf{A} - 3\mathbf{I} = \begin{bmatrix} 1 & 1 & 1 & 2 & 2 \\ -1 & -1 & 1 & 3 & 0 \\ 0 & 0 & 0 & 0 & 0 \\ 0 & 0 & 0 & -1 & 1 \\ 0 & 0 & 0 & 1 & -1 \end{bmatrix}$$

has rank 3, while

$$(\mathbf{A}-3\mathbf{I})^2 = \begin{bmatrix} 0 & 0 & 2 & 5 & 2 \\ 0 & 0 & -2 & -8 & 1 \\ 0 & 0 & 0 & 0 & 0 \\ 0 & 0 & 0 & 2 & -2 \\ 0 & 0 & 0 & -2 & 2 \end{bmatrix}$$

has rank 2, and

$$(\mathbf{A}-3\mathbf{I})^3 = \begin{bmatrix} 0 & 0 & 0 & -3 & 3 \\ 0 & 0 & 0 & 9 & -9 \\ 0 & 0 & 0 & 0 & 0 \\ 0 & 0 & 0 & -4 & 4 \\ 0 & 0 & 0 & 4 & -4 \end{bmatrix}$$

has rank 1. Thus, $p = 3$, and $N_3 = 2 - 1 = 1$, $N_2 = 3 - 2 = 1$, and $N_1 = 5 - 3 = 2$.

A generalized eigenvector of rank $p = 3$ for $\lambda = 3$ is $\mathbf{X}_3 = [0, 0, 1, 0, 0]^T$, which generates

$$\mathbf{X}_2 = (\mathbf{A}-3\mathbf{I})\mathbf{X}_3 = \begin{bmatrix} 1 & 1 & 1 & 2 & 2 \\ -1 & -1 & 1 & 3 & 0 \\ 0 & 0 & 0 & 0 & 0 \\ 0 & 0 & 0 & -1 & 1 \\ 0 & 0 & 0 & 1 & -1 \end{bmatrix}\begin{bmatrix} 0 \\ 0 \\ 1 \\ 0 \\ 0 \end{bmatrix} = \begin{bmatrix} 1 \\ 1 \\ 0 \\ 0 \\ 0 \end{bmatrix}$$

and

$$\mathbf{X}_1 = (\mathbf{A}-3\mathbf{I})\mathbf{X}_2 = \begin{bmatrix} 1 & 1 & 1 & 2 & 2 \\ -1 & -1 & 1 & 3 & 0 \\ 0 & 0 & 0 & 0 & 0 \\ 0 & 0 & 0 & -1 & 1 \\ 0 & 0 & 0 & 1 & -1 \end{bmatrix}\begin{bmatrix} 1 \\ 1 \\ 0 \\ 0 \\ 0 \end{bmatrix} = \begin{bmatrix} 2 \\ -2 \\ 0 \\ 0 \\ 0 \end{bmatrix}$$

We now reduce the N_k by 1 to obtain $N_3 = N_2 = 0$ and $N_1 = 1$. Thus, a canonical basis for $\mathbf{A}$ will contain generalized eigenvector of rank 1 for $\lambda = 3$. This is an eigenvector, and it must be linearly independent of $\mathbf{X}_1, \mathbf{X}_2$, and $\mathbf{X}_3$; to find it we solve $(\mathbf{A}-3\mathbf{I})\mathbf{Y}_1 = \mathbf{0}$ to obtain, as one possibility, $\mathbf{Y}_1 = [0, -1, -7, 2, 2]^T$.

Since $\lambda = 1$ is an eigenvalue of multiplicity one, its contribution to a canonical basis is any eigenvector corresponding to it. One choice is $\mathbf{Z}_1 = [-3, 9, 0, -4, 4]^T$. A canonical basis for the matrix $\mathbf{A}$ is, then, $\{\mathbf{X}_3, \mathbf{X}_2, \mathbf{X}_1, \mathbf{Y}_1, \mathbf{Z}_1\}$, comprised of one chain of length three and two chains of length one.

9.14 Find the minimum polynomial for the matrix given in Problem 9.12.

In Problem 9.12 we found that the matrix has the single distinct eigenvalue $\lambda = 3$, with $p = 2$. Its minimum polynomial is then

$$m(\lambda) = (\lambda - 3)^2 = \lambda^2 - 6\lambda + 9$$

9.15 Find the minimum polynomial for the matrix given in Problem 9.13.

From Problem 9.13 we know that the matrix has two distinct eigenvalues: $\lambda_1 = 3$ with multiplicity four and $p_1 = 3$, and $\lambda_2 = 1$ with multiplicity one and thus $p_2 = 1$. Then

$$m(\lambda) = (\lambda - 3)^3(\lambda - 1) = \lambda^4 - 10\lambda^3 + 36\lambda^2 - 54\lambda + 27$$

Supplementary Problems

9.16 Determine which of the following are generalized eigenvectors of rank 3 corresponding to $\lambda = 1$ for the matrix

$$\mathbf{A} = \begin{bmatrix} 1 & 0 & 1 & 0 & -1 \\ 0 & 1 & 0 & 0 & 0 \\ 0 & 0 & 1 & -1 & 2 \\ 0 & 0 & 0 & 1 & 1 \\ 0 & 0 & 0 & 0 & 1 \end{bmatrix}$$

(*a*) $[1, 1, 0, 0, 0]^T$ (*b*) $[0, 0, 0, 0, 2]^T$ (*c*) $[0, 0, 0, 1, 0]^T$
(*d*) $[1, 1, 0, 2, 0]^T$ (*e*) $[0, 0, 0, 0, 0]^T$ (*f*) $[1, 1, 1, 1, 1]^T$

9.17 Find the chain generated by $\mathbf{X}_4 = [0, 0, 0, 0, 1]^T$, a generalized eigenvector of rank 4 corresponding to $\lambda = 1$ for the matrix given in Problem 9.16.

9.18 Find a generalized eigenvector of rank 2 corresponding to $\lambda = 5$ for the matrix

$$\mathbf{A} = \begin{bmatrix} 5 & 1 & 0 \\ 0 & 5 & 1 \\ 0 & 0 & 5 \end{bmatrix}$$

9.19 Find the chain generated by $\mathbf{X}_3 = [0, 0, 1]^T$, a generalized eigenvector of rank 3 corresponding to $\lambda = 5$ for the matrix given in Problem 9.18.

9.20 Determine the lengths of the chains associated with an eigenvalue λ that are included in a canonical basis if

(*a*) $N_2 = N_1 = 2$ (*b*) $N_2 = 1, N_1 = 3$
(*c*) $N_3 = N_2 = 1, N_1 = 2$ (*d*) $N_3 = 2, N_2 = 3, N_1 = 3$
(*e*) $N_3 = 1, N_2 = 2, N_1 = 5$ (*f*) $N_2 = 3, N_1 = 5$
(*g*) $N_3 = 2, N_2 = 2, N_1 = 1$

In Problems 9.21 through 9.31, (*a*) determine the number of generalized eigenvectors of each rank that will form a canonical basis for the given matrix, and (*b*) find such a basis.

9.21 The matrix in Problem 9.16. **9.22** The matrix in Problem 9.18.

9.23 $\begin{bmatrix} -5 & -3 \\ 3 & 1 \end{bmatrix}$ **9.24** $\begin{bmatrix} 1 & 3 \\ 1 & 3 \end{bmatrix}$ **9.25** $\begin{bmatrix} 2 & 2 & 1 \\ 0 & 2 & 0 \\ 0 & 0 & 2 \end{bmatrix}$

9.26 $\begin{bmatrix} 2 & 2 & 2 \\ 0 & 4 & 0 \\ 3 & -3 & 1 \end{bmatrix}$ **9.27** $\begin{bmatrix} 2 & 1 & 1 \\ 1 & 2 & 1 \\ -2 & -2 & -1 \end{bmatrix}$ **9.28** $\begin{bmatrix} 7 & 1 & 3 \\ 2 & 4 & 2 \\ -7 & -3 & -3 \end{bmatrix}$

9.29 $\begin{bmatrix} 3 & 2 & 1 & 0 \\ 0 & 3 & 0 & 0 \\ -1 & 1 & 1 & 0 \\ 0 & 1 & 1 & 3 \end{bmatrix}$ **9.30** $\begin{bmatrix} 3 & 1 & 1 & 0 \\ 0 & 3 & 0 & 0 \\ -1 & 1 & 1 & 0 \\ 0 & 1 & 1 & 3 \end{bmatrix}$ **9.31** $\begin{bmatrix} 5 & 1 & 2 & 0 & 0 \\ 0 & 5 & -1 & 0 & 0 \\ 0 & 0 & 5 & 0 & 0 \\ 0 & 0 & 0 & 3 & 2 \\ 0 & 0 & 0 & 1 & 4 \end{bmatrix}$

9.32 Find the minimum polynomial for the matrix in
(*a*) Problem 9.25. (*b*) Problem 9.26. (*c*) Problem 9.27.
(*d*) Problem 9.28. (*e*) Problem 9.29. (*f*) Problem 9.30.

Chapter 10

Similarity

SIMILAR MATRICES

A matrix $\mathbf{A}$ is *similar* to a matrix $\mathbf{B}$ if there exists an invertible matrix $\mathbf{S}$ such that

$$\mathbf{A} = \mathbf{S}^{-1}\mathbf{B}\mathbf{S} \tag{10.1}$$

If $\mathbf{A}$ is similar to $\mathbf{B}$, then $\mathbf{B}$ is also similar to $\mathbf{A}$ and both matrices must be of the same order and square.

Property 10.1: Similar matrices have the same characteristic equation and, therefore, the same eigenvalues and the same trace.

Property 10.2: If $\mathbf{X}$ is an eigenvector of $\mathbf{A}$ associated with eigenvalue λ and (*10.1*) holds, then $\mathbf{Y} = \mathbf{S}\mathbf{X}$ is an eigenvector of $\mathbf{B}$ associated with the same eigenvalue.

(See Problems 10.1 through 10.3 and 10.43.)

MODAL MATRIX

Associated with every square matrix $\mathbf{A}$ is a canonical basis (see Chapter 9). A *modal matrix* $\mathbf{M}$ for $\mathbf{A}$ is a matrix of the same order as $\mathbf{A}$ having as its columns all the vectors of a canonical basis for $\mathbf{A}$. A canonical basis is a set of linearly independent vectors, so $\mathbf{M}$ has an inverse.

A modal matrix $\mathbf{M}$ is not unique. To standardize $\mathbf{M}$ somewhat, we shall always assume it has been constructed as follows:

(M1): All chains of length one precede all longer chains (if any exist).

(M2): The vectors of each chain of length two or more are contiguous, with rank increasing from left to right.

(See Problems 10.4 through 10.6.)

JORDAN CANONICAL FORM

A *Jordan block* is a square matrix whose diagonal elements are all equal, whose superdiagonal elements (those immediately above the main diagonal) all equal 1, and whose other elements are all zero. It has the form

$$\begin{bmatrix} \lambda & 1 & 0 & \cdots & 0 & 0 \\ 0 & \lambda & 1 & \cdots & 0 & 0 \\ 0 & 0 & \lambda & \cdots & 0 & 0 \\ \cdots & \cdots & \cdots & \cdots & \cdots & \cdots \\ 0 & 0 & 0 & \cdots & \lambda & 1 \\ 0 & 0 & 0 & \cdots & 0 & \lambda \end{bmatrix}$$

A Jordan block is completely determined by its order and the value of its diagonal elements.

A matrix is in *Jordan canonical form* if it is a diagonal matrix or if it has one of the following two partitioned forms:

$$\begin{bmatrix} \mathbf{D} & & & 0 \\ & \mathbf{J}_1 & & \\ & & \mathbf{J}_2 & \\ & & & \ddots \\ 0 & & & \mathbf{J}_k \end{bmatrix} \qquad \begin{bmatrix} \mathbf{J}_1 & & & 0 \\ & \mathbf{J}_2 & & \\ & & \ddots & \\ 0 & & & \mathbf{J}_k \end{bmatrix}$$

where **D** denotes a diagonal matrix (whose diagonal elements need not be equal) and $\mathbf{J}_i$ ($i = 1, 2, \ldots, k$) represents a Jordan block. Although the diagonal elements in any one Jordan block must be equal, different Jordan blocks within the Jordan canonical form may have different diagonals. (See Problem 10.7.)

SIMILARITY AND JORDAN CANONICAL FORM

Every square matrix **A** is similar to a matrix **J** in Jordan canonical form. If **M** is a modal matrix for **A**, then

$$\mathbf{A} = \mathbf{MJM}^{-1} \tag{10.2}$$

Equation (*10.2*) has the form of (*10.1*) with $\mathbf{S} = \mathbf{M}^{-1}$.

The matrix **J** in (*10.2*) is uniquely determined by **M**. Each chain of length r appearing in **M** and corresponding to eigenvalue λ generates an $r \times r$ Jordan block in **J** with λ on the diagonal. The chains of length one (if they exist) give rise collectively to the diagonal submatrix of **J**; the diagonal elements of this submatrix are the eigenvalues associated with the chains of length one, in the same order as their corresponding eigenvectors in **M**. If **M** consists solely of (generalized) eigenvectors (of rank 1), then **J** is simply a diagonal matrix; if **M** contains no chains of length one, then **J** is a partitioned matrix of Jordan blocks. (See Problems 10.8 through 10.11.)

The Jordan canonical form of a matrix is unique as regards the individual Jordan blocks it contains, as well as the diagonal elements associated with chains of length one, if they exist. However, the positions of Jordan blocks and diagonal elements are not unique. Each chain must appear as contiguous columns in a modal matrix, but there is no criterion for ordering the chains. Different orderings of entire chains will produce different permutations of the associated Jordan blocks in **J**, or of the diagonal elements of **D**.

Jordan blocks are defined on occasion to have 0s on the superdiagonal and 1s on the subdiagonal. Such forms are obtained easily by changing rule M2 so that rank decreases from left to right in **M**.

FUNCTIONS OF MATRICES

Functions of matrices are easily computed for matrices in Jordan canonical form. If **J** is the diagonal matrix

$$\mathbf{J} = \begin{bmatrix} \lambda_1 & 0 & \cdots & 0 \\ 0 & \lambda_2 & \cdots & 0 \\ \cdots & \cdots & \cdots & \cdots \\ 0 & 0 & \cdots & \lambda_n \end{bmatrix}$$

then

$$f(\mathbf{J}) = \begin{bmatrix} f(\lambda_1) & 0 & \cdots & 0 \\ 0 & f(\lambda_2) & \cdots & 0 \\ \cdots & \cdots & \cdots & \cdots \\ 0 & 0 & \cdots & f(\lambda_n) \end{bmatrix} \tag{10.3}$$

(See Problem 10.14.) If **J** is the $r \times r$ Jordan block

$$\mathbf{J} = \begin{bmatrix} \lambda & 1 & 0 & \cdots & 0 & 0 \\ 0 & \lambda & 1 & \cdots & 0 & 0 \\ 0 & 0 & \lambda & \cdots & 0 & 0 \\ \cdots & \cdots & \cdots & \cdots & \cdots & \cdots \\ 0 & 0 & 0 & \cdots & \lambda & 1 \\ 0 & 0 & 0 & \cdots & 0 & \lambda \end{bmatrix}$$

then

$$f(\mathbf{J}) = \begin{bmatrix} \frac{f(\lambda)}{0!} & \frac{f'(\lambda)}{1!} & \frac{f''(\lambda)}{2!} & \cdots & \frac{f^{(r-2)}(\lambda)}{(r-2)!} & \frac{f^{(r-1)}(\lambda)}{(r-1)!} \\ 0 & \frac{f(\lambda)}{0!} & \frac{f'(\lambda)}{1!} & \cdots & \frac{f^{(r-3)}(\lambda)}{(r-3)!} & \frac{f^{(r-2)}(\lambda)}{(r-2)!} \\ \cdots & \cdots & \cdots & \cdots & \cdots & \cdots \\ 0 & 0 & 0 & \cdots & \frac{f(\lambda)}{0!} & \frac{f'(\lambda)}{1!} \\ 0 & 0 & 0 & \cdots & 0 & \frac{f(\lambda)}{0!} \end{bmatrix} \tag{10.4}$$

where all derivatives are taken with respect to λ. (See Problem 10.15.) If **J** is a partitioned matrix of Jordan blocks and (perhaps) a diagonal matrix such that

$$\mathbf{J} = \begin{bmatrix} \mathbf{D} & & & \\ & \mathbf{J}_1 & & 0 \\ & & \cdots & \\ 0 & & & \mathbf{J}_k \end{bmatrix}$$

then

$$f(\mathbf{J}) = \begin{bmatrix} f(\mathbf{D}) & & & \\ & f(\mathbf{J}_1) & & 0 \\ & & \cdots & \\ 0 & & & f(\mathbf{J}_k) \end{bmatrix}$$

where $f(\mathbf{D})$ and $f(\mathbf{J}_i)$ $(i = 1, \ldots, k)$ are defined by (*10.3*) and (*10.4*), respectively. (See Problem 10.16.) If **A** is similar to the matrix **J** in Jordan canonical form, then

$$f(\mathbf{A}) = \mathbf{M}f(\mathbf{J})\mathbf{M}^{-1} \tag{10.5}$$

(See Problems 10.17 through 10.19.) This formula is computationally efficient only when **M** is known or easily found; otherwise the procedure given in Chapter 8 is preferred. Since $f(\mathbf{J})$ is upper triangular, (*10.5*) and Property 10.1 imply:

Property 10.3: If λ is an eigenvalue of **A**, the $f(\lambda)$ is an eigenvalue of $f(\mathbf{A})$.

Solved Problems

10.1 Determine whether

$$\mathbf{A} = \begin{bmatrix} 2 & 0 \\ 3 & 2 \end{bmatrix} \quad \text{is similar to} \quad \mathbf{B} = \begin{bmatrix} 2 & 0 \\ 0 & 2 \end{bmatrix}$$

The matrices are similar if and only if there exists a matrix **S** such that $\mathbf{A} = \mathbf{S}^{-1}\mathbf{B}\mathbf{S}$ or, equivalently, such that

$$\mathbf{SA} = \mathbf{BS} \tag{1}$$

Set

$$\mathbf{S} = \begin{bmatrix} a & b \\ c & d \end{bmatrix}$$

Then (*1*) becomes

$$\begin{bmatrix} a & b \\ c & d \end{bmatrix}\begin{bmatrix} 2 & 0 \\ 3 & 2 \end{bmatrix} = \begin{bmatrix} 2 & 0 \\ 0 & 2 \end{bmatrix}\begin{bmatrix} a & b \\ c & d \end{bmatrix}$$

or

$$\begin{bmatrix} 2a+3b & 2b \\ 2c+3d & 2d \end{bmatrix} = \begin{bmatrix} 2a & 2b \\ 2c & 2d \end{bmatrix}$$

The solution to this matrix equality is $b = d = 0$, with a and c arbitrary. For (*1*) to be valid, **S** must thus have the form

$$\mathbf{S} = \begin{bmatrix} a & 0 \\ c & 0 \end{bmatrix}$$

which is nonsingular for any choice of a and c. Thus, there is no invertible matrix which satisfies (*1*), and the matrices **A** and **B** are not similar. Observe that **A** and **B** have the same eigenvalues, so *matrices may have the same eigenvalues and not be similar.*

10.2 Determine whether

$$\mathbf{A} = \begin{bmatrix} 3 & 1 & 1 \\ 2 & 2 & 4 \\ 1 & 1 & 1 \end{bmatrix} \quad \text{is similar to} \quad \mathbf{B} = \begin{bmatrix} 2 & 1 & 1 \\ 3 & 0 & 1 \\ 4 & 2 & 1 \end{bmatrix}$$

The trace of **A** is $3 + 2 + 1 = 6$, while that of **B** is $2 + 0 + 1 = 3$. Since the traces are not equal, **A** and **B** must have different sets of eigenvalues and, therefore, cannot be similar.

10.3 Prove that similar matrices have the same characteristic polynomial.

If **A** and **B** are similar, there exists a matrix **S** such that $\mathbf{A} = \mathbf{S}^{-1}\mathbf{B}\mathbf{S}$. Therefore,

$$|\mathbf{A} - \lambda\mathbf{I}| = |\mathbf{S}^{-1}\mathbf{B}\mathbf{S} - \lambda\mathbf{S}^{-1}\mathbf{S}| = |\mathbf{S}^{-1}(\mathbf{B} - \lambda\mathbf{I})\mathbf{S}|$$

$$= |\mathbf{S}^{-1}||\mathbf{B} - \lambda\mathbf{I}||\mathbf{S}| = \frac{1}{|\mathbf{S}|}|\mathbf{B} - \lambda\mathbf{I}||\mathbf{S}| = |\mathbf{B} - \lambda\mathbf{I}|$$

10.4 Construct a modal matrix for

$$\mathbf{A} = \begin{bmatrix} 4 & 1 & 1 & 2 & 2 \\ -1 & 2 & 1 & 3 & 0 \\ 0 & 0 & 3 & 0 & 0 \\ 0 & 0 & 0 & 2 & 1 \\ 0 & 0 & 0 & 1 & 2 \end{bmatrix}$$

A canonical basis for **A** was found in Problem 9.13. It consists of one chain of length three,

$$\mathbf{X}_3 = [0, 0, 1, 0, 0]^T \qquad \mathbf{X}_2 = [1, 1, 0, 0, 0]^T \qquad \mathbf{X}_1 = [2, -2, 0, 0, 0]^T$$

and two chains of length one,

$$\mathbf{Y}_1 = [0, -1, -7, 2, 2]^T \qquad \text{and} \qquad \mathbf{Z}_1 = [-3, 9, 0, -4, 4]^T$$

Thus,

$$\mathbf{M} = [\mathbf{Y}_1, \mathbf{Z}_1, \mathbf{X}_1, \mathbf{X}_2, \mathbf{X}_3] = \begin{bmatrix} 0 & -3 & 2 & 1 & 0 \\ -1 & 9 & -2 & 1 & 0 \\ -7 & 0 & 0 & 0 & 1 \\ 2 & -4 & 0 & 0 & 0 \\ 2 & 4 & 0 & 0 & 0 \end{bmatrix}$$

A second modal matrix may be obtained by interchanging the first two columns of **M**.

10.5 Construct a modal matrix for

$$\mathbf{A} = \begin{bmatrix} 4 & 2 & 1 & 0 & 0 & 0 \\ 0 & 4 & -1 & 0 & 0 & 0 \\ 0 & 0 & 4 & 0 & 0 & 0 \\ 0 & 0 & 0 & 4 & 2 & 0 \\ 0 & 0 & 0 & 0 & 4 & 0 \\ 0 & 0 & 0 & 0 & 0 & 7 \end{bmatrix}$$

A canonical basis for this matrix was determined in Problem 9.11 to consist of one chain of length three,

$$\mathbf{X}_3 = [0, 0, 1, 0, 0, 0]^T \qquad \mathbf{X}_2 = [1, -1, 0, 0, 0, 0]^T \qquad \mathbf{X}_1 = [-2, 0, 0, 0, 0, 0]^T$$

one chain of length two,

$$\mathbf{Y}_2 = [0, 0, 0, 0, 1, 0]^T \qquad \mathbf{Y}_1 = [0, 0, 0, 2, 0, 0]^T$$

and one chain of length one,

$$\mathbf{Z}_1 = [0, 0, 0, 0, 0, 0\,1]^T$$

A modal matrix for **A** is either $\mathbf{M} = [\mathbf{Z}_1, \mathbf{Y}_1, \mathbf{Y}_2, \mathbf{X}_1, \mathbf{X}_2, \mathbf{X}_3]$ or

$$\mathbf{M} = [\mathbf{Z}_1, \mathbf{X}_1, \mathbf{X}_2, \mathbf{X}_3, \mathbf{Y}_1, \mathbf{Y}_2] = \begin{bmatrix} 0 & -2 & 1 & 0 & 0 & 0 \\ 0 & 0 & -1 & 0 & 0 & 0 \\ 0 & 0 & 0 & 1 & 0 & 0 \\ 0 & 0 & 0 & 0 & 2 & 0 \\ 0 & 0 & 0 & 0 & 0 & 1 \\ 1 & 0 & 0 & 0 & 0 & 0 \end{bmatrix}$$

10.6 Construct a modal matrix for

$$\mathbf{A} = \begin{bmatrix} 5 & 2 & 2 \\ 3 & 6 & 3 \\ 6 & 6 & 9 \end{bmatrix}$$

A set of three linearly independent eigenvectors for **A** was determined in Problem 7.4 to consist of

$[-1, 1, 0]^T$, $[-1, 0, 1]^T$, and $[2, 3, 6]^T$. Since these three vectors form a full complement of generalized eigenvectors of rank 1, they are a canonical basis for **A**. A modal matrix for **A** is then

$$\mathbf{M} = \begin{bmatrix} -1 & -1 & 2 \\ 1 & 0 & 3 \\ 0 & 1 & 6 \end{bmatrix}$$

Any permutation of the columns of **M** will produce another equally acceptable modal matrix.

10.7 Determine which of the following matrices are in Jordan canonical form:

$$\mathbf{A} = \begin{bmatrix} 3 & 0 & 0 & 0 \\ 0 & 3 & 0 & 0 \\ 0 & 0 & 1 & 0 \\ 0 & 0 & 0 & 2 \end{bmatrix} \qquad \mathbf{B} = \begin{bmatrix} 3 & 1 & 0 & 0 & 0 \\ 0 & 3 & 1 & 0 & 0 \\ 0 & 0 & 3 & 0 & 0 \\ 0 & 0 & 0 & 4 & 1 \\ 0 & 0 & 0 & 0 & 4 \end{bmatrix} \qquad \mathbf{C} = \begin{bmatrix} 0 & 0 & 0 & 0 & 0 & 0 \\ 0 & 1 & 0 & 0 & 0 & 0 \\ 0 & 0 & 1 & 1 & 0 & 0 \\ 0 & 0 & 0 & 1 & 0 & 0 \\ 0 & 0 & 0 & 0 & 1 & 1 \\ 0 & 0 & 0 & 0 & 0 & 1 \end{bmatrix}$$

All three matrices are in Jordan canonical form: **A**, because it is a diagonal matrix; **B**, because it is in the form

$$\mathbf{B} = \begin{bmatrix} \mathbf{J}_1 & \mathbf{0} \\ \mathbf{0} & \mathbf{J}_2 \end{bmatrix} \quad \text{with} \quad \mathbf{J}_1 = \begin{bmatrix} 3 & 1 & 0 \\ 0 & 3 & 1 \\ 0 & 0 & 3 \end{bmatrix} \quad \text{and} \quad \mathbf{J}_2 = \begin{bmatrix} 4 & 1 \\ 0 & 4 \end{bmatrix}$$

and **C**, because it is in the form

$$\mathbf{C} = \begin{bmatrix} \mathbf{D} & & 0 \\ & \mathbf{J}_1 & \\ 0 & & \mathbf{J}_2 \end{bmatrix} \quad \text{with} \quad \mathbf{D} = \begin{bmatrix} 0 & 0 \\ 0 & 1 \end{bmatrix} \quad \text{and} \quad \mathbf{J}_1 = \mathbf{J}_2 = \begin{bmatrix} 1 & 1 \\ 0 & 1 \end{bmatrix}$$

10.8 Find a matrix in Jordan canonical form that is similar to the matrix **A** of Problem 10.6.

Using the results of Problem 10.6, we note that the columns of **M** are all eigenvectors (chains of length one) corresponding, respectively, to the eigenvalues 3, 3, and 14 (as found in Problem 7.2). Thus, **A** is similar to the diagonal matrix

$$\mathbf{J} = \begin{bmatrix} 3 & 0 & 0 \\ 0 & 3 & 0 \\ 0 & 0 & 14 \end{bmatrix}$$

To see that it is, note that

$$\mathbf{MJM}^{-1} = \begin{bmatrix} -1 & -1 & 2 \\ 1 & 0 & 3 \\ 0 & 1 & 6 \end{bmatrix} \begin{bmatrix} 3 & 0 & 0 \\ 0 & 3 & 0 \\ 0 & 0 & 14 \end{bmatrix} \begin{bmatrix} -3/11 & 8/11 & -3/11 \\ -6/11 & -6/11 & 5/11 \\ 1/11 & 1/11 & 1/11 \end{bmatrix}$$

$$= \begin{bmatrix} 5 & 2 & 2 \\ 3 & 6 & 3 \\ 6 & 6 & 9 \end{bmatrix} = \mathbf{A}$$

10.9 Find a matrix **J** in Jordan canonical form that is similar to the matrix **A** of Problem 10.4.

In Problem 10.4 we found that $\mathbf{M} = [\mathbf{Y}_1, \mathbf{Z}_1, \mathbf{X}_1, \mathbf{X}_2, \mathbf{X}_3]$. The two generalized eigenvectors of rank 1, $\mathbf{Y}_1$ and $\mathbf{Z}_1$, correspond to the eigenvalues 3 and 1, respectively, and generate the diagonal submatrix of **J**:

$$\mathbf{D} = \begin{bmatrix} 3 & 0 \\ 0 & 1 \end{bmatrix}$$

The chain of length three, $\{\mathbf{X}_1, \mathbf{X}_2, \mathbf{X}_3\}$, corresponds to the eigenvalue $\lambda = 3$, so it generates the Jordan block

$$\mathbf{J}_1 = \begin{bmatrix} 3 & 1 & 0 \\ 0 & 3 & 1 \\ 0 & 0 & 3 \end{bmatrix}$$

A is thus similar to

$$\mathbf{J} = \begin{bmatrix} \mathbf{D} & \mathbf{0} \\ \mathbf{0} & \mathbf{J}_1 \end{bmatrix} = \begin{bmatrix} 3 & 0 & 0 & 0 & 0 \\ 0 & 1 & 0 & 0 & 0 \\ 0 & 0 & 3 & 1 & 0 \\ 0 & 0 & 0 & 3 & 1 \\ 0 & 0 & 0 & 0 & 3 \end{bmatrix}$$

10.10 Find a matrix **J** in Jordan canonical form that is similar to the matrix **A** of Problem 10.5.

In Problem 10.5, we found that $\mathbf{M} = [\mathbf{Z}_1, \mathbf{X}_1, \mathbf{X}_2, \mathbf{X}_3, \mathbf{Y}_1, \mathbf{Y}_2]$. The single generalized eigenvector of rank 1, $\mathbf{Z}_1$, corresponds to the eigenvalue 7 and generates the 1×1 diagonal submatrix of **J** comprised of this eigenvalue. The chain of length three, $\{\mathbf{X}_1, \mathbf{X}_2, \mathbf{X}_3\}$, corresponds to the eigenvalue 4 and generates the Jordan block

$$\mathbf{J}_1 = \begin{bmatrix} 4 & 1 & 0 \\ 0 & 4 & 1 \\ 0 & 0 & 4 \end{bmatrix}$$

The chain of length two, $\{\mathbf{Y}_1, \mathbf{Y}_2\}$, also corresponds to the eigenvalue 4 and generates the Jordan block

$$\mathbf{J}_2 = \begin{bmatrix} 4 & 1 \\ 0 & 4 \end{bmatrix}$$

Thus **A** is similar to

$$\mathbf{J} = \begin{bmatrix} [7] & & 0 \\ & \mathbf{J}_1 & \\ 0 & & \mathbf{J}_2 \end{bmatrix} = \begin{bmatrix} 7 & 0 & 0 & 0 & 0 & 0 \\ 0 & 4 & 1 & 0 & 0 & 0 \\ 0 & 0 & 4 & 1 & 0 & 0 \\ 0 & 0 & 0 & 4 & 0 & 0 \\ 0 & 0 & 0 & 0 & 4 & 1 \\ 0 & 0 & 0 & 0 & 0 & 4 \end{bmatrix}$$

10.11 Find a matrix **J** in Jordan canonical form that is similar to a matrix **A** whose characteristic equation is $(\lambda - 2)^5(\lambda - 3)^5 = 0$, and that has eigenvalue rank numbers $N_3 = N_2 = 1$ and $N_1 = 3$ associated with eigenvalue $\lambda = 2$, and $N_2 = 2$ and $N_1 = 3$ associated with $\lambda = 3$.

A must be 10×10 matrix, and so too is **J**. A canonical basis for **A** will contain one chain of length three and two chains of length one corresponding to $\lambda = 2$. Denote these vectors, respectively, as $\mathbf{X}_3(2)$, $\mathbf{X}_2(2)$, $\mathbf{X}_1(2)$, $\mathbf{Y}_1(2)$, $\mathbf{Z}_1(2)$, where the integer in parenthesis denotes the associated eigenvalue. This same canonical basis must also contain two chains of length two and one chain of length one corresponding to $\lambda = 3$. Denote these vectors, respectively, as $\mathbf{U}_2(3)$, $\mathbf{U}_1(3)$, $\mathbf{V}_2(3)$, $\mathbf{V}_1(3)$, $\mathbf{W}_1(3)$. We use this canonical basis to construct the modal matrix

$$\mathbf{M} = [\mathbf{Y}_1(2), \mathbf{Z}_1(2), \mathbf{W}_1(3), \mathbf{U}_1(3), \mathbf{U}_2(3), \mathbf{X}_1(2), \mathbf{X}_2(2), \mathbf{X}_3(2), \mathbf{V}_1(3), \mathbf{V}_2(3)]$$

which determines the matrix in Jordan canonical form

$$\mathbf{J} = \begin{bmatrix} 2 & 0 & 0 & 0 & 0 & 0 & 0 & 0 & 0 & 0 \\ 0 & 2 & 0 & 0 & 0 & 0 & 0 & 0 & 0 & 0 \\ 0 & 0 & 3 & 0 & 0 & 0 & 0 & 0 & 0 & 0 \\ 0 & 0 & 0 & 3 & 1 & 0 & 0 & 0 & 0 & 0 \\ 0 & 0 & 0 & 0 & 3 & 0 & 0 & 0 & 0 & 0 \\ 0 & 0 & 0 & 0 & 0 & 2 & 1 & 0 & 0 & 0 \\ 0 & 0 & 0 & 0 & 0 & 0 & 2 & 1 & 0 & 0 \\ 0 & 0 & 0 & 0 & 0 & 0 & 0 & 2 & 0 & 0 \\ 0 & 0 & 0 & 0 & 0 & 0 & 0 & 0 & 3 & 1 \\ 0 & 0 & 0 & 0 & 0 & 0 & 0 & 0 & 0 & 3 \end{bmatrix}$$

10.12 Verify (*10.2*) for a modal matrix consisting solely of generalized eigenvectors of rank 1.

Denote the columns of **M** as $\mathbf{E}_1, \mathbf{E}_2, \ldots, \mathbf{E}_n$, where each $\mathbf{E}_i$ $(i = 1, 2, \ldots, n)$ is an eigenvector of **A**. Thus, $\mathbf{AE}_i = \lambda_i\mathbf{E}_i$. The eigenvalues $\lambda_1, \lambda_2, \ldots, \lambda_n$ of **A** need not be distinct. Now define

$$\mathbf{J} = \begin{bmatrix} \lambda_1 & 0 & \cdots & 0 \\ 0 & \lambda_2 & \cdots & 0 \\ \cdots & \cdots & \cdots & \cdots \\ 0 & 0 & \cdots & \lambda_n \end{bmatrix}$$

and note that

$$\begin{aligned} \mathbf{AM} &= \mathbf{A}[\mathbf{E}_1, \mathbf{E}_2, \ldots, \mathbf{E}_n] = [\mathbf{AE}_1, \mathbf{AE}_2, \ldots, \mathbf{AE}_n] \\ &= [\lambda_1\mathbf{E}_1, \lambda_2\mathbf{E}_2, \ldots, \lambda_n\mathbf{E}_n] = [\mathbf{E}_1, \mathbf{E}_2, \ldots, \mathbf{E}_n]\mathbf{J} = \mathbf{MJ} \end{aligned}$$

from which (*10.2*) immediately follows.

10.13 Verify (*10.2*) for a modal matrix consisting of a single chain of length r.

Denote the columns of **M** as $\mathbf{X}_1, \mathbf{X}_2, \ldots, \mathbf{X}_r$, where each $\mathbf{X}_i$ $(i = 1, 2, \ldots, r)$ is a generalized eigenvector of rank i for **A** and all $\mathbf{X}_i$ correspond to the same eigenvalue λ. Now

$$\begin{aligned} \mathbf{X}_{r-1} &= (\mathbf{A} - \lambda\mathbf{I})\mathbf{X}_r = \mathbf{AX}_r - \lambda\mathbf{X}_r \\ \mathbf{X}_{r-2} &= (\mathbf{A} - \lambda\mathbf{I})\mathbf{X}_{r-1} = \mathbf{AX}_{r-1} - \lambda\mathbf{X}_{r-1} \\ &\cdots\cdots\cdots\cdots\cdots\cdots \\ \mathbf{X}_2 &= (\mathbf{A} - \lambda\mathbf{I})\mathbf{X}_3 = \mathbf{AX}_3 - \lambda\mathbf{X}_3 \\ \mathbf{X}_1 &= (\mathbf{A} - \lambda\mathbf{I})\mathbf{X}_2 = \mathbf{AX}_2 - \lambda\mathbf{X}_2 \end{aligned}$$

which may be rewritten in reverse order as

$$\begin{aligned} \mathbf{AX}_2 &= \lambda\mathbf{X}_2 + \mathbf{X}_1 \\ \mathbf{AX}_3 &= \lambda\mathbf{X}_3 + \mathbf{X}_2 \\ &\cdots\cdots\cdots \\ \mathbf{AX}_r &= \lambda\mathbf{X}_r + \mathbf{X}_{r-1} \end{aligned}$$

In addition, since $\mathbf{X}_1$ is an eigenvector, $\mathbf{AX}_1 = \lambda\mathbf{X}_1$.

Define **J** to be the $r \times r$ Jordan block with λ on the diagonal. Then

$$\begin{aligned} \mathbf{AM} &= \mathbf{A}[\mathbf{X}_1, \mathbf{X}_2, \mathbf{X}_3, \ldots, \mathbf{X}_r] = [\mathbf{AX}_1, \mathbf{AX}_2, \mathbf{AX}_3, \ldots, \mathbf{AX}_r] \\ &= [\lambda\mathbf{X}_1, \lambda\mathbf{X}_2 + \mathbf{X}_1, \lambda\mathbf{X}_3 + \mathbf{X}_2, \ldots, \lambda\mathbf{X}_r + \mathbf{X}_{r-1}] \\ &= [\mathbf{X}_1, \mathbf{X}_2, \mathbf{X}_3, \ldots, \mathbf{X}_r]\mathbf{J} = \mathbf{MJ} \end{aligned}$$

and (*10.2*) follows immediately.

10.14 Calculate sin **J** for the diagonal matrix

$$\mathbf{J} = \begin{bmatrix} 3 & 0 & 0 \\ 0 & 3 & 0 \\ 0 & 0 & 14 \end{bmatrix}$$

Here $f(\lambda) = \sin \lambda$, so it follows from (*10.3*) that

$$f(\mathbf{J}) = \begin{bmatrix} \sin 3 & 0 & 0 \\ 0 & \sin 3 & 0 \\ 0 & 0 & \sin 14 \end{bmatrix} = \begin{bmatrix} 0.141120 & 0 & 0 \\ 0 & 0.141120 & 0 \\ 0 & 0 & 0.990607 \end{bmatrix}$$

10.15 Calculate $\sin \mathbf{J}$ for the Jordan block

$$\mathbf{J}=\begin{bmatrix}2&1&0&0\\0&2&1&0\\0&0&2&1\\0&0&0&2\end{bmatrix}$$

Here $f(\lambda)=\sin\lambda$, $f'(\lambda)=\cos\lambda$, $f''(\lambda)=-\sin\lambda$, and $f'''(\lambda)=-\cos\lambda$, so $f(2)=\sin 2$, $f'(2)=\cos 2$, $f''(2)=-\sin 2$, and $f'''(2)=-\cos 2$. It follows from (*10.4*) that

$$f(\mathbf{J})=\begin{bmatrix}\dfrac{\sin 2}{1}&\dfrac{\cos 2}{1}&\dfrac{-\sin 2}{2}&\dfrac{-\cos 2}{6}\\0&\dfrac{\sin 2}{1}&\dfrac{\cos 2}{1}&\dfrac{-\sin 2}{2}\\0&0&\dfrac{\sin 2}{1}&\dfrac{\cos 2}{1}\\0&0&0&\dfrac{\sin 2}{1}\end{bmatrix}=\begin{bmatrix}0.909297&-0.416147&-0.454649&0.0693578\\0&0.909297&-0.416147&-0.454649\\0&0&0.909297&-0.416147\\0&0&0&0.909297\end{bmatrix}$$

10.16 Calculate $\cos \mathbf{J}$ for

$$\mathbf{J}=\begin{bmatrix}0&0&0&0&0&0&0\\0&0&0&0&0&0&0\\0&0&0&1&0&0&0\\0&0&0&0&0&0&0\\0&0&0&0&0&1&0\\0&0&0&0&0&0&1\\0&0&0&0&0&0&0\end{bmatrix}$$

This matrix is in Jordan canonical form, with all eigenvalues equal to zero and with the blocks

$$\mathbf{D}=\begin{bmatrix}0&0\\0&0\end{bmatrix}\qquad \mathbf{J}_1=\begin{bmatrix}0&1\\0&0\end{bmatrix}\qquad \mathbf{J}_2=\begin{bmatrix}0&1&0\\0&0&1\\0&0&0\end{bmatrix}$$

Since $f(\lambda)=\cos\lambda$, we have $f(0)=\cos 0$, $f'(0)=-\sin 0$, and $f''(0)=-\cos 0$. Therefore,

$$\cos\mathbf{D}=\begin{bmatrix}\cos 0&0\\0&\cos 0\end{bmatrix}=\begin{bmatrix}1&0\\0&1\end{bmatrix}$$

$$\cos\mathbf{J}_1=\begin{bmatrix}\dfrac{\cos 0}{0!}&\dfrac{-\sin 0}{1!}\\0&\dfrac{\cos 0}{0!}\end{bmatrix}=\begin{bmatrix}1&0\\0&1\end{bmatrix}$$

$$\cos\mathbf{J}_2=\begin{bmatrix}\dfrac{\cos 0}{0!}&\dfrac{-\sin 0}{1!}&\dfrac{-\cos 0}{2!}\\0&\dfrac{\cos 0}{0!}&\dfrac{-\sin 0}{1!}\\0&0&\dfrac{\cos 0}{0!}\end{bmatrix}=\begin{bmatrix}1&0&-1/2\\0&1&0\\0&0&1\end{bmatrix}$$

and
$$\cos\mathbf{J}=\begin{bmatrix}\cos\mathbf{D}&&\mathbf{0}\\&\cos\mathbf{J}_1&\\\mathbf{0}&&\cos\mathbf{J}_2\end{bmatrix}=\begin{bmatrix}1&0&0&0&0&0&0\\0&1&0&0&0&0&0\\0&0&1&0&0&0&0\\0&0&0&1&0&0&0\\0&0&0&0&1&0&-1/2\\0&0&0&0&0&1&0\\0&0&0&0&0&0&1\end{bmatrix}$$

10.17 Calculate sin **A** for the matrix given in Problem 10.6.

Using the results of Problems 10.6, 10.8, and 10.14 along with (*10.2*), we have

$$\sin \mathbf{A} = \mathbf{M}(\sin \mathbf{J})\mathbf{M}^{-1} = \begin{bmatrix} -1 & -1 & 2 \\ 1 & 0 & 3 \\ 0 & 1 & 6 \end{bmatrix} \begin{bmatrix} 0.141120 & 0 & 0 \\ 0 & 0.141120 & 0 \\ 0 & 0 & 0.990607 \end{bmatrix} \begin{bmatrix} -3/11 & 8/11 & -3/11 \\ -6/11 & -6/11 & 5/11 \\ 1/11 & 1/11 & 1/11 \end{bmatrix}$$

$$= \begin{bmatrix} 0.295572 & 0.154452 & 0.154452 \\ 0.231678 & 0.372798 & 0.231678 \\ 0.463357 & 0.463357 & 0.604477 \end{bmatrix}$$

10.18 Calculate $e^{\mathbf{A}t}$ for

$$\mathbf{A} = \begin{bmatrix} 2 & 1 & 0 \\ 0 & 2 & 1 \\ 0 & 0 & 2 \end{bmatrix}$$

We set

$$\mathbf{B} = \mathbf{A}t = \begin{bmatrix} 2t & t & 0 \\ 0 & 2t & t \\ 0 & 0 & 2t \end{bmatrix}$$

(see Chapter 8) and calculate $e^{\mathbf{B}}$. Even though **A** is a Jordan block, **B** is not because it no longer has 1s on the superdiagonal. We find that modal matrix for **B** is

$$\mathbf{M} = \begin{bmatrix} t^2 & 0 & 0 \\ 0 & t & 0 \\ 0 & 0 & 1 \end{bmatrix}$$

so **B** is similar to the matrix in Jordan canonical form

$$\mathbf{J} = \mathbf{M}^{-1}\mathbf{B}\mathbf{M} = \begin{bmatrix} 2t & 1 & 0 \\ 0 & 2t & 1 \\ 0 & 0 & 2t \end{bmatrix}$$

We have $f(\lambda) = f'(\lambda) = f''(\lambda) = e^{\lambda}$, so

$$f(\mathbf{J}) = e^{\mathbf{J}} = \begin{bmatrix} e^{2t} & e^{2t} & e^{2t}/2 \\ 0 & e^{2t} & e^{2t} \\ 0 & 0 & e^{2t} \end{bmatrix}$$

and

$$e^{\mathbf{A}t} = e^{\mathbf{B}} = f(\mathbf{B}) = \mathbf{M}f(\mathbf{J})\mathbf{M}^{-1} = e^{2t}\begin{bmatrix} 1 & t & t^2/2 \\ 0 & 1 & t \\ 0 & 0 & 1 \end{bmatrix}$$

10.19 Find $e^{\mathbf{A}t}$ for

$$\mathbf{A} = \begin{bmatrix} 0 & 1 \\ -1 & 0 \end{bmatrix}$$

We set

$$\mathbf{B} = \mathbf{A}t = \begin{bmatrix} 0 & t \\ -t & 0 \end{bmatrix}$$

and compute $e^{\mathbf{B}}$. The eigenvalues for **B** are the complex conjugates it and $-it$, so **B** is similar to

$$\mathbf{J} = \begin{bmatrix} it & 0 \\ 0 & -it \end{bmatrix}$$

A modal matrix for **B**, consisting of two generalized eigenvectors of rank 1, is

$$\mathbf{M} = \begin{bmatrix} 1 & i \\ i & 1 \end{bmatrix}$$

Thus,
$$e^{\mathbf{A}t} = e^{\mathbf{B}} = \mathbf{M}e^{\mathbf{J}}\mathbf{M}^{-1} = \begin{bmatrix} 1 & i \\ i & 1 \end{bmatrix}\begin{bmatrix} e^{it} & 0 \\ 0 & e^{-it} \end{bmatrix}\begin{bmatrix} 1/2 & -i/2 \\ -i/2 & 1/2 \end{bmatrix}$$

$$= \begin{bmatrix} \dfrac{e^{it} + e^{-it}}{2} & \dfrac{-ie^{it} + ie^{-it}}{2} \\ \dfrac{ie^{it} - ie^{-it}}{2} & \dfrac{e^{it} + e^{-it}}{2} \end{bmatrix} = \begin{bmatrix} \cos t & \sin t \\ -\sin t & \cos t \end{bmatrix}$$

(Compare with Problem 8.7.)

Supplementary Problems

10.20 Determine which of the following pairs of matrices are similar matrices:

(*a*) $\begin{bmatrix} 2 & 3 \\ 5 & 7 \end{bmatrix}$ and $\begin{bmatrix} 3 & 7 \\ 7 & 5 \end{bmatrix}$

(*b*) $\begin{bmatrix} 2 & 1 \\ 0 & 2 \end{bmatrix}$ and $\begin{bmatrix} 2 & 0 \\ 0 & 2 \end{bmatrix}$

(*c*) $\begin{bmatrix} 1 & 2 & 3 \\ 4 & 5 & 6 \\ 7 & 8 & 9 \end{bmatrix}$ and $\begin{bmatrix} 2 & 2 & 1 \\ 1 & 1 & 2 \\ 1 & 2 & 2 \end{bmatrix}$

In Problems 10.21 through 10.25, find a modal matrix associated with the given matrix.

10.21 $\begin{bmatrix} -5 & -3 \\ 3 & 1 \end{bmatrix}$

(*Hint:* See Problem 9.23.)

10.22 $\begin{bmatrix} 1 & 3 \\ 1 & 3 \end{bmatrix}$

(*Hint:* See Problem 9.24.)

10.23 $\begin{bmatrix} 2 & 2 & 1 \\ 0 & 2 & 0 \\ 0 & 0 & 2 \end{bmatrix}$

(*Hint:* See Problem 9.25.)

10.24 $\begin{bmatrix} 2 & 2 & 2 \\ 0 & 4 & 0 \\ 3 & -3 & 1 \end{bmatrix}$

(*Hint:* See Problem 9.26.)

10.25 $\begin{bmatrix} 3 & 2 & 1 & 0 \\ 0 & 3 & 0 & 0 \\ -1 & 1 & 1 & 0 \\ 0 & 1 & 1 & 3 \end{bmatrix}$

(*Hint:* See Problem 9.29.)

10.26 Each of the following is a complete set of eigenvalue rank numbers for a matrix whose only eigenvalue is $\lambda = 2$. Find, in each case, a matrix in Jordan canonical form which is similar to that matrix.

(*a*) $N_2 = N_1 = 2$ (*b*) $N_2 = 1, N_1 = 3$
(*c*) $N_3 = N_2 = 1, N_1 = 2$ (*d*) $N_3 = 2, N_2 = 3, N_1 = 3$
(*e*) $N_3 = 1, N_2 = 2, N_1 = 5$ (*f*) $N_2 = 3, N_1 = 5$

In Problems 10.27 through 10.33, find a matrix in Jordan canonical form that is similar to the given matrix.

10.27 The matrix in Problem 10.21. **10.28** The matrix in Problem 10.22.

10.29 The matrix in Problem 10.23.

10.30 The matrix in Problem 10.24.

10.31 The matrix in Problem 10.25.

10.32 $\begin{bmatrix} 3 & 1 & 1 & 0 \\ 0 & 3 & 0 & 0 \\ -1 & 1 & 1 & 0 \\ 0 & 1 & 1 & 3 \end{bmatrix}$

(*Hint:* See Problem 9.30.)

10.33 $\begin{bmatrix} 5 & 1 & 2 & 0 & 0 \\ 0 & 5 & -1 & 0 & 0 \\ 0 & 0 & 5 & 0 & 0 \\ 0 & 0 & 0 & 3 & 2 \\ 0 & 0 & 0 & 1 & 4 \end{bmatrix}$

(*Hint:* See Problem 9.31.)

10.34 Find $\cos \mathbf{A}$ for the following matrices:

(*a*) $\mathbf{A} = \begin{bmatrix} 2 & 0 & 0 \\ 0 & 2 & 0 \\ 0 & 0 & 2 \end{bmatrix}$ (*b*) $\mathbf{A} = \begin{bmatrix} 2 & 1 & 0 \\ 0 & 2 & 1 \\ 0 & 0 & 2 \end{bmatrix}$

(*c*) $\mathbf{A} = \begin{bmatrix} 2 & 1 & 0 & 0 \\ 0 & 2 & 0 & 0 \\ 0 & 0 & 2 & 1 \\ 0 & 0 & 0 & 2 \end{bmatrix}$ (*d*) $\mathbf{A} = \begin{bmatrix} 2 & 0 & 0 & 0 & 0 & 0 \\ 0 & 2 & 0 & 0 & 0 & 0 \\ 0 & 0 & 2 & 1 & 0 & 0 \\ 0 & 0 & 0 & 2 & 1 & 0 \\ 0 & 0 & 0 & 0 & 2 & 1 \\ 0 & 0 & 0 & 0 & 0 & 2 \end{bmatrix}$

10.35 Find $e^{\mathbf{A}}$ for the matrix $\mathbf{A}$ given in Problem 10.21.

10.36 Find $e^{\mathbf{A}}$ for the matrix $\mathbf{A}$ given in Problem 10.23.

10.37 Find $\mathbf{A}^{31}$ for the matrix $\mathbf{A}$ given in Problem 10.23.

In Problems 10.38 through 10.41, determine $e^{\mathbf{A}t}$.

10.38 $\mathbf{A} = \begin{bmatrix} 2 & 1 \\ 0 & 2 \end{bmatrix}$

10.39 $\mathbf{A} = \begin{bmatrix} 5 & 6 \\ -4 & -5 \end{bmatrix}$

10.40 $\mathbf{A} = \begin{bmatrix} 2 & 0 & 0 \\ 0 & 2 & 1 \\ 0 & 0 & 2 \end{bmatrix}$

10.41 $\mathbf{A} = \begin{bmatrix} -1 & 1 & 0 \\ 0 & 2 & 1 \\ 0 & 0 & 2 \end{bmatrix}$

10.42 Prove that if $\mathbf{A}$ is similar to $\mathbf{B}$, then $\mathbf{B}$ is similar to $\mathbf{A}$.

10.43 Prove that if (*10.1*) is valid for $\mathbf{A}$ and $\mathbf{B}$ and $\mathbf{X}$ is an eigenvector for $\mathbf{A}$ corresponding to eigenvalue λ, then $\mathbf{Y} = \mathbf{SX}$ is an eigenvector for $\mathbf{B}$ also corresponding to λ.

Chapter 11

Inner Products

COMPLEX CONJUGATES

The complex conjugate of a scalar $z = a + ib$ (where a and b are real) is $\bar{z} = a - ib$; the *complex conjugate of a matrix* **A** is the matrix $\bar{\mathbf{A}}$ whose elements are the complex conjugates of the elements of **A**. The following properties are valid for scalars x and y and matrices **A** and **B**:

(C1): $\bar{\bar{x}} = x$; and $\bar{\bar{\mathbf{A}}} = \mathbf{A}$.

(C2): x is real if and only if $\bar{x} = x$; and **A** is a real matrix if and only if $\bar{\mathbf{A}} = \mathbf{A}$.

(C3): $x + \bar{x}$ is a real scalar; and $\mathbf{A} + \bar{\mathbf{A}}$ is a real matrix.

(C4): $\overline{xy} = (\bar{x})(\bar{y})$; and $\overline{\mathbf{AB}} = (\bar{\mathbf{A}})(\bar{\mathbf{B}})$ if the latter product is defined.

(C5): $\overline{(x + y)} = \bar{x} + \bar{y}$; and $\overline{(\mathbf{A} + \mathbf{B})} = \bar{\mathbf{A}} + \bar{\mathbf{B}}$ if the latter sum is defined.

(C6): $x\bar{x} = |x|^2$ is always real and positive, except that $x\bar{x} = 0$ when $x = 0$.

THE INNER PRODUCT

Let **W** denote a nonsingular $n \times n$ matrix. The *inner product* of n-dimensional column vectors **X** and **Y** with respect to **W**, denoted by $\langle \mathbf{X}, \mathbf{Y} \rangle_{\mathbf{W}}$, is the dot product (see Chapter 1)

$$\langle \mathbf{X}, \mathbf{Y} \rangle_{\mathbf{W}} = (\mathbf{WX}) \cdot (\overline{\mathbf{WY}}) \tag{11.1}$$

If $\mathbf{W} = \mathbf{I}$, then the subscript in *(11.1)* is dropped, and the inner product

$$\langle \mathbf{X}, \mathbf{Y} \rangle = \mathbf{X} \cdot \bar{\mathbf{Y}} \tag{11.2}$$

is called the *Euclidean inner product*. If **X** and **Y** are also real, then the Euclidean inner product reduces to the dot product of the two vectors. (See Problems 11.1 through 11.5.)

PROPERTIES OF INNER PRODUCTS

Property 11.1: $\langle \mathbf{X}, \mathbf{X} \rangle_{\mathbf{W}}$ is real and positive if $\mathbf{X} \neq \mathbf{0}$.

Property 11.2: $\langle \mathbf{X}, \mathbf{X} \rangle_{\mathbf{W}} = \mathbf{0}$ if and only if $\mathbf{X} = \mathbf{0}$.

Property 11.3: $\langle \mathbf{X}, \mathbf{Y} \rangle_{\mathbf{W}} = \overline{\langle \mathbf{Y}, \mathbf{X} \rangle_{\mathbf{W}}}$.

Property 11.4: $\langle c\mathbf{X}, \mathbf{Y} \rangle_{\mathbf{W}} = c\langle \mathbf{X}, \mathbf{Y} \rangle_{\mathbf{W}}$ and $\langle \mathbf{X}, c\mathbf{Y} \rangle_{\mathbf{W}} = \bar{c}\langle \mathbf{X}, \mathbf{Y} \rangle_{\mathbf{W}}$ for any scalar c, real or complex.

Property 11.5: $\langle \mathbf{X} + \mathbf{Y}, \mathbf{Z} \rangle_{\mathbf{W}} = \langle \mathbf{X}, \mathbf{Z} \rangle_{\mathbf{W}} + \langle \mathbf{Y}, \mathbf{Z} \rangle_{\mathbf{W}}$ and $\langle \mathbf{X}, \mathbf{Y} + \mathbf{Z} \rangle_{\mathbf{W}} = \langle \mathbf{X}, \mathbf{Y} \rangle_{\mathbf{W}} + \langle \mathbf{X}, \mathbf{Z} \rangle_{\mathbf{W}}$.

Property 11.6 (Schwarz inequality): $|\langle \mathbf{X}, \mathbf{Y} \rangle_{\mathbf{W}}|^2 \leq \langle \mathbf{X}, \mathbf{X} \rangle_{\mathbf{W}} \langle \mathbf{Y}, \mathbf{Y} \rangle_{\mathbf{W}}$.

(See Problems 11.9, 11.10, 11.12, 11.28, and 11.29.)

ORTHOGONALITY

Two vectors are *orthogonal* if their inner product is zero. Since different matrices **W** in *(11.1)* generate different inner products, two vectors may be orthogonal under one inner product and not orthogonal under another inner product. (See Problem 11.5.) Orthogonality reduces to the geometric concept of perpendicularity under the Euclidean inner product when the vectors are real and restricted to two or three dimensions.

A *set* of vectors is orthogonal if each vector in the set is orthogonal to every other vector in that set. Such a set is linearly independent when the vectors are all nonzero. (See Problem 11.27.)

GRAM-SCHMIDT ORTHOGONALIZATION

Every finite set of linearly independent vectors $\{\mathbf{X}_1, \mathbf{X}_2, \ldots, \mathbf{X}_n\}$ has associated with it an orthogonal set of nonzero vectors $\{\mathbf{Q}_1, \mathbf{Q}_2, \ldots, \mathbf{Q}_n\}$ with respect to a specified inner product, such that each vector $\mathbf{Q}_j$ $(j = 1, 2, \ldots, n)$ is a linear combination of $\mathbf{X}_1$ through $\mathbf{X}_{j-1}$. The following algorithm for producing the vectors $\mathbf{Q}_j$ is called the *Gram-Schmidt orthogonalization process*.

STEP 11.1: Set

$$\mathbf{Q}_1 = \frac{1}{\sqrt{\langle \mathbf{X}_1, \mathbf{X}_1 \rangle_\mathbf{W}}} \mathbf{X}_1 \qquad \text{and } j = 1$$

STEP 11.2: If $j = n$, stop; the algorithm is complete. Otherwise, increase j by 1 and continue.

STEP 11.3: Calculate

$$\mathbf{Y}_j = \mathbf{X}_j - \sum_{i=1}^{j-1} \langle \mathbf{X}_j, \mathbf{Q}_i \rangle_\mathbf{W} \mathbf{Q}_i$$

STEP 11.4: Set

$$\mathbf{Q}_j = \frac{1}{\sqrt{\langle \mathbf{Y}_j, \mathbf{Y}_j \rangle_\mathbf{W}}} \mathbf{Y}_j$$

STEP 11.5: Return to Step 11.2.

(See Problems 11.6 through 11.8) A modification of this algorithm, which is less susceptible to roundoff error, is presented in Chapter 20.

Besides producing orthogonal vectors, the Gram-Schmidt process generates vectors having the property that the inner product of each vector with itself is unity. This property is discussed further in the next chapter.

Solved Problems

11.1 Calculate $\langle \mathbf{X}, \mathbf{Y} \rangle_\mathbf{W}$ if

$$\mathbf{X} = \begin{bmatrix} 1 \\ 2 \\ 3 \end{bmatrix} \qquad \mathbf{Y} = \begin{bmatrix} 4 \\ 5 \\ 6 \end{bmatrix} \qquad \mathbf{W} = \begin{bmatrix} 1 & 1 & 0 \\ 0 & 1 & 1 \\ 1 & 0 & 1 \end{bmatrix}$$

All elements are real, so the conjugate notation in (*11.1*) can be suppressed. Thus,

$$\langle \mathbf{X}, \mathbf{Y} \rangle_\mathbf{W} = (\mathbf{WX}) \cdot (\mathbf{WY}) = \begin{bmatrix} 3 \\ 5 \\ 4 \end{bmatrix} \cdot \begin{bmatrix} 9 \\ 11 \\ 10 \end{bmatrix} = 3(9) + 5(11) + 4(10) = 122$$

11.2 Calculate the Euclidean inner product for the vectors given in Problem 11.1.

Since both vectors are real, the Euclidean inner product is the dot product of those vectors: $\langle \mathbf{X}, \mathbf{Y} \rangle = 1(4) + 2(5) + 3(6) = 32$.

11.3 Calculate $\langle \mathbf{X}, \mathbf{X} \rangle_{\mathbf{W}}$ if

$$\mathbf{X} = \begin{bmatrix} 1-i \\ i \end{bmatrix} \quad \text{and} \quad \mathbf{W} = \begin{bmatrix} 1 & i \\ 2+i5 & 0 \end{bmatrix}$$

Here
$$\mathbf{WX} = \begin{bmatrix} -i \\ 7+i3 \end{bmatrix} \quad \text{so} \quad \overline{\mathbf{WX}} = \begin{bmatrix} i \\ 7-i3 \end{bmatrix}$$

and
$$\langle \mathbf{X}, \mathbf{X} \rangle_{\mathbf{W}} = (\mathbf{WX}) \cdot (\overline{\mathbf{WX}}) = \begin{bmatrix} -i \\ 7+i3 \end{bmatrix} \cdot \begin{bmatrix} i \\ 7-i3 \end{bmatrix} = (-i)(i) + (7+i3)(7-i3) = 59$$

11.4 Calculate $\langle \mathbf{X}, \mathbf{Y} \rangle_{\mathbf{W}}$ if

$$\mathbf{X} = \begin{bmatrix} -i2 \\ 3-i2 \\ i \end{bmatrix} \quad \mathbf{Y} = \begin{bmatrix} i2 \\ 2+i \\ 7 \end{bmatrix} \quad \mathbf{W} = \begin{bmatrix} 1 & 2 & 3 \\ 4 & 5 & 6 \\ 7 & 8 & 9 \end{bmatrix}$$

W is a singular matrix, so the inner product $\langle \mathbf{X}, \mathbf{Y} \rangle_{\mathbf{W}}$ is not defined, even though the matrix operations on the right side of Eq. (*11.1*) can be performed. When the matrix **W** is singular, it is always possible to find a nonzero vector **Z** (in this case, $\mathbf{Z} = [1, -2, 1]^T$ will do) for which $\langle \mathbf{Z}, \mathbf{Z} \rangle_{\mathbf{W}} = 0$, thereby violating Property 11.2.

11.5 Calculate the inner product $\langle \mathbf{X}, \mathbf{Y} \rangle_{\mathbf{W}}$ for the vectors given in Problem 11.4 when (*a*) $\mathbf{W} = \mathbf{I}$ and (*b*) **W** is as given in Problem 11.1.

(*a*) The Euclidean inner product is

$$\langle \mathbf{X}, \mathbf{Y} \rangle = \mathbf{X} \cdot \bar{\mathbf{Y}} = \begin{bmatrix} -i2 \\ 3-i2 \\ i \end{bmatrix} \cdot \begin{bmatrix} -i2 \\ 2-i \\ 7 \end{bmatrix} = (-i2)(-i2) + (3-i2)(2-i) + (i)(7) = 0$$

(*b*) With **W** as given in Problem 11.1, we have

$$\mathbf{WX} = [3-i4,\ \ 3-i,\ \ -i]^T \quad \text{and} \quad \mathbf{WY} = [2+i3,\ \ 9+i,\ \ 7+i2]^T$$

so
$$\langle \mathbf{X}, \mathbf{Y} \rangle_{\mathbf{W}} = (\mathbf{WX}) \cdot (\overline{\mathbf{WY}}) = (3-i4)(2-i3) + (3-i)(9-i) + (-i)(7-i2) = 18-i36$$

Thus, **X** and **Y** are orthogonal under the Euclidean inner product but not orthogonal under the inner product in part *b*.

11.6 Use the Gram-Schmidt orthogonalization process with the Euclidean inner product to construct an orthogonal set of vectors associated with $\{\mathbf{X}_1, \mathbf{X}_2\}$ when

$$\mathbf{X}_1 = \begin{bmatrix} 2 \\ 1+i2 \end{bmatrix} \quad \text{and} \quad \mathbf{X}_2 = \begin{bmatrix} i \\ 1+i \end{bmatrix}$$

These two vectors are linearly independent, so Steps 11.1 through 11.5 may be used to find

$$\langle \mathbf{X}_1, \mathbf{X}_1 \rangle = \mathbf{X}_1 \cdot \bar{\mathbf{X}}_1 = 9$$

and
$$\mathbf{Q}_1 = \frac{1}{\sqrt{9}} \begin{bmatrix} 2 \\ 1+i2 \end{bmatrix} = \begin{bmatrix} 2/3 \\ (1+i2)/3 \end{bmatrix}$$

Also,
$$\mathbf{Y}_2 = \mathbf{X}_2 - \langle \mathbf{X}_2, \mathbf{Q}_1 \rangle \mathbf{Q}_1 = \begin{bmatrix} i \\ 1+i \end{bmatrix} - \left(1 + \frac{i}{3}\right) \begin{bmatrix} 2/3 \\ (1+i2)/3 \end{bmatrix} = \begin{bmatrix} (-6+i7)/9 \\ (8+i2)/9 \end{bmatrix}$$

so
$$\langle \mathbf{Y}_2, \mathbf{Y}_2 \rangle = \mathbf{Y}_2 \cdot \bar{\mathbf{Y}}_2 = 153/81$$

$$\text{and}\quad \mathbf{Q}_2 = \frac{1}{\sqrt{153/81}}\begin{bmatrix}(-6+i7)/9\\(8+i2)/9\end{bmatrix} = \begin{bmatrix}(-6+i7)/\sqrt{153}\\(8+i2)/\sqrt{153}\end{bmatrix}.$$

The orthogonal set is $\{\mathbf{Q}_1, \mathbf{Q}_2\}$.

11.7 Use the Gram-Schmidt orthogonalization process with the Euclidean inner product to construct an orthogonal set of vectors associated with $\{\mathbf{X}_1, \mathbf{X}_2, \mathbf{X}_3, \mathbf{X}_4\}$ when

$$\mathbf{X}_1 = \begin{bmatrix}0\\1\\1\\1\end{bmatrix} \qquad \mathbf{X}_2 = \begin{bmatrix}1\\0\\1\\1\end{bmatrix} \qquad \mathbf{X}_3 = \begin{bmatrix}1\\1\\0\\1\end{bmatrix} \qquad \mathbf{X}_4 = \begin{bmatrix}1\\1\\1\\0\end{bmatrix}$$

These vectors can be shown to be linearly independent (see Chapter 6). Using Steps 11.1 through 11.5, we find

$$\langle \mathbf{X}_1, \mathbf{X}_1 \rangle = \mathbf{X}_1 \cdot \mathbf{X}_1 = 3$$

and $\mathbf{Q}_1 = \frac{1}{\sqrt{3}}\mathbf{Y}_1 = [0,\ 1/\sqrt{3},\ 1/\sqrt{3},\ 1/\sqrt{3}]^T$

Then
$$\mathbf{Y}_2 = \mathbf{X}_2 - \langle \mathbf{X}_2, \mathbf{0}_1 \rangle \mathbf{0}_1 = \mathbf{X}_2 - \frac{2}{\sqrt{3}}\mathbf{Q}_1$$
$$= [1,\ -2/3,\ 1/3,\ 1/3]^T$$

so
$$\langle \mathbf{Y}_2, \mathbf{Y}_2 \rangle = \mathbf{Y}_2 \cdot \mathbf{Y}_2 = 15/9$$

and $\mathbf{Q}_2 = \sqrt{9/15}\mathbf{Y}_2 = [3/\sqrt{15},\ -2/\sqrt{15},\ 1/\sqrt{15},\ 1/\sqrt{15}]^T$

Also,
$$\mathbf{Y}_3 = \mathbf{X}_3 - \langle \mathbf{X}_3, \mathbf{Q}_1 \rangle \mathbf{Q}_1 - \langle \mathbf{X}_3, \mathbf{Q}_2 \rangle \mathbf{Q}_2$$
$$= \mathbf{X}_3 - \frac{2}{\sqrt{3}}\mathbf{Q}_1 - \frac{2}{\sqrt{15}}\mathbf{Q}_2 = [3/5,\ 3/5,\ -4/5,\ 1/5]^T$$

so
$$\langle \mathbf{Y}_3, \mathbf{Y}_3 \rangle = \mathbf{Y}_3 \cdot \mathbf{Y}_3 = 35/25$$

and $\mathbf{Q}_3 = \sqrt{25/35}\mathbf{Y}_3 = [3/\sqrt{35},\ 3/\sqrt{35},\ -4/\sqrt{35},\ 1/\sqrt{35}]^T$

Lastly,
$$\mathbf{Y}_4 = \mathbf{X}_4 - \langle \mathbf{X}_4, \mathbf{Q}_1 \rangle \mathbf{Q}_1 - \langle \mathbf{X}_4, \mathbf{Q}_2 \rangle \mathbf{Q}_2 - \langle \mathbf{X}_4, \mathbf{Q}_3 \rangle \mathbf{Q}_3$$
$$= \mathbf{X}_4 - \frac{2}{\sqrt{3}}\mathbf{Q}_1 - \frac{2}{\sqrt{15}}\mathbf{Q}_2 - \frac{2}{\sqrt{35}}\mathbf{Q}_3$$
$$= [3/7,\ 3/7,\ 3/7,\ -6/7]^T$$

so
$$\langle \mathbf{Y}_4, \mathbf{Y}_4 \rangle = \mathbf{Y}_4 \cdot \mathbf{Y}_4 = 63/49$$

and $\mathbf{Q}_4 = \sqrt{49/63}\mathbf{Y}_4 = [3/\sqrt{63},\ 3/\sqrt{63},\ 3/\sqrt{63},\ -6/\sqrt{63}]^T$
$= [1/\sqrt{7},\ 1/\sqrt{7},\ 1/\sqrt{7},\ -2/\sqrt{7}]^T$

The orthogonal set is $\{\mathbf{Q}_1, \mathbf{Q}_2, \mathbf{Q}_3, \mathbf{Q}_4\}$.

11.8 Use the Gram-Schmidt orthogonalization process to construct an orthogonal set of vectors associated with the set $\{\mathbf{X}_1, \mathbf{X}_2, \mathbf{X}_3\}$ from Problem 11.7 and with

$$\mathbf{W} = \begin{bmatrix} 1 & -1 & 2 & 0 \\ 0 & 1 & 2 & 3 \\ 2 & 1 & 1 & 2 \\ 2 & -1 & -2 & 0 \end{bmatrix}$$

Using Steps 11.1 through 11.5, we calculate

$$\mathbf{WX}_1 = [1, 6, 4, -3]^T$$

so $$\langle \mathbf{X}_1, \mathbf{X}_1 \rangle_\mathbf{W} = (\mathbf{WX}_1) \cdot (\mathbf{WX}_1) = 62$$

and $$\mathbf{Q}_1 = \frac{1}{\sqrt{62}}\mathbf{X}_1 = [0,\ 1/\sqrt{62},\ 1/\sqrt{62},\ 1/\sqrt{62}]^T$$

Then $$\mathbf{WQ}_1 = [1/\sqrt{62},\ 6/\sqrt{62},\ 4/\sqrt{62},\ -3/\sqrt{62}]^T$$

and $$\mathbf{WX}_2 = [3,\ 5,\ 5,\ 0]^T$$

so $$\langle \mathbf{X}_2, \mathbf{Q}_1 \rangle_\mathbf{W} = (\mathbf{WX}_2) \cdot (\mathbf{WQ}_1) = 53/\sqrt{62}$$

and $$\mathbf{Y}_2 = \mathbf{X}_2 - \langle \mathbf{X}_2, \mathbf{Q}_1 \rangle_\mathbf{W}\mathbf{Q}_1 = \mathbf{X}_2 - \frac{53}{\sqrt{62}}\mathbf{Q}_1$$

$$= [1,\ -53/62,\ 9/62,\ 9/62]^T$$

Now $$\mathbf{WY}_2 = [133/62,\ -8/62,\ 98/62,\ 159/62]^T$$

so $$\langle \mathbf{Y}_2, \mathbf{Y}_2 \rangle_\mathbf{W} = (\mathbf{WY}_2) \cdot (\mathbf{WY}_2) = 52{,}638/(62)^2$$

and $$\mathbf{Q}_2 = \frac{62}{\sqrt{52{,}638}}\mathbf{Y}_2 = \frac{1}{\sqrt{52{,}638}}[62,\ -53,\ 9,\ 9]^T$$

Lastly, $$\mathbf{WQ}_2 = \frac{1}{\sqrt{52{,}638}}[133,\ -8,\ 98,\ 159]^T$$

and $$\mathbf{WX}_3 = [0,\ 4,\ 5,\ 1]^T$$

so $$\langle \mathbf{X}_3, \mathbf{Q}_1 \rangle_\mathbf{W} = (\mathbf{WX}_3) \cdot (\mathbf{WQ}_1) = 41\sqrt{62}$$

and $$\langle \mathbf{X}_3, \mathbf{Q}_2 \rangle_\mathbf{W} = (\mathbf{WX}_3) \cdot (\mathbf{WQ}_2) = 617/\sqrt{52{,}638}$$

giving $$\mathbf{Y}_3 = \mathbf{X}_3 - \langle \mathbf{X}_3, \mathbf{Q}_1 \rangle_\mathbf{W}\mathbf{Q}_1 - \langle \mathbf{X}_3, \mathbf{Q}_2 \rangle_\mathbf{W}\mathbf{Q}_2$$

$$= \mathbf{X}_3 - \frac{41}{\sqrt{62}}\mathbf{Q}_1 - \frac{617}{\sqrt{52{,}638}}\mathbf{Q}_2$$

$$= \frac{1}{26{,}319}[7{,}192,\ 25{,}265,\ -20{,}181,\ 6{,}138]^T$$

Now $$\mathbf{WY}_3 = \frac{1}{26{,}319}[-58{,}435,\ 3{,}317,\ 31{,}744,\ 29{,}481]^T$$

so $$\langle \mathbf{Y}_3, \mathbf{Y}_3 \rangle_\mathbf{W} = (\mathbf{WY}_3) \cdot (\mathbf{WY}_3) = \frac{5{,}302{,}462{,}611}{(26{,}319)^2}$$

and $$\mathbf{Q}_3 = \frac{1}{\sqrt{5{,}302{,}462{,}611}}[7{,}192,\ 25,\ 265,\ -20{,}181,\ 6{,}138]^T$$

The orthogonal set is $\{\mathbf{Q}_1, \mathbf{Q}_2, \mathbf{Q}_3\}$.

11.9 Prove Property 11.3: $\langle \mathbf{X}, \mathbf{Y} \rangle_\mathbf{W} = \overline{\langle \mathbf{Y}, \mathbf{X} \rangle_\mathbf{W}}$.

$$\langle \mathbf{X}, \mathbf{Y} \rangle_\mathbf{W} = (\mathbf{WX}) \cdot (\overline{\mathbf{WY}}) = (\overline{\mathbf{WY}}) \cdot (\mathbf{WX}) = \overline{(\mathbf{WY}) \cdot (\overline{\mathbf{WX}})} = \overline{\langle \mathbf{Y}, \mathbf{X} \rangle_\mathbf{W}}$$

11.10 Prove Property 11.4.

$$\langle c\mathbf{X}, \mathbf{Y}\rangle_{\mathbf{W}} = (\mathbf{W}c\mathbf{X})\cdot(\overline{\mathbf{WY}}) = \{c(\mathbf{WX})\}\cdot(\overline{\mathbf{WY}}) = c\{(\mathbf{WX})\cdot(\overline{\mathbf{WY}})\} = c\langle \mathbf{X}, \mathbf{Y}\rangle_{\mathbf{W}}$$

and

$$\langle \mathbf{X}, c\mathbf{Y}\rangle_{\mathbf{W}} = (\mathbf{WX})\cdot(\overline{\mathbf{W}c\mathbf{Y}}) = (\mathbf{WX})\cdot\{\overline{c(\mathbf{WY})}\} = (\mathbf{WX})\cdot\{\bar{c}(\overline{\mathbf{WY}})\}$$
$$= \bar{c}\{(\mathbf{WX})\cdot(\overline{\mathbf{WY}})\} = \bar{c}\langle \mathbf{X}, \mathbf{Y}\rangle_{\mathbf{W}}$$

11.11 Prove that $\langle \mathbf{0}, \mathbf{Y}\rangle_{\mathbf{W}} = 0$ for any $\mathbf{Y}$ of appropriate dimension.

$$\langle \mathbf{0}, \mathbf{Y}\rangle_{\mathbf{W}} = \langle 0\mathbf{0}, \mathbf{Y}\rangle_{\mathbf{W}} = 0\langle \mathbf{0}, \mathbf{Y}\rangle_{\mathbf{W}} = 0$$

because the inner product is a scalar, and zero times any scalar is zero.

11.12 Prove the Schwarz inequality.

When $\mathbf{X} = \mathbf{0}$, both sides of the inequality are zero (see Problem 11.11), and the inequality is satisfied. If $\mathbf{X} \neq \mathbf{0}$, then $\langle \mathbf{X}, \mathbf{X}\rangle_{\mathbf{W}} \neq 0$ (Property 11.2), and for any vectors $\mathbf{X}$ and $\mathbf{Y}$ and any scalar c, we have

$$0 \le \langle c\mathbf{X} - \mathbf{Y}, c\mathbf{X} - \mathbf{Y}\rangle_{\mathbf{W}} \qquad \text{(Property 11.1)}$$
$$= \langle c\mathbf{X}, c\mathbf{X}\rangle_{\mathbf{W}} - \langle c\mathbf{X}, \mathbf{Y}\rangle_{\mathbf{W}} - \langle \mathbf{Y}, c\mathbf{X}\rangle_{\mathbf{W}} + \langle \mathbf{Y}, \mathbf{Y}\rangle_{\mathbf{W}} \qquad \text{(Property 11.5)}$$
$$= c\bar{c}\langle \mathbf{X}, \mathbf{X}\rangle_{\mathbf{W}} - c\langle \mathbf{X}, \mathbf{Y}\rangle_{\mathbf{W}} - \bar{c}\langle \mathbf{Y}, \mathbf{X}\rangle_{\mathbf{W}} + \langle \mathbf{Y}, \mathbf{Y}\rangle_{\mathbf{W}} \qquad \text{(Property 11.4)}$$

Setting $c = \overline{\langle \mathbf{X}, \mathbf{Y}\rangle_{\mathbf{W}}}/\langle \mathbf{X}, \mathbf{X}\rangle_{\mathbf{W}}$ and noting that $\overline{\langle \mathbf{X}, \mathbf{X}\rangle_{\mathbf{W}}} = \langle \mathbf{X}, \mathbf{X}\rangle_{\mathbf{W}}$ by Property 11.1, we cancel the first two terms on the right side of the last equality and obtain

$$0 \le \frac{-\langle \mathbf{X}, \mathbf{Y}\rangle_{\mathbf{W}}\langle \mathbf{Y}, \mathbf{X}\rangle_{\mathbf{W}}}{\langle \mathbf{X}, \mathbf{X}\rangle_{\mathbf{W}}} + \langle \mathbf{Y}, \mathbf{Y}\rangle_{\mathbf{W}} = \frac{-\langle \mathbf{X}, \mathbf{Y}\rangle_{\mathbf{W}}\overline{\langle \mathbf{X}, \mathbf{Y}\rangle_{\mathbf{W}}}}{\langle \mathbf{X}, \mathbf{X}\rangle_{\mathbf{W}}} + \langle \mathbf{Y}, \mathbf{Y}\rangle_{\mathbf{W}} \qquad \text{(Property 11.3)}$$
$$= \frac{-|\langle \mathbf{X}, \mathbf{Y}\rangle_{\mathbf{W}}|^2}{\langle \mathbf{X}, \mathbf{X}\rangle_{\mathbf{W}}} + \langle \mathbf{Y}, \mathbf{Y}\rangle_{\mathbf{W}} \qquad \text{(Property C6)}$$

Thus,

$$\frac{|\langle \mathbf{X}, \mathbf{Y}\rangle_{\mathbf{W}}|^2}{\langle \mathbf{X}, \mathbf{X}\rangle_{\mathbf{W}}} \le \langle \mathbf{Y}, \mathbf{Y}\rangle$$

from which the Schwarz inequality immediately follows.

Supplementary Problems

11.13 Calculate (*a*) $\langle \mathbf{X}, \mathbf{Y}\rangle$, (*b*) $\langle \mathbf{X}, \mathbf{Z}\rangle$, (*c*) $\langle \mathbf{Y}, \mathbf{Z}\rangle$, (*d*) $\langle \mathbf{X}, \mathbf{Y}\rangle_{\mathbf{A}}$, (*e*) $\langle \mathbf{X}, \mathbf{Z}\rangle_{\mathbf{A}}$, (*f*) $\langle \mathbf{X}, \mathbf{Y}\rangle_{\mathbf{B}}$, and (*g*) $\langle \mathbf{X}, \mathbf{Z}\rangle_{\mathbf{B}}$ when

$$\mathbf{X} = \begin{bmatrix} 1 \\ 2 \\ 1 \end{bmatrix} \quad \mathbf{Y} = \begin{bmatrix} 2 \\ -1 \\ 0 \end{bmatrix} \quad \mathbf{Z} = \begin{bmatrix} 1 \\ -1 \\ 1 \end{bmatrix} \quad \mathbf{A} = \begin{bmatrix} 1 & 2 & 3 \\ -1 & -1 & 1 \\ 1 & 2 & 1 \end{bmatrix} \quad \mathbf{B} = \begin{bmatrix} 1 & 0 & 0 \\ 0 & 2 & 0 \\ 0 & 0 & 3 \end{bmatrix}$$

11.14 Calculate (*a*) $\langle \mathbf{X}, \mathbf{Y}\rangle$, (*b*) $\langle \mathbf{Y}, \mathbf{X}\rangle$, (*c*) $\langle \mathbf{X}, \mathbf{Z}\rangle$, (*d*) $\langle \mathbf{Y}, \mathbf{Z}\rangle$, (*e*) $\langle \mathbf{X}, \mathbf{Y}\rangle_{\mathbf{W}}$, and (*f*) $\langle \mathbf{X}, \mathbf{Z}\rangle_{\mathbf{W}}$ when

$$\mathbf{X} = \begin{bmatrix} i \\ 1 \end{bmatrix} \quad \mathbf{Y} = \begin{bmatrix} i \\ -i \end{bmatrix} \quad \mathbf{Z} = \begin{bmatrix} 1+i2 \\ 2+i3 \end{bmatrix} \quad \mathbf{W} = \begin{bmatrix} 1 & 3 \\ 2 & 1 \end{bmatrix}$$

In Problems 11.15 through 11.24, use the Gram-Schmidt orthogonalization algorithm with the Euclidean inner product to produce an orthogonal set of vectors from the given set.

11.15 $\mathbf{X}_1 = \begin{bmatrix} 1 \\ 1 \end{bmatrix}$ $\mathbf{X}_2 = \begin{bmatrix} 2 \\ 3 \end{bmatrix}$ **11.16** $\mathbf{X}_1 = \begin{bmatrix} 3 \\ 2 \end{bmatrix}$ $\mathbf{X}_2 = \begin{bmatrix} 4 \\ -5 \end{bmatrix}$

11.17 $\mathbf{X}_1 = \begin{bmatrix} 1 \\ 2 \end{bmatrix}$ $\mathbf{X}_2 = \begin{bmatrix} 2 \\ 4 \end{bmatrix}$ **11.18** $\mathbf{X}_1 = \begin{bmatrix} i \\ 1 \end{bmatrix}$ $\mathbf{X}_2 = \begin{bmatrix} 1 \\ 1 \end{bmatrix}$

11.19 $\mathbf{X}_1 = \begin{bmatrix} 1 \\ 1 \\ 0 \end{bmatrix}$ $\mathbf{X}_2 = \begin{bmatrix} 0 \\ 1 \\ 1 \end{bmatrix}$ $\mathbf{X}_3 = \begin{bmatrix} 1 \\ 0 \\ 1 \end{bmatrix}$

11.20 $\mathbf{X}_1 = \begin{bmatrix} 2 \\ 1 \\ -1 \end{bmatrix}$ $\mathbf{X}_2 = \begin{bmatrix} 3 \\ 6 \\ 2 \end{bmatrix}$ $\mathbf{X}_3 = \begin{bmatrix} -1 \\ 1 \\ 3 \end{bmatrix}$

11.21 $\mathbf{X}_1 = \begin{bmatrix} 1 \\ 1 \\ 1 \end{bmatrix}$ $\mathbf{X}_2 = \begin{bmatrix} 1 \\ 1 \\ -1 \end{bmatrix}$ $\mathbf{X}_3 = \begin{bmatrix} -1 \\ 1 \\ 1 \end{bmatrix}$

11.22 $\mathbf{X}_1 = \begin{bmatrix} 1 \\ 0 \\ i \end{bmatrix}$ $\mathbf{X}_2 = \begin{bmatrix} 1+i \\ 1 \\ 1 \end{bmatrix}$ $\mathbf{X}_3 = \begin{bmatrix} 0 \\ 1-i \\ -i \end{bmatrix}$

11.23 $\mathbf{X}_1 = \begin{bmatrix} 1 \\ 1 \\ 0 \\ 0 \end{bmatrix}$ $\mathbf{X}_2 = \begin{bmatrix} 0 \\ 1 \\ -1 \\ 0 \end{bmatrix}$ $\mathbf{X}_3 = \begin{bmatrix} 1 \\ 0 \\ -1 \\ 0 \end{bmatrix}$ $\mathbf{X}_4 = \begin{bmatrix} 1 \\ 0 \\ 0 \\ -1 \end{bmatrix}$

11.24 $\mathbf{X}_1 = \begin{bmatrix} 1 \\ i \\ 0 \\ 0 \end{bmatrix}$ $\mathbf{X}_2 = \begin{bmatrix} i \\ 0 \\ 2 \\ 0 \end{bmatrix}$ $\mathbf{X}_3 = \begin{bmatrix} 0 \\ 1 \\ 0 \\ 1 \end{bmatrix}$ $\mathbf{X}_4 = \begin{bmatrix} 0 \\ 0 \\ 0 \\ i \end{bmatrix}$

11.25 Use the Gram-Schmidt orthogonalization algorithm to construct an orthogonal set of vectors associated with

$$\mathbf{X}_1 = \begin{bmatrix} 1 \\ 0 \end{bmatrix} \quad \text{and} \quad \mathbf{X}_2 = \begin{bmatrix} 0 \\ 1 \end{bmatrix} \quad \text{for} \quad \mathbf{W} = \begin{bmatrix} 1 & 2 \\ 3 & 4 \end{bmatrix}$$

11.26 Use the Gram-Schmidt orthogonalization algorithm to construct an orthogonal set of vectors associated with $\{\mathbf{X}_1, \mathbf{X}_2, \mathbf{X}_3\}$, where

$$\mathbf{X}_1 = \begin{bmatrix} 1 \\ 1 \\ 0 \end{bmatrix} \quad \mathbf{X}_2 = \begin{bmatrix} 0 \\ 1 \\ 1 \end{bmatrix} \quad \mathbf{X}_3 = \begin{bmatrix} 1 \\ 0 \\ 1 \end{bmatrix} \quad \text{and} \quad \mathbf{W} = \begin{bmatrix} 1 & 1 & 1 \\ 1 & 1 & -1 \\ -1 & 1 & 1 \end{bmatrix}$$

11.27 Prove that an orthogonal set of nonzero vectors is linearly independent.

11.28 Prove Property 11.1.

11.29 Prove Property 11.2.

Chapter 12

Norms

VECTOR NORMS

A *norm* for an arbitrary finite-dimensional vector $\mathbf{X}$, denoted $\|\mathbf{X}\|$, is a real-valued function satisfying the following four conditions for all vectors $\mathbf{X}$ and $\mathbf{Y}$ of the same dimension:

(N1): $\|\mathbf{X}\| \geq 0$.

(N2): $\|\mathbf{X}\| = 0$ if and only if $\mathbf{X} = \mathbf{0}$.

(N3): $\|c\mathbf{X}\| = |c|\|\mathbf{X}\|$ for any scalar c.

(N4) (Triangle inequality): $\|\mathbf{X} + \mathbf{Y}\| \leq \|\mathbf{X}\| + \|\mathbf{Y}\|$.

A vector norm is a measure of the length or magnitude of a vector. Just as there are various bases for measuring scalar length—such as feet and meters—there are alternative norms for measuring the magnitude of a vector. Some of the more common vector norms for $\mathbf{X} = [x_1, x_2, \ldots, x_n]^T$ are:

- *The inner-product-generated norm:* $\|\mathbf{X}\|_\mathbf{W} = \sqrt{\langle \mathbf{X}, \mathbf{X} \rangle_\mathbf{W}}$
- *The Euclidean (or l_2) norm:* $\|\mathbf{X}\|_2 = \sqrt{\langle \mathbf{X}, \mathbf{X} \rangle}$
- *The l_1 norm:* $\|\mathbf{X}\|_1 = |x_1| + |x_2| + \cdots + |x_n|$
- *The l_∞ norm:* $\|\mathbf{X}\|_\infty = \max\,(|x_1|, |x_2|, \ldots, |x_n|)$
- *The l_p norm $(p \geq 1)$:* $\|\mathbf{X}\|_p = (|x_1|^p + |x_2|^p + \cdots + |x_n|^p)^{1/p}$

The Euclidean norm is the most popular, and it is a special case of the inner-product norm when $\mathbf{W} = \mathbf{I}$; the Euclidean norm and the l_1 norm are special cases of the l_p norm for $p = 2$ and $p = 1$, respectively. (See Problems 12.1 through 12.3.) Finally, in the limit as $p \to \infty$, the l_p norm yields the l_∞ norm.

NORMALIZED VECTORS AND DISTANCE

A *unit vector* is a vector having norm equal to unity. A nonzero vector is *normalized* when it is multiplied by the reciprocal of its norm; consequently, normalized vectors are unit vectors. A set of vectors is *orthonormal* if the set is orthogonal and if each vector in the set is a unit vector.

The *distance* between two vectors $\mathbf{X}$ and $\mathbf{Y}$ is $\|\mathbf{X} - \mathbf{Y}\|$. Its value, as well as the designation of a vector as a unit vector, depends on the particular norm selected. (See Problems 12.4 and 12.5.)

MATRIX NORMS

A norm for a square matrix $\mathbf{A}$, denoted $\|\mathbf{A}\|$, is a real-valued function satisfying the following conditions for all $n \times n$ matrices $\mathbf{A}$ and $\mathbf{B}$:

(M1): $\|\mathbf{A}\| \geq 0$.

(M2): $\|\mathbf{A}\| = 0$ if and only if $\mathbf{A} = \mathbf{0}$.

(M3): $\|c\mathbf{A}\| = |c|\|\mathbf{A}\|$ for any scalar c.

(M4) (Triangle inequality): $\|\mathbf{A} + \mathbf{B}\| \leq \|\mathbf{A}\| + \|\mathbf{B}\|$

(M5) (Consistency condition): $\|\mathbf{AB}\| \leq \|\mathbf{A}\|\|\mathbf{B}\|$

Because of the added consistency condition (M5), not all vector norms can be extended to become matrix norms. (See Problem 12.6.) Two that can be extended are the l_1 norm (see Problem 12.7) and the Euclidean norm. For the $n \times n$ matrix $\mathbf{A} = [a_{ij}]$, the Euclidean norm becomes

- *The Frobenius (or Euclidean) norm:* $\|\mathbf{A}\|_F = \left(\sum_{i=1}^{n} \sum_{j=1}^{n} |a_{ij}|^2\right)^{1/2}$

INDUCED NORMS

Each vector norm induces (or generates) the matrix norm

$$\|\mathbf{A}\| = \max_{\|\mathbf{X}\|=1} (\|\mathbf{AX}\|) \tag{12.1}$$

on an arbitrary $n \times n$ matrix $\mathbf{A}$ where the maximum is taken over all n-dimensional vectors $\mathbf{X}$ having vector norm equal to unity. Some induced norms for $\mathbf{A} = [a_{ij}]$ are:

- *The L_1 norm (induced by the l_1 norm):*

$$\|\mathbf{A}\|_1 = \max_{j=1,2,\ldots,n} \left(\sum_{i=1}^{n} |a_{ij}|\right)$$

which is the largest column sum of absolute value.

- *The L_∞ norm (induced by the l_∞ norm):*

$$\|\mathbf{A}\|_\infty = \max_{i=1,2,\ldots,n} \left(\sum_{j=1}^{n} |a_{ij}|\right)$$

which is the largest row sum of absolute values.

- *The spectral norm (induced by the Euclidean norm):*

$$\|\mathbf{A}\|_S = \max(\sqrt{\lambda}\colon \lambda \text{ is an eigenvalue of } \bar{\mathbf{A}}^T\mathbf{A})$$

(See Problems 12.8 through 12.12 and Problem 15.12.)

COMPATIBILITY

A vector norm is *compatible* with a matrix norm if

$$\|\mathbf{AY}\| \le \|\mathbf{A}\|\|\mathbf{Y}\| \tag{12.2}$$

for every $n \times n$ matrix $\mathbf{A}$ and every n-dimensional vector $\mathbf{Y}$. Induced norms are always compatible with the vector norms that generated them, and in those cases there always exists at least one vector $\mathbf{Y}$ for which (*12.2*) is an equality. (See Problem 12.13.) Compatibility is not restricted to induced norms; the Frobenius norm, for example, is compatible with the Euclidean vector norm even though the former is not induced by the latter. (See Problems 12.15 and 12.16.)

SPECTRAL RADIUS

The *spectral radius* of a square matrix $\mathbf{A}$, denoted by $\sigma(\mathbf{A})$, is the largest absolute value of any eigenvalue of $\mathbf{A}$. That is, $\sigma(\mathbf{A}) = \max(|\lambda|\colon \lambda \text{ is an eigenvalue of } \mathbf{A})$. If λ is any eigenvalue of a matrix $\mathbf{A}$, then $|\lambda| \le \sigma(\mathbf{A})$, and there is at least one eigenvalue for which this inequality is an equality. For any matrix norm,

$$\sigma(\mathbf{A}) \le \|\mathbf{A}\| \tag{12.3}$$

Inequality (*12.3*) provides bounds on the eigenvalues of a matrix. (See Problems 12.17 and 12.18.) An equivalent expression for the spectral radius is

$$\sigma(\mathbf{A}) = \lim_{m\to\infty} \|\mathbf{A}^m\|^{1/m} \tag{12.4}$$

Solved Problems

12.1 Determine $\|\mathbf{X}\|_\mathbf{W}$ and $\|\mathbf{Y}\|_\mathbf{W}$ for

$$\mathbf{X} = \begin{bmatrix} 1 \\ 2 \\ 3 \end{bmatrix} \qquad \mathbf{Y} = \begin{bmatrix} 4 \\ 5 \\ 6 \end{bmatrix} \qquad \mathbf{W} = \begin{bmatrix} 1 & 1 & 0 \\ 0 & 1 & 1 \\ 1 & 0 & 1 \end{bmatrix}$$

For the given vectors, we have

$$\|\mathbf{X}\|_\mathbf{W} = \sqrt{\langle \mathbf{X}, \mathbf{X}\rangle_\mathbf{W}} = \sqrt{(\mathbf{WX})\cdot(\mathbf{WX})} = \sqrt{3(3) + 5(5) + 4(4)} = 50$$

and

$$\|\mathbf{Y}\|_\mathbf{W} = \sqrt{\langle \mathbf{Y}, \mathbf{Y}\rangle_\mathbf{W}} = \sqrt{(\mathbf{WY})\cdot(\mathbf{WY})} = \sqrt{9(9) + 11(11) + 10(10)} = \sqrt{302}$$

12.2 Find (*a*) $\|\mathbf{X}\|_2$, (*b*) $\|\mathbf{X}\|_1$, (*c*) $\|\mathbf{X}\|_\infty$, and (*d*) $\|\mathbf{X}\|_5$ for the vector **X** of Problem 12.1.

(*a*) $\|\mathbf{X}\|_2 = \sqrt{\mathbf{X}\cdot\mathbf{X}} = \sqrt{1(1) + 2(2) + 3(3)} = \sqrt{14}$.
(*b*) $\|\mathbf{X}\|_1 = |1| + |2| + |3| = 6$.
(*c*) $\|\mathbf{X}\|_\infty = \max(|1|, |2|, |3|) = 3$.
(*d*) $\|\mathbf{X}\|_5 = (|1|^5 + |2|^5 + |3|^5)^{1/5} = (276)^{1/5} = 3.077$.

12.3 Find (*a*) $\|\mathbf{X}\|_\mathbf{W}$, (*b*) $\|\mathbf{X}\|_2$, (*c*) $\|\mathbf{X}\|_1$, (*d*) $\|\mathbf{X}\|_\infty$, and (*e*) $\|\mathbf{X}\|_4$ when

$$\mathbf{X} = \begin{bmatrix} 1-i \\ i \end{bmatrix} \qquad \text{and} \qquad \mathbf{W} = \begin{bmatrix} 1 & i \\ 2+i5 & 0 \end{bmatrix}$$

(*a*) From Problem 11.3, we have $\|\mathbf{X}\|_\mathbf{W} = \sqrt{\langle \mathbf{X}, \mathbf{X}\rangle_\mathbf{W}} = \sqrt{59}$.
(*b*) $\|\mathbf{X}\|_2 = \sqrt{\mathbf{X}\cdot\bar{\mathbf{X}}} = \sqrt{(1-i)(1+i) + i(-i)} = \sqrt{3}$.
(*c*) $\|\mathbf{X}\|_1 = |1-i| + |i| = \sqrt{2} + 1 = 2.414$.
(*d*) $\|\mathbf{X}\|_\infty = \max(|1-i|, |i|) = \max(\sqrt{2}, 1) = \sqrt{2}$.
(*e*) $\|\mathbf{X}\|_4 = (|1-i|^4 + |i|^4)^{1/4} = (5)^{1/4} = 1.495$.

12.4 Find the distance between **X** and **Y** with respect to (*a*) the Euclidean norm, (*b*) the inner-product norm with respect to **W**, and (*c*) the l_3 norm when **X**, **Y**, and **W** are as given in Problem 12.1.

For these vectors, $\mathbf{X} - \mathbf{Y} = [-3, -3, -3]^T$, so

(*a*) $\|\mathbf{X} - \mathbf{Y}\|_2 = \sqrt{(-3)^2 + (-3)^2 + (-3)^2} = \sqrt{27}$.
(*b*) $\mathbf{W}(\mathbf{X} - \mathbf{Y}) = [-6, -6, -6]^T$, and

$$\|\mathbf{X} - \mathbf{Y}\|_\mathbf{W} = \sqrt{\mathbf{W}(\mathbf{X}-\mathbf{Y})\cdot\mathbf{W}(\mathbf{X}-\mathbf{Y})} = \sqrt{(-6)(-6) + (-6)(-6) + (-6)(-6)} = 10.392$$

(*c*) $\|\mathbf{X} - \mathbf{Y}\|_3 = (|-3|^3 + |-3|^3 + |-3|^3)^{1/3} = (81)^{1/3} = 4.327$.

12.5 Normalize the vector **X** given in Problem 12.1 with respect to (*a*) the l_2 norm, (*b*), the l_1 norm, and (*c*) the l_∞ norm.

Using the results of Problem 12.2, we obtain the normalized vectors (*a*) $[1/\sqrt{14}, 2/\sqrt{14}, 3/\sqrt{14}]^T$; (*b*) $[1/6, 2/6, 3/6]^T$; and (*c*) $[1/3, 2/3, 1]^T$. Each of these vectors is a unit vector with respect to its associated norm.

12.6 Show that the l_∞ norm for vectors does not extend to a matrix norm.

The l_∞ norm is simply the largest component of a vector in absolute value, and its extension to matrices would be the largest element of a matrix in absolute value. That is,

$$\|\mathbf{A}\| = \max_{\substack{i=1,\ldots,n \\ j=1,\ldots,n}} (|a_{ij}|)$$

Consider the matrices

$$\mathbf{A} = \mathbf{B} = \begin{bmatrix} 1 & 1 \\ 1 & 1 \end{bmatrix} \quad \text{for which} \quad \mathbf{AB} = \begin{bmatrix} 2 & 2 \\ 2 & 2 \end{bmatrix}$$

We have $\|\mathbf{A}\| = \|\mathbf{B}\| = 1$, but $\|\mathbf{AB}\| = 2$. Since condition M5 is violated, the proposed norm is not a norm.

12.7 Extend the l_1 vector norm to a matrix norm, and use it to compute the norm of

$$\mathbf{A} = \begin{bmatrix} 4 + i3 & -7 \\ 3 & i4 \end{bmatrix}$$

The l_1 norm is the sum of the absolute values of all the components of the vector; its extension to the matrix $\mathbf{A} = [a_{ij}]$ is the sum of the absolute values of all the elements of the matrix. That is, $\|\mathbf{A}\|_1 = \Sigma_{i=1}^n \Sigma_{j=1}^n |a_{ij}|$. This norm automatically satisfies conditions M1 through M4 because they are identical to conditions N1 through N4. In addition, for two $n \times n$ matrices **A** and **B**, the extended norm gives us

$$\|\mathbf{AB}\|_1 + \sum_{i=1}^n \sum_{j=1}^n \left| \sum_{k=1}^n a_{ik} b_{kj} \right| \le \sum_{i=1}^n \sum_{j=1}^n \sum_{k=1}^n |a_{ik}||b_{kj}|$$

$$\le \sum_{i=1}^n \sum_{j=1}^n \sum_{k=1}^n \sum_{m=1}^n |a_{ik}||b_{mj}| = \left(\sum_{i=1}^n \sum_{k=1}^n |a_{ik}| \right) \left(\sum_{m=1}^n \sum_{j=1}^n |b_{mj}| \right) = \|\mathbf{A}\|_1 \|\mathbf{B}\|_1$$

Thus, condition M5 is satisfied and this extension is a matrix norm. Applying it to the given matrix, we calculate

$$\|\mathbf{A}\| = |4 + i3| + |-7| + |3| + |i4| = 5 + 7 + 3 + 4 = 19$$

12.8 Calculate the (*a*) Frobenius norm, (*b*) L_1 norm, (*c*) L_∞ norm, and (*d*) spectral norm for the matrix **A** of Problem 12.7.

(*a*) $\|\mathbf{A}\|_F = (|4 + i3|^2 + |-7|^2 + |3|^2 + |i4|^2)^{1/2} = \sqrt{25 + 49 + 9 + 16} = 9.950$.

(*b*) $\|\mathbf{A}\|_1 = \max(|4 + i3| + |3|, |-7| + |i4|) = \max(5 + 3, 7 + 4) = 11$.

(*c*) $\|\mathbf{A}\|_\infty = \max(|4 + i3| + |-7|, |3| + |i4|) = \max(5 + 7, 3 + 4) = 12$.

(*d*) We compute

$$\bar{\mathbf{A}}^T\mathbf{A} = \begin{bmatrix} 4 - i3 & 3 \\ -7 & -i4 \end{bmatrix} \begin{bmatrix} 4 + i3 & -7 \\ 3 & i4 \end{bmatrix} = \begin{bmatrix} 34 & -28 + i33 \\ -28 - i33 & 65 \end{bmatrix}$$

which has the characteristic equation $\lambda^2 - 99\lambda + 337 = 0$ and the eigenvalues $\lambda_1 = 95.470$ and $\lambda_2 = 3.530$. Thus, $\|\mathbf{A}\|_S = \max(\sqrt{95.470}, \sqrt{3.530}) = 9.771$.

12.9 Calculate the (*a*) Frobenius norm, (*b*) L_1 norm, (*c*) L_∞ norm, and (*d*) spectral norm for

$$\mathbf{A} = \begin{bmatrix} 7 & -2 & 0 \\ -4 & -6 & 0 \\ 0 & 0 & -9 \end{bmatrix}$$

(*a*) $\|\mathbf{A}\|_F = \{(7)^2 + (-4)^2 + (0)^2 + (-2)^2 + (-6)^2 + (0)^2 + (0)^2 + (0)^2 + (-9)^2\}^{1/2} = 13.638.$

(*b*) $\|\mathbf{A}\|_1 = \max(|7| + |-2| + |0|, |-4| + |-6| + |0|, |0| + |0| + |-9|) = \max(9, 10, 9) = 10.$

(c) $\|\mathbf{A}\|_\infty = \max(|7| + |-4| + |0|, |-2| + |-6| + |0|, |0| + |0| + |-9|) = \max(11, 8, 9) = 11.$

(*d*) Here we have

$$\bar{\mathbf{A}}^T\mathbf{A} = \mathbf{A}^T\mathbf{A} = \begin{bmatrix} 65 & 10 & 0 \\ 10 & 40 & 0 \\ 0 & 0 & 81 \end{bmatrix}$$

which has eigenvalues 68.5078, 36.4922, and 81. Thus, $\|\mathbf{A}\|_S = \max(\sqrt{68.5078}, \sqrt{36.4922}, \sqrt{81}) = 9.$

12.10 Prove that an induced norm generated by (12.1) is a matrix norm.

Let $\mathbf{A}$ and $\mathbf{B}$ denote arbitrary $n \times n$ matrices, and $\mathbf{X}$ and $\mathbf{Y}$ arbitrary n-dimensional vectors. We are given a vector norm that satisfies conditions N1 through N4, and we wish to show that a proposed matrix norm defined by (*12.1*) satisfies conditions M1 through M5. For clarity we subscript the vector norm with V and the matrix norm with M.

(M1): $\|\mathbf{A}\|_M$ is the maximum of nonnegative quantities $\|\mathbf{AX}\|_V$ and must be nonnegative.

(M2): If $\mathbf{A} = \mathbf{0}$, then $\mathbf{AX} = \mathbf{0}$ for all vectors $\mathbf{X}$, and $\|\mathbf{A}\|_M = \max(\|\mathbf{AX}\|_V) = \max(0) = 0$. If $\mathbf{A} \neq \mathbf{0}$, then $\mathbf{A}$ must contain at least one nonzero column. Designate the first such column as column i, and construct $\mathbf{Y}$ by setting its ith component equal to 1 and all other components equal to 0. Now $\|\mathbf{AY}\|_V$ and $\|\mathbf{Y}\|_V$ are positive, and since $\mathbf{Y}/\|\mathbf{Y}\|_V$ is a unit vector,

$$\|\mathbf{A}\|_M = \max_{\|\mathbf{X}\|_V = 1} (\|\mathbf{AX}\|_V) \geq \left\| \mathbf{A} \frac{\mathbf{Y}}{\|\mathbf{Y}\|_V} \right\|_V = \frac{\|\mathbf{AY}\|_V}{\|\mathbf{Y}\|_V} > 0$$

(M3): $\|c\mathbf{A}\|_M = \max(\|c\mathbf{AX}\|_V) = \max(|c|\|\mathbf{AX}\|_V) = |c| \max(\|\mathbf{AX}\|_V) = |c|\|\mathbf{A}\|_M.$

(M4):

$$\begin{aligned} \|\mathbf{A} + \mathbf{B}\|_M &= \max\{\|(\mathbf{A} + \mathbf{B})\mathbf{X}\|_V\} = \max(\|\mathbf{AX} + \mathbf{BX}\|_V) \\ &\leq \max(\|\mathbf{AX}\|_V + \|\mathbf{BX}\|_V) \\ &\leq \max(\|\mathbf{AX}\|_V) + \max(\|\mathbf{BX}\|_V) = \|\mathbf{A}\|_M + \|\mathbf{B}\|_M \end{aligned}$$

(M5): If $\mathbf{B} = \mathbf{0}$, then $\mathbf{AB} = \mathbf{0}$, $\|\mathbf{AB}\|_M = \|\mathbf{B}\|_M = \|\mathbf{0}\|_M = 0$, and the desired inequality is trivially true. If $\mathbf{B} \neq \mathbf{0}$, we can restrict our attention to those unit vectors $\mathbf{X}$ for which $\mathbf{BX} \neq \mathbf{0}$, since all others have no influence on the norm. Then

$$\begin{aligned} \|\mathbf{AB}\|_M &= \max_{\|\mathbf{X}\|_V=1} \{\|(\mathbf{AB})\mathbf{X}\|_V\} = \max_{\|\mathbf{X}\|_V=1} \{\|\mathbf{A}(\mathbf{BX})\|_V\} = \max_{\|\mathbf{X}\|_V=1} \left\{ \frac{\|\mathbf{A}(\mathbf{BX})\|_V}{\|\mathbf{BX}\|_V} \|\mathbf{BX}\|_V \right\} \\ &\leq \max_{\|\mathbf{X}\|_V=1} \left\{ \frac{\|\mathbf{A}(\mathbf{BX})\|_V}{\|\mathbf{BX}\|_V} \right\} \max_{\|\mathbf{X}\|_V=1} (\|\mathbf{BX}\|_V) = \max_{\|\mathbf{X}\|_V=1} \left(\left\| \mathbf{A} \frac{\mathbf{BX}}{\|\mathbf{BX}\|_V} \right\|_V \right) \max_{\|\mathbf{X}\|_V=1} (\|\mathbf{BX}\|_V) \\ &= \max_{\mathbf{Y}=\mathbf{BX}/\|\mathbf{BX}\|_V} (\|\mathbf{AY}\|_V) \max_{\|\mathbf{X}\|_V=1} (\|\mathbf{BX}\|_V) \\ &\leq \max_{\|\mathbf{Y}\|_V=1} (\|\mathbf{AY}\|_V) \max_{\|\mathbf{X}\|_V=1} (\|\mathbf{BX}\|_V) = \|\mathbf{A}\|_M \|\mathbf{B}\|_M \end{aligned}$$

12.11 Show that the l_1 vector norm induces the L_1 matrix norm under Eq. *(12.1)*.

Set $\|\mathbf{A}\|_1 = \max_{\|\mathbf{X}\|_1=1} (\|\mathbf{AX}\|_1)$. Then $\|\mathbf{A}\|_1$ is a matrix norm as a result of Problem 12.10. Denote the columns of $\mathbf{A}$ as vectors $\mathbf{A}_1, \mathbf{A}_2, \ldots, \mathbf{A}_n$, respectively, and set

$$H = \max_{j=1,\ldots,n} \left(\sum_{i=1}^{n} |a_{ij}|\right) = \max_{j=1,\ldots,n} (\|\mathbf{A}_j\|_1)$$

We wish to show that $\|\mathbf{A}\|_1 = H$.

For any unit vector $\mathbf{X} = [x_1, x_2, \ldots, x_n]^T$,

$$\begin{aligned}\|\mathbf{AX}\|_1 &= \|x_1\mathbf{A}_1 + x_2\mathbf{A}_2 + \cdots + x_n\mathbf{A}_n\|_1 \\ &\le \|x_1\mathbf{A}_1\|_1 + \|x_2\mathbf{A}_2\|_1 + \cdots + \|x_n\mathbf{A}_n\|_1 = |x_1|\|\mathbf{A}_1\|_1 + |x_2|\|\mathbf{A}_2\|_1 + \cdots + |x_n|\|\mathbf{A}_n\|_1 \\ &\le |x_1|H + |x_2|H + \cdots + |x_n|H = H(|x_1| + |x_2| + \cdots + |x_n|) = H\|\mathbf{X}\|_1 = H\end{aligned}$$

Thus, $$\|\mathbf{A}\|_1 = \max_{\|\mathbf{X}\|_1=1} (\|\mathbf{AX}\|_1) \le \max_{\|\mathbf{X}\|_1=1} (H) = H \qquad (1)$$

But for unit vectors $\mathbf{Y}_k$ $(k = 1, 2, \ldots, n)$ having a 1 as the kth component and 0s as all other components,

$$\|\mathbf{A}\|_1 = \max_{\|\mathbf{X}\|_1=1} (\|\mathbf{AX}\|_1) \ge \|\mathbf{AY}_k\|_1 = \|\mathbf{A}_k\|_1$$

so $$\|\mathbf{A}\|_1 \ge \max_{k=1,\ldots,n} (\|\mathbf{A}_k\|_1) = H \qquad (2)$$

Together, *(1)* and *(2)* imply the desired equality.

12.12 Show that the l_∞ vector norm induces the L_∞ matrix norm under *(12.1)*.

Set $\|\mathbf{A}\|_\infty = \max_{\|\mathbf{X}\|_\infty=1} (\|\mathbf{AX}\|_\infty)$. Then $\|\mathbf{A}\|_\infty$ is a matrix norm as a result of Problem 12.10. Now

$$H = \max_{i=1,\ldots,n} \left(\sum_{j=1}^{n} |a_{ij}|\right)$$

We wish to show that $\|\mathbf{A}\|_\infty = H$. This equality obviously holds if $\mathbf{A} = \mathbf{0}$, so we consider only nonzero matrices in what follows. For any unit vector $\mathbf{X} = [x_1, x_2, \ldots, x_n]^T$, we have

$$\begin{aligned}\|\mathbf{AX}\|_\infty &= \max_{i=1,\ldots,n} \left(\left|\sum_{j=1}^{n} a_{ij}x_j\right|\right) \le \max_{i=1,\ldots,n} \left(\sum_{j=1}^{n} |a_{ij}||x_j|\right) \\ &\le \max_{i=1,\ldots,n} \left(\sum_{j=1}^{n} |a_{ij}|\|\mathbf{X}\|_\infty\right) = \max_{i=1,\ldots,n} \left(\sum_{j=1}^{n} |a_{ij}|\right) = H\end{aligned}$$

Thus, $$\|\mathbf{A}\|_\infty = \max_{\|\mathbf{X}\|_\infty=1} (\|\mathbf{AX}\|_\infty) \le \max_{\|\mathbf{X}\|_\infty=1} (H) = H \qquad (1)$$

Corresponding to the kth row of $\mathbf{A}$ $(k = 1, 2, \ldots, n)$, define a vector $\mathbf{Y}_k = [y_1^{(k)}, y_2^{(k)}, \ldots, y_n^{(k)}]^T$ by setting

$$y_j^{(k)} = \begin{cases} 1 & \text{when } a_{kj} = 0 \\ \dfrac{\bar{a}_{kj}}{|a_{kj}|} & \text{when } a_{kj} \ne 0 \end{cases}$$

Then $\|\mathbf{Y}_k\|_\infty = 1$, and the kth component of $\mathbf{AY}_k$ is $\Sigma_{j=1}^{n} |a_{kj}|$. Also for each k,

$$\|\mathbf{A}\|_\infty = \max_{\|\mathbf{X}\|_\infty=1} (\|\mathbf{AX}\|_\infty) \ge \|\mathbf{AY}_k\|_\infty$$

The last term on the right is the maximum component in absolute value of $\mathbf{AY}_k$, which is at least as large as the absolute value of the kth component of $\mathbf{AY}_k$. Hence, $\|\mathbf{A}\|_\infty \ge \Sigma_{j=1}^{n} |a_{kj}|$ for all k, and

$$\|\mathbf{A}\|_\infty \geq \max_{k=1,\ldots,n} \left(\sum_{j=1}^{n} |a_{kj}| \right) = H \tag{2}$$

Together, (*1*) and (*2*) imply the desired equality.

12.13 Show that an induced matrix norm with its associated vector norm satisfy the compatibility condition $\|\mathbf{AY}\| \leq \|\mathbf{A}\|\|\mathbf{Y}\|$.

The inequality is immediate when $\mathbf{Y} = \mathbf{0}$. For any nonzero vector $\mathbf{Y}, \mathbf{Y}/\|\mathbf{Y}\|$ is a unit vector, and

$$\|\mathbf{A}\| = \max_{\|\mathbf{X}\|=1} (\|\mathbf{AX}\|) \geq \left\| \mathbf{A}\frac{\mathbf{Y}}{\|\mathbf{Y}\|} \right\| = \frac{\|\mathbf{AY}\|}{\|\mathbf{Y}\|}$$

12.14 Show that $\|\mathbf{A}\| = \max_{\|\mathbf{X}\|=1} (\|\mathbf{AX}\|) = \max_{\mathbf{X}\neq\mathbf{0}} (\|\mathbf{AX}\|/\|\mathbf{X}\|)$.

Set $H = \max_{\mathbf{X}\neq\mathbf{0}} (\|\mathbf{AX}\|/\|\mathbf{X}\|)$. We must show that $\|\mathbf{A}\| = H$. First, we have

$$\|\mathbf{A}\| = \max_{\|\mathbf{X}\|=1} (\|\mathbf{AX}\|) = \max_{\|\mathbf{X}\|=1} \left(\frac{\|\mathbf{AX}\|}{\|\mathbf{X}\|} \right) \leq \max_{\mathbf{X}\neq\mathbf{0}} \left(\frac{\|\mathbf{AX}\|}{\|\mathbf{X}\|} \right) = H$$

where the inequality follows from taking the maximum over a larger set of vectors. Thus,

$$\|\mathbf{A}\| \leq H \tag{1}$$

It follows from Problem 12.13 that

$$H = \max_{\mathbf{X}\neq\mathbf{0}} \left(\frac{\|\mathbf{AX}\|}{\|\mathbf{X}\|} \right) \leq \max_{\mathbf{X}\neq\mathbf{0}} \left(\frac{\|\mathbf{A}\|\|\mathbf{X}\|}{\|\mathbf{X}\|} \right) = \max_{\mathbf{X}\neq\mathbf{0}} (\|\mathbf{A}\|) = \|\mathbf{A}\|$$

so

$$\|\mathbf{A}\| \geq H \tag{2}$$

Together, (*1*) and (*2*) imply the desired inequality.

12.15 Show that the Frobenius matrix norm is compatible with the Euclidean vector norm.

For any $n \times n$ matrix $\mathbf{A}$ and n-dimensional vector $\mathbf{X}$,

$$\|\mathbf{AX}\|_2^2 = \sum_{i=1}^{n} |i\text{th component of } \mathbf{AX}|^2 = \sum_{i=1}^{n} \left| \sum_{j=1}^{n} a_{ij}x_j \right|^2$$

Using the subscript i to designate the ith row of a matrix and employing the Schwarz inequality (see Chapter 11), we have

$$\left| \sum_{j=1}^{n} a_{ij}x_j \right|^2 = |\mathbf{A}_i^T \cdot \mathbf{X}|^2 = |\langle \mathbf{A}_i^T, \bar{\mathbf{X}} \rangle|^2$$

$$\leq \langle \mathbf{A}_i^T, \mathbf{A}_i^T \rangle \langle \bar{\mathbf{X}}, \bar{\mathbf{X}} \rangle = (\mathbf{A}_i^T \cdot \bar{\mathbf{A}}_i^T)(\bar{\mathbf{X}} \cdot \mathbf{X}) = \left(\sum_{j=1}^{n} |a_{ij}|^2 \right) \left(\sum_{j=1}^{n} |x_j|^2 \right) = \left(\sum_{j=1}^{n} |a_{ij}|^2 \right) \|\mathbf{X}\|_2^2$$

Therefore,

$$\|\mathbf{AX}\|_2^2 \leq \left(\sum_{i=1}^{n} \sum_{j=1}^{n} |a_{ij}|^2 \right) \|\mathbf{X}\|_2^2 = \|\mathbf{A}\|_F^2 \|\mathbf{X}\|_2^2$$

We obtain the required inequality by taking square roots.

12.16 Show that any matrix norm has a vector norm with which it is compatible.

Let $\|\mathbf{A}\|_M$ designate an arbitrary matrix norm on $n \times n$ matrices. If $\mathbf{Y}$ is an arbitrary but fixed n-dimensional, nonzero column vector, then the function $f(\mathbf{X}) = \|\mathbf{XY}^T\|_m$ satisfies all the properties of a vector norm on the set of all n-dimensional column vectors $\mathbf{X}$. Furthermore,

$$f(\mathbf{AX}) = \|(\mathbf{AX})\mathbf{Y}^T\|_M = \|\mathbf{A}(\mathbf{XY}^T)\|_M \leq \|\mathbf{A}\|_M \|\mathbf{XY}^T\|_M = \|\mathbf{A}\|_M f(\mathbf{X})$$

so $\|\mathbf{A}\|_m$ is compatible with the vector norm $f(\mathbf{X})$.

12.17 Determine bounds on the eigenvalues of

$$\mathbf{A}=\begin{bmatrix}10 & 7 & 8 & 7\\ 7 & 5 & 6 & 5\\ 8 & 6 & 10 & 9\\ 7 & 5 & 9 & 10\end{bmatrix}$$

The row sums and column sums are both 32, 23, 33, and 31, so $\|\mathbf{A}\|_1 = \|\mathbf{A}\|_\infty = 33$. The Frobenius norm is $\|\mathbf{A}\|_F = 30.5450$. It follows from (*12.3*) that $\sigma(\mathbf{A}) \le 33$ and $\sigma(\mathbf{A}) \le 30.5450$, from which we conclude that every eigenvalue must be no greater than 30.5450 in absolute value. (See also Problem 20.8.) Of course, other norms not considered here might place a still lower bound on the eigenvalues of **A**.

12.18 Prove that $\sigma(\mathbf{A}) \le \|\mathbf{A}\|$ for any matrix norm.

Let λ be an eigenvalue of **A** for which $|\lambda| = \sigma(\mathbf{A})$, and let **X** denote a corresponding eigenvector. Construct a matrix **B** having each of its columns equal to **X**. Then $\mathbf{AB} = \lambda\mathbf{B}$, and for any matrix norm

$$|\lambda|\|\mathbf{B}\| = \|\lambda\mathbf{B}\| = \|\mathbf{AB}\| \le \|\mathbf{A}\|\|\mathbf{B}\|$$

Since **B** is not a zero matrix, it follows that $|\lambda| \le \|\mathbf{A}\|$. But $|\lambda| = \sigma(\mathbf{A})$, so $\sigma(\mathbf{A}) \le \|\mathbf{A}\|$.

Supplementary Problems

12.19 Determine the l_2 norm of each of the following vectors:

(*a*) $\mathbf{X}=\begin{bmatrix}1\\0\\0\end{bmatrix}$ (*b*) $\mathbf{Y}=\begin{bmatrix}1\\2\\-3\end{bmatrix}$ (*c*) $\mathbf{Z}=\begin{bmatrix}2\\1\\-4\end{bmatrix}$ (*d*) $\mathbf{U}=\begin{bmatrix}1\\8\\0\end{bmatrix}$ (*e*) $\mathbf{V}=\begin{bmatrix}4\\4\\4\end{bmatrix}$

12.20 Determine the l_1 norms of the vectors in Problem 12.19.

12.21 Determine the l_∞ norms of the vectors in Problem 12.19.

12.22 For each of the vectors in Problem 12.19, determine the inner-product-generated norm with respect to

$$\mathbf{W}=\begin{bmatrix}0 & 1 & 1\\ 1 & -2 & 1\\ 3 & 0 & 2\end{bmatrix}$$

12.23 For the vectors given in Problem 12.9, determine (*a*) $\|\mathbf{Z}-\mathbf{Y}\|_4$; (*b*) $\|\mathbf{Z}-\mathbf{Y}\|_2$; (*c*) $\|\mathbf{Z}-\mathbf{Y}\|_1$; (*d*) $\|\mathbf{U}-\mathbf{V}\|_3$; (*e*) $\|\mathbf{U}-\mathbf{V}\|_1$; and (*f*) $\|\mathbf{U}-\mathbf{V}\|_\infty$

12.24 Determine the Euclidean norm of

(*a*) $\mathbf{X}=\begin{bmatrix}i\\0\end{bmatrix}$ (*b*) $\mathbf{Y}=\begin{bmatrix}-5\\i2\end{bmatrix}$ (*c*) $\mathbf{Z}=\begin{bmatrix}0\\8\end{bmatrix}$ (*d*) $\mathbf{U}=\begin{bmatrix}1+i2\\3+i4\end{bmatrix}$

12.25 Determine the l_1 norms of the vectors in Problem 12.24.

12.26 Determine the l_∞ norms of the vectors in Problem 12.24.

12.27 Determine the l_3 norms of the vectors in Problem 12.24.

12.28 Determine the Frobenius norms for the following matrices:

$$(a)\begin{bmatrix} 1 & 2 \\ 3 & 4 \end{bmatrix} \quad (b)\begin{bmatrix} -1 & 7 \\ 2 & 5 \end{bmatrix} \quad (c)\begin{bmatrix} 3 & 4 \\ 4 & 5 \end{bmatrix} \quad (d)\begin{bmatrix} i & i4 \\ 2 & -i \end{bmatrix} \quad (e)\begin{bmatrix} 1+i3 & 2-i4 \\ 3-i4 & 4+i2 \end{bmatrix}$$

12.29 Determine the L_1 norms of the matrices in Problem 12.28.

12.30 Determine the L_∞ norms of the matrices in Problem 12.28.

12.31 Determine the spectral norms of the matrices in Problem 12.28.

12.32 Prove that for any induced matrix norm, $\|\mathbf{I}\| = 1$.

12.33 Show that the Frobenius matrix norm satisfies condition M5.

12.34 Prove the Pythagorean theorem for an inner-product-generated vector norm; that is, prove that if $\langle \mathbf{X}, \mathbf{Y} \rangle_{\mathbf{W}} = 0$, then $\|\mathbf{X}+\mathbf{Y}\|_{\mathbf{W}}^2 = \|\mathbf{X}\|_{\mathbf{W}}^2 + \|\mathbf{Y}\|_{\mathbf{W}}^2$.

12.35 Using the L_1, L_∞, and Frobenius norms, determine an upper bound on the spectral radius for each of the matrices in Problem 12.28.

12.36 Determine the spectral radii of matrices (a), (b), and (c) of Problem 12.28.

12.37 Determine the spectral radius of

$$\mathbf{A} = \begin{bmatrix} -3 & 0 & 0 \\ 2 & 1 & 5 \\ 1 & -4 & 4 \end{bmatrix}$$

12.38 Prove that $\sigma(\mathbf{A}^T) = \sigma(\mathbf{A})$.

12.39 The *condition number* of a square matrix **A** with respect to a matrix norm is

$$c(\mathbf{A}) = \begin{cases} \|\mathbf{A}\|\|\mathbf{A}^{-1}\| & \text{if } \mathbf{A} \text{ is nonsingular} \\ \infty & \text{if } \mathbf{A} \text{ is singular} \end{cases}$$

Determine the condition numbers of the matrices in Problem 12.28 with respect to the Frobenius norm.

12.40 Show that the condition number of an identity matrix is unity for all induced matrix norms.

12.41 Show that the condition number as defined in Problem 12.39 cannot be less than 1.

Chapter 13

Hermitian Matrices

NORMAL MATRICES

The *Hermitian transpose* of a matrix $\mathbf{A}$, denoted $\mathbf{A}^H$, is the complex conjugate transpose of $\mathbf{A}$; that is, $\mathbf{A}^H = \bar{\mathbf{A}}^T$. A matrix $\mathbf{A}$ is *normal* if

$$\mathbf{A}\mathbf{A}^H = \mathbf{A}^H\mathbf{A} \tag{13.1}$$

(See Problem 13.1.) Normal matrices have the following properties:

Property 13.1: Every normal matrix is similar to a diagonal matrix.

Property 13.2: Every normal matrix possesses a canonical basis of eigenvectors which can be arranged to form an orthonormal set.

(See Problems 13.7 through 13.9.)

HERMITIAN MATRICES

A matrix is *Hermitian* if it equals its own Hermitian transpose (or complex conjugate transpose); that is, $\mathbf{A}$ is Hermitian if

$$\mathbf{A} = \mathbf{A}^H \tag{13.2}$$

The sum of Hermitian matrices is Hermitian, as is the product of a Hermitian matrix with a real scalar. A Hermitian matrix is also normal, because $\mathbf{A}\mathbf{A}^H = \mathbf{A}\mathbf{A} = \mathbf{A}^H\mathbf{A}$. Therefore, Hermitian matrices possess Properties 13.1 and 13.2. In addition,

Property 13.3: The eigenvalues of a Hermitian matrix are real.

Property 13.4: If a Hermitian matrix $\mathbf{A}$ can be reduced to upper triangular form $\mathbf{U}$ using only elementary row operations of the third kind (E3), then the diagonal of $\mathbf{U}$ contains the same number of zeros, the same number of positive values, and the same number of negative values as the eigenvalues of $\mathbf{A}$.

Property 13.5: An $n \times n$ matrix $\mathbf{A}$ is Hermitian if and only if $\langle \mathbf{AX}, \mathbf{X} \rangle$ is real for all (real and complex) n-dimensional vectors $\mathbf{X}$.

(See Problems 13.4, 13.11, and 13.19.)

REAL SYMMETRIC MATRICES

A matrix is *symmetric* if it equals its own transpose. A symmetric matrix that contains only real elements is Hermitian and, therefore, normal. Consequently, real symmetric matrices possess Properties 13.1 through 13.5 as well as the following:

Property 13.6: The eigenvectors of a real symmetric matrix can be chosen to be real.

(See Problem 13.15.)

THE ADJOINT

The *adjoint* of an $n \times m$ matrix $\mathbf{A}$ is an $m \times n$ matrix $\mathbf{A}^*$ having the property that

$$\langle \mathbf{X}, \mathbf{AY} \rangle_{\mathbf{W}} = \langle \mathbf{A}^*\mathbf{X}, \mathbf{Y} \rangle_{\mathbf{W}} \tag{13.3}$$

for all m-dimensional vectors **Y** and n-dimensional vectors **X**, where the inner product is as defined in Chapter 11. The adjoint always exists and it is

$$\mathbf{A}^* = (\mathbf{W}^H\mathbf{W})^{-1}\mathbf{A}^H(\mathbf{W}^H\mathbf{W}) \tag{13.4}$$

For the special case $\mathbf{W} = \mathbf{I}$ (the Euclidean inner product), (*13.4*) reduces to

$$\mathbf{A}^* = \mathbf{A}^H \tag{13.5}$$

(See Problems 13.16 through 13.18.) Adjoints satisfy the following identities:

(A1): $(\mathbf{A}^*)^* = \mathbf{A}$.

(A2): $(\mathbf{A} + \mathbf{B})^* = \mathbf{A}^* + \mathbf{B}^*$.

(A3): $(\mathbf{AB})^* = \mathbf{B}^*\mathbf{A}^*$.

(A4): $(c\mathbf{A})^* = \bar{c}\mathbf{A}^*$ for any scalar c.

SELF-ADJOINT MATRICES

A matrix **A** is *self-adjoint* if it equals its own adjoint. Such a matrix is necessarily square, and it satisfies the identity

$$\langle \mathbf{X}, \mathbf{AY} \rangle_{\mathbf{W}} = \langle \mathbf{AX}, \mathbf{Y} \rangle_{\mathbf{W}} \tag{13.6}$$

for all vectors **X** and **Y** of appropriate dimension. A matrix is self-adjoint with respect to the Euclidean inner product if and only if it is Hermitian.

Solved Problems

13.1 Determine which of the following matrices are normal:

$$\mathbf{A} = \begin{bmatrix} 1 & 2 & 3 \\ 4 & 5 & 6 \end{bmatrix} \quad \mathbf{B} = \begin{bmatrix} 1 & 2 & 3 \\ 2 & 4 & -5 \\ 3 & -5 & 0 \end{bmatrix} \quad \mathbf{C} = \begin{bmatrix} 2 & 3+i4 \\ 3-i4 & -5 \end{bmatrix} \quad \mathbf{D} = \begin{bmatrix} 2 & 6 & -3 \\ 3 & 2 & 6 \\ -6 & 3 & 2 \end{bmatrix}$$

A is not square, so it cannot be normal. **B** is real and symmetric and, therefore, normal. **C** is Hermitian and, therefore, normal. **D** is normal because

$$\mathbf{DD}^H = \begin{bmatrix} 2 & 6 & -3 \\ 3 & 2 & 6 \\ -6 & 3 & 2 \end{bmatrix} \begin{bmatrix} 2 & 3 & -6 \\ 6 & 2 & 3 \\ -3 & 6 & 2 \end{bmatrix} = \begin{bmatrix} 49 & 0 & 0 \\ 0 & 49 & 0 \\ 0 & 0 & 49 \end{bmatrix} = \begin{bmatrix} 2 & 3 & -6 \\ 6 & 2 & 3 \\ -3 & 6 & 2 \end{bmatrix} \begin{bmatrix} 2 & 6 & -3 \\ 3 & 2 & 6 \\ -6 & 3 & 2 \end{bmatrix} = \mathbf{D}^H\mathbf{D}$$

13.2 Show that $\mathbf{A}^H\mathbf{A}$ and $\mathbf{AA}^H$ are normal for any matrix **A**.

$$(\mathbf{A}^H\mathbf{A})^H = (\overline{\bar{\mathbf{A}}^T\mathbf{A}})^T = (\mathbf{A}^T\bar{\mathbf{A}})^T = (\bar{\mathbf{A}})^T(\mathbf{A}^T)^T = \bar{\mathbf{A}}^T\mathbf{A} = \mathbf{A}^H\mathbf{A}$$

and

$$(\mathbf{AA}^H)^H = (\overline{\mathbf{A}\bar{\mathbf{A}}^T})^T = (\bar{\mathbf{A}}\mathbf{A}^T)^T = (\mathbf{A}^T)^T(\bar{\mathbf{A}})^T = \mathbf{A}\bar{\mathbf{A}}^T = \mathbf{AA}^H$$

Both $\mathbf{A}^H\mathbf{A}$ and $\mathbf{AA}^H$ equal their own Hermitian transposes, so they are Hermitian and, therefore, normal.

13.3 Prove that the eigenvalues of $\mathbf{A}^H\mathbf{A}$ are nonnegative.

If λ is an eigenvalue of $\mathbf{A}^H\mathbf{A}$, then there must exist a nonzero eigenvector $\mathbf{X}$ associated with λ satisfying the equality $\mathbf{A}^H\mathbf{A}\mathbf{X} = \lambda\mathbf{X}$. For the Euclidean inner product, it follows from Property 11.1 and Eqs. (*13.3*) and (*13.5*) that

$$0 \leq \langle \mathbf{AX}, \mathbf{AX} \rangle = \langle \mathbf{A}^*\mathbf{AX}, \mathbf{X} \rangle = \langle \mathbf{A}^H\mathbf{AX}, \mathbf{X} \rangle = \langle \lambda\mathbf{X}, \mathbf{X} \rangle = \lambda\langle \mathbf{X}, \mathbf{X} \rangle \qquad (1)$$

Since $\mathbf{X}$ is an eigenvector, it is nonzero and we may infer from Property 11.2 that $\langle \mathbf{X}, \mathbf{X} \rangle$ is positive. Dividing (*1*) by $\langle \mathbf{X}, \mathbf{X} \rangle$ yields $\lambda \geq 0$.

13.4 Prove that the eigenvalues of a Hermitian matrix are real.

Let λ denote an eigenvalue of a Hermitian matrix $\mathbf{A}$, and let $\mathbf{X}$ denote a corresponding eigenvector. Then, under the Euclidean inner product,

$$\lambda\langle \mathbf{X}, \mathbf{X} \rangle = \langle \lambda\mathbf{X}, \mathbf{X} \rangle = \langle \mathbf{AX}, \mathbf{X} \rangle = \langle \mathbf{X}, \mathbf{A}^*\mathbf{X} \rangle = \langle \mathbf{X}, \mathbf{A}^H\mathbf{X} \rangle = \langle \mathbf{X}, \mathbf{AX} \rangle = \langle \mathbf{X}, \lambda\mathbf{X} \rangle = \bar{\lambda}\langle \mathbf{X}, \mathbf{X} \rangle \qquad (1)$$

Since $\mathbf{X}$ is an eigenvector, it is nonzero and so too is $\langle \mathbf{X}, \mathbf{X} \rangle$. Dividing (*1*) by $\langle \mathbf{X}, \mathbf{X} \rangle$ gives us $\lambda = \bar{\lambda}$, which implies that λ is real.

13.5 Show that if $\mathbf{X}$ is an eigenvector of a normal matrix $\mathbf{A}$ corresponding to eigenvalue λ, then $\mathbf{X}$ is an eigenvector of $\mathbf{A}^H$ corresponding to $\bar{\lambda}$.

Using the Euclidean inner product and (*13.1*), we obtain

$$\langle \mathbf{AX}, \mathbf{AX} \rangle = \langle \mathbf{A}^*\mathbf{AX}, \mathbf{X} \rangle = \langle \mathbf{A}^H\mathbf{AX}, \mathbf{X} \rangle = \langle \mathbf{AA}^H\mathbf{X}, \mathbf{X} \rangle = \langle \mathbf{A}^H\mathbf{X}, \mathbf{A}^*\mathbf{X} \rangle = \langle \mathbf{A}^H\mathbf{X}, \mathbf{A}^H\mathbf{X} \rangle$$

It then follows that

$$\begin{aligned} 0 = \langle \mathbf{0}, \mathbf{0} \rangle &= \langle \mathbf{AX} - \lambda\mathbf{X}, \mathbf{AX} - \lambda\mathbf{X} \rangle \\ &= \langle \mathbf{AX}, \mathbf{AX} \rangle - \bar{\lambda}\langle \mathbf{AX}, \mathbf{X} \rangle - \lambda\langle \mathbf{X}, \mathbf{AX} \rangle + \langle \lambda\mathbf{X}, \lambda\mathbf{X} \rangle \\ &= \langle \mathbf{A}^H\mathbf{X}, \mathbf{A}^H\mathbf{X} \rangle - \bar{\lambda}\langle \mathbf{X}, \mathbf{A}^*\mathbf{X} \rangle - \lambda\langle \mathbf{A}^*\mathbf{X}, \mathbf{X} \rangle + \lambda\bar{\lambda}\langle \mathbf{X}, \mathbf{X} \rangle \\ &= \langle \mathbf{A}^H\mathbf{X}, \mathbf{A}^H\mathbf{X} \rangle - \bar{\lambda}\langle \mathbf{X}, \mathbf{A}^H\mathbf{X} \rangle - \lambda\langle \mathbf{A}^H\mathbf{X}, \mathbf{X} \rangle + \langle \bar{\lambda}\mathbf{X}, \bar{\lambda}\mathbf{X} \rangle \\ &= \langle \mathbf{A}^H\mathbf{X} - \bar{\lambda}\mathbf{X}, \mathbf{A}^H\mathbf{X} - \bar{\lambda}\mathbf{X} \rangle \end{aligned}$$

Thus, $\mathbf{A}^H\mathbf{X} - \bar{\lambda}\mathbf{X} = \mathbf{0}$, which implies that $\mathbf{X}$ is an eigenvector of $\mathbf{A}^H$ corresponding to $\bar{\lambda}$.

13.6 Show that eigenvectors corresponding to distinct eigenvalues of a normal matrix are orthogonal with respect to the Euclidean inner product.

Let λ_1 and λ_2 be two distinct eigenvalues of a normal matrix $\mathbf{A}$ with corresponding eigenvectors $\mathbf{X}_1$ and $\mathbf{X}_2$. Then $\mathbf{AX}_1 = \lambda_1\mathbf{X}_1$ and $\mathbf{AX}_2 = \lambda_2\mathbf{X}_2$, and $\mathbf{A}^H\mathbf{X}_2 = \bar{\lambda}_2\mathbf{X}_2$ as a result of Problem 13.5. Furthermore,

$$\begin{aligned} (\lambda_1 - \lambda_2)\langle \mathbf{X}_1, \mathbf{X}_2 \rangle &= \lambda_1\langle \mathbf{X}_1, \mathbf{X}_2 \rangle - \lambda_2\langle \mathbf{X}_1, \mathbf{X}_2 \rangle = \langle \lambda_1\mathbf{X}_1, \mathbf{X}_2 \rangle - \langle \mathbf{X}_1, \bar{\lambda}_2\mathbf{X}_2 \rangle \\ &= \langle \mathbf{AX}_1, \mathbf{X}_2 \rangle - \langle \mathbf{X}_1, \mathbf{A}^H\mathbf{X}_2 \rangle = \langle \mathbf{AX}_1, \mathbf{X}_2 \rangle - \langle \mathbf{X}_1, \mathbf{A}^*\mathbf{X}_2 \rangle \\ &= \langle \mathbf{AX}_1, \mathbf{X}_2 \rangle - \langle \mathbf{AX}_1, \mathbf{X}_2 \rangle = 0 \end{aligned}$$

Since $\lambda_1 \neq \lambda_2$, it follows that $\langle \mathbf{X}_1, \mathbf{X}_2 \rangle = 0$.

13.7 Show that a set of linearly independent eigenvectors of a normal matrix can be arranged to form an orthonormal set of eigenvectors.

Eigenvectors corresponding to distinct eigenvalues of a normal matrix are orthogonal by Problem 13.6, and they remain orthogonal eigenvectors if each is normalized. Therefore, we need only show that linearly independent eigenvectors corresponding to the same eigenvalue can be so arranged. But this is easily accomplished by the Gram-Schmidt orthogonalization process. Because this process forms linear combinations from a given set (in such a way as to produce orthonormal vectors), it follows from Problem 7.12 that the resulting vectors will remain eigenvectors.

13.8 Determine a canonical basis of orthonormal eigenvectors with respect to the Euclidean inner product for

$$\mathbf{A} = \begin{bmatrix} 2 & 2 & -2 \\ 2 & 2 & -2 \\ -2 & -2 & 6 \end{bmatrix}$$

The matrix is real and symmetric and, therefore, normal. The eigenvalues for **A** are 0, 2, and 8, and a corresponding set of eigenvectors is

$$\mathbf{X}_1 = \begin{bmatrix} -1 \\ 1 \\ 0 \end{bmatrix} \quad \mathbf{X}_2 = \begin{bmatrix} 1 \\ 1 \\ 1 \end{bmatrix} \quad \mathbf{X}_3 = \begin{bmatrix} -1 \\ -1 \\ 2 \end{bmatrix}$$

Since each eigenvector corresponds to a different eigenvalue, the vectors are guaranteed to be orthogonal with respect to the Euclidean inner product. Dividing each vector by its Euclidean norm, we obtain the orthonormal set of eigenvectors

$$\mathbf{Q}_1 = \begin{bmatrix} -1/\sqrt{2} \\ 1/\sqrt{2} \\ 0 \end{bmatrix} \quad \mathbf{Q}_2 = \begin{bmatrix} 1/\sqrt{3} \\ 1/\sqrt{3} \\ 1/\sqrt{3} \end{bmatrix} \quad \mathbf{Q}_3 = \begin{bmatrix} -1/\sqrt{6} \\ -1/\sqrt{6} \\ 2/\sqrt{6} \end{bmatrix}$$

13.9 Determine a canonical basis of orthonormal vectors with respect to the Euclidean inner product for

$$\mathbf{A} = \begin{bmatrix} 3 & -i2 & 0 & i2 \\ i2 & 1 & -2 & 0 \\ 0 & -2 & 3 & -2 \\ -i2 & 0 & -2 & 1 \end{bmatrix}$$

A is Hermitian and, therefore, normal. Its eigenvalues are 5, 5, −1, and −1, with corresponding eigenvectors

$$\mathbf{X}_1 = \begin{bmatrix} i \\ -1 \\ 1 \\ 0 \end{bmatrix} \quad \mathbf{X}_2 = \begin{bmatrix} i2 \\ -1 \\ 0 \\ 1 \end{bmatrix} \quad \mathbf{X}_3 = \begin{bmatrix} i \\ 2 \\ 1 \\ 0 \end{bmatrix} \quad \mathbf{X}_4 = \begin{bmatrix} -i \\ -1 \\ 0 \\ 1 \end{bmatrix}$$

Since $\mathbf{X}_1$ and $\mathbf{X}_2$ correspond to one eigenvalue, and $\mathbf{X}_3$ and $\mathbf{X}_4$ to another, each of the first two vectors is guaranteed to be orthogonal to the latter two. Applying the Gram-Schmidt orthogonalization process to the first two vectors, we obtain

$$\mathbf{Q}_1 = \begin{bmatrix} i/\sqrt{3} \\ -1/\sqrt{3} \\ 1/\sqrt{3} \\ 0 \end{bmatrix} \quad \mathbf{Q}_2 = \begin{bmatrix} i/\sqrt{3} \\ 0 \\ -1/\sqrt{3} \\ 1/\sqrt{3} \end{bmatrix}$$

Applying the Gram-Schmidt orthogonalization process to the latter two vectors, we calculate

$$\mathbf{Q}_3 = \begin{bmatrix} i/\sqrt{6} \\ 2/\sqrt{6} \\ 1/\sqrt{6} \\ 0 \end{bmatrix} \quad \mathbf{Q}_4 = \begin{bmatrix} -i/\sqrt{6} \\ 0 \\ 1/\sqrt{6} \\ 2/\sqrt{6} \end{bmatrix}$$

The set $\{\mathbf{Q}_1, \mathbf{Q}_2, \mathbf{Q}_3, \mathbf{Q}_4\}$ is a canonical basis of orthonormal eigenvectors for **A**.

13.10 Verify Property 13.4 for the matrix in Problem 13.8.

The eigenvalues for the matrix are 0, 2, and 8, so it has one zero eigenvalue and two positive eigenvalues. Reducing the matrix to upper triangular form using only elementary row operations of the third kind, we obtain

$$\begin{bmatrix} 2 & 2 & -2 \\ 2 & 2 & -2 \\ -2 & -2 & 6 \end{bmatrix}$$

$$\rightarrow \begin{bmatrix} 2 & 2 & -2 \\ 0 & 0 & 0 \\ -2 & -2 & 6 \end{bmatrix} \quad \text{Adding } -1 \text{ times the first row to the second row}$$

$$\rightarrow \begin{bmatrix} 2 & 2 & -2 \\ 0 & 0 & 0 \\ 0 & 0 & 4 \end{bmatrix} \quad \text{Adding the first row to the third row}$$

This last matrix is in upper triangular form, and the diagonal elements consist of one zero and two positive numbers.

13.11 Verify Property 13.4 for the matrix in Problem 13.9.

The eigenvalues for that matrix are 5, 5, -1, and -1, which consist of two positive and two negative numbers. Reduced to upper triangular form via elementary row operations of the third kind, the matrix becomes

$$\begin{bmatrix} 3 & -i2 & 0 & i2 \\ 0 & -1/3 & -2 & 4/3 \\ 0 & 0 & 15 & -10 \\ 0 & 0 & 0 & -5/3 \end{bmatrix}$$

The diagonal elements of this matrix also consist of two positive and two negative numbers.

13.12 Show that if $\mathbf{A}$ is Hermitian, then $\mathbf{A} - c\mathbf{I}$ is also Hermitian for any real scalar c.

$$(\overline{\mathbf{A} - c\mathbf{I}})^T = (\bar{\mathbf{A}} - \overline{c\mathbf{I}})^T = (\bar{\mathbf{A}} - c\mathbf{I})^T = \bar{\mathbf{A}}^T - (c\mathbf{I})^T = \mathbf{A} - c\mathbf{I}$$

13.13 Prove that a Hermitian matrix is similar to a diagonal matrix.

We need only show that a Hermitian matrix $\mathbf{A}$ does not possess any generalized eigenvectors of rank 2. This, in turn, implies that it possesses no generalized eigenvectors of rank greater than 2, because otherwise we could form a chain and obtain, as part of the chain, a generalized eigenvector of rank 2. Thus, all generalized eigenvectors have rank 1, and it follows from Chapter 10 that the Jordan canonical form of $\mathbf{A}$ is a diagonal matrix.

Assume that $\mathbf{X}$ is a generalized eigenvector of rank 2 corresponding to the eigenvalue λ. Then

$$(\mathbf{A} - \lambda\mathbf{I})^2\mathbf{X} = \mathbf{0}$$

and

$$(\mathbf{A} - \lambda\mathbf{I})\mathbf{X} \neq \mathbf{0} \tag{1}$$

We may infer from Problem 13.4 that λ is real and from Problem 13.12 that $\mathbf{A} - \lambda\mathbf{I}$ is also Hermitian. Thus,

$$\begin{aligned} 0 = \langle \mathbf{X}, \mathbf{0} \rangle &= \langle \mathbf{X}, (\mathbf{A} - \lambda\mathbf{I})^2\mathbf{X} \rangle = \langle (\mathbf{A} - \lambda\mathbf{I})^*\mathbf{X}, (\mathbf{A} - \lambda\mathbf{I})\mathbf{X} \rangle \\ &= \langle (\mathbf{A} - \lambda\mathbf{I})^H\mathbf{X}, (\mathbf{A} - \lambda\mathbf{I})\mathbf{X} \rangle = \langle (\mathbf{A} - \lambda\mathbf{I})\mathbf{X}, (\mathbf{A} - \lambda\mathbf{I})\mathbf{X} \rangle \end{aligned} \tag{2}$$

We conclude from (2) and Property 11.2 that $(\mathbf{A} - \lambda\mathbf{I})\mathbf{X} = 0$. But this contradicts (1), so $\mathbf{X}$ cannot be a generalized eigenvector of rank 2. (See Problem 15.11 for the generalization of this result to all normal matrices.)

13.14 Show if a matrix is upper triangular and normal, then it must be a diagonal matrix.

Let $\mathbf{A} = [a_{ij}]$ be an $n \times n$ upper triangular matrix that is also normal. Then $a_{ij} = 0$ for $i > j$. We show sequentially, for $i = 1, 2, \ldots, n-1$, that $a_{ij} = 0$ when $i < j$. Since

$$\mathbf{A}^H\mathbf{A} = \mathbf{A}\mathbf{A}^H \tag{1}$$

it follows from equating the (1,1) elements of the two products in (*1*) that

$$\bar{a}_{11}a_{11} = a_{11}\bar{a}_{11} + \sum_{j=2}^{n} a_{1j}\bar{a}_{1j}$$

so that

$$0 = \sum_{j=2}^{n} |a_{1j}|^2$$

Thus,

$$a_{1j} = 0 \qquad (j = 2, 3, \ldots, n) \tag{2}$$

Next, equating the (2,2) elements of the two products in (*1*) and using (*2*), we obtain

$$\bar{a}_{22}a_{22} = a_{22}\bar{a}_{22} + \sum_{j=3}^{n} a_{2j}\bar{a}_{2j}$$

so that

$$0 = \sum_{j=3}^{n} |a_{2j}|^2$$

for which we infer that

$$a_{2j} = 0 \qquad (j = 3, 4, \ldots, n)$$

Continuing in this manner—working with each successive diagonal element in turn—we find that all elements above the diagonal of **A** must be zero. Thus, all nondiagonal elements of **A** are zero, and **A** is a diagonal matrix.

13.15 Show that the eigenvectors of a real symmetric matrix can always be chosen to be real.

Let **X** be an eigenvector of a real symmetric matrix **A** corresponding to the eigenvalue λ (which must be real as a result of Problem 13.4). If the components of **X** are all pure imaginary, then $\mathbf{Y} = i\mathbf{X}$ is real, and

$$\mathbf{AY} = \mathbf{A}(i\mathbf{X}) = i(\mathbf{AX}) = i(\lambda\mathbf{X}) = \lambda(i\mathbf{X}) = \lambda\mathbf{Y}$$

so **Y** is also an eigenvector of **A** corresponding to λ. If the components of **X** are not all pure imaginary, then $\mathbf{Y} = \mathbf{X} + \bar{\mathbf{X}}$ is not zero but is real, and

$$\mathbf{AY} = \mathbf{A}(\mathbf{X} + \bar{\mathbf{X}}) = \mathbf{AX} + \mathbf{A}\bar{\mathbf{X}} = \lambda\mathbf{X} + \overline{\mathbf{AX}} = \lambda\mathbf{X} + \overline{\lambda\mathbf{X}} = \lambda\mathbf{X} + \lambda\bar{\mathbf{X}} = \lambda(\mathbf{X} + \bar{\mathbf{X}}) = \lambda\mathbf{Y}$$

and **Y** is a real eigenvector of **A** corresponding to λ.

13.16 Determine the adjoints of the following matrices with respect to the Euclidean inner product:

$$\mathbf{A} = \begin{bmatrix} 1 & 2 \\ 3 & 4 \\ 5 & 6 \end{bmatrix} \qquad \mathbf{B} = [i5,\ -5,\ 2 + i3] \qquad \mathbf{C} = \begin{bmatrix} 3 & 4 + i5 \\ 4 - i5 & 6 \end{bmatrix}$$

$$\mathbf{D} = \begin{bmatrix} i2 & -i \\ 4 & 3 \end{bmatrix} \qquad \mathbf{E} = \begin{bmatrix} -1 & 2 & 3 \\ 2 & 4 & -5 \\ 3 & -5 & 6 \end{bmatrix}$$

In each case, the adjoint is the Hermitian transpose of the given matrix, as provided by (*13.5*); hence,

$$\mathbf{A}^* = \begin{bmatrix} 1 & 3 & 5 \\ 2 & 4 & 6 \end{bmatrix} \qquad \mathbf{B}^* = \begin{bmatrix} -i5 \\ -5 \\ 2-i3 \end{bmatrix} \qquad \mathbf{C}^* = \begin{bmatrix} 3 & 4+i5 \\ 4-i5 & 6 \end{bmatrix}$$

$$\mathbf{D}^* = \begin{bmatrix} -i2 & 4 \\ i & 3 \end{bmatrix} \qquad \mathbf{E}^* = \begin{bmatrix} -1 & 2 & 3 \\ 2 & 4 & -5 \\ 3 & -5 & 6 \end{bmatrix}$$

C and **E** are self-adjoint because both are Hermitian.

13.17 Determine the adjoint of **A** under an inner product with respect to **W**, where

$$\mathbf{A} = \begin{bmatrix} 2 & 0 \\ -i3 & 3+i4 \end{bmatrix} \quad \text{and} \quad \mathbf{W} = \begin{bmatrix} i4 & 5 \\ -1 & i \end{bmatrix}$$

Using (*13.4*), we calculate

$$\mathbf{W}^H\mathbf{W} = \begin{bmatrix} -i4 & -1 \\ 5 & -i \end{bmatrix}\begin{bmatrix} i4 & 5 \\ -1 & i \end{bmatrix} = \begin{bmatrix} 17 & -i21 \\ i21 & 26 \end{bmatrix}$$

and

$$\mathbf{A}^* = (\mathbf{W}^H\mathbf{W})^{-1}\mathbf{A}^H(\mathbf{W}^H\mathbf{W}) = \begin{bmatrix} 26 & i21 \\ -i21 & 17 \end{bmatrix}\begin{bmatrix} 2 & i3 \\ 0 & 3-i4 \end{bmatrix}\begin{bmatrix} 17 & -i21 \\ i21 & 26 \end{bmatrix}$$

$$= \begin{bmatrix} -2{,}077+i1{,}764 & 2{,}184+i2{,}574 \\ 1{,}428+i1{,}680 & 2{,}082-i1{,}768 \end{bmatrix}$$

13.18 Derive (*13.4*).

For an arbitrary inner product defined with respect to a nonsingular matrix **W**, we have

$$\langle \mathbf{X}, \mathbf{AY}\rangle_\mathbf{W} = (\mathbf{WX})\cdot(\overline{\mathbf{WAY}}) = (\mathbf{WX})^T(\overline{\mathbf{WAY}}) = \mathbf{X}^T\mathbf{W}^T\overline{\mathbf{WAY}}$$

and

$$\langle \mathbf{A}^*\mathbf{X}, \mathbf{Y}\rangle_\mathbf{W} = (\mathbf{WA}^*\mathbf{X})\cdot(\overline{\mathbf{WY}}) = (\mathbf{WA}^*\mathbf{X})^T(\overline{\mathbf{WY}}) = \mathbf{X}^T(\mathbf{A}^*)^T\mathbf{W}^T\overline{\mathbf{WY}}$$

The two inner products are equal by (*13.3*), which implies that

$$\mathbf{X}^T\mathbf{W}^T\overline{\mathbf{WAY}} = \mathbf{X}^T(\mathbf{A}^*)^T\mathbf{W}^T\overline{\mathbf{WY}}$$

or

$$\mathbf{X}^T\{\mathbf{W}^T\overline{\mathbf{WA}} - (\mathbf{A}^*)^T\mathbf{W}^T\bar{\mathbf{W}}\}\bar{\mathbf{Y}} = \mathbf{0}$$

This last equation is valid for all vectors **X** and **Y** if and only if $\mathbf{W}^T\overline{\mathbf{WA}} - (\mathbf{A}^*)^T\mathbf{W}^T\bar{\mathbf{W}} = \mathbf{0}$, from which we infer that $(\mathbf{W}^T\bar{\mathbf{W}})\bar{\mathbf{A}}(\mathbf{W}^T\bar{\mathbf{W}})^{-1} = (\mathbf{A}^*)^T$, and

$$\mathbf{A}^* = \{(\mathbf{W}^T\bar{\mathbf{W}})\bar{\mathbf{A}}(\mathbf{W}^T\bar{\mathbf{W}})^{-1}\}^T = \{(\mathbf{W}^T\bar{\mathbf{W}})^{-1}\}^T\bar{\mathbf{A}}^T(\mathbf{W}^T\bar{\mathbf{W}})^T = \{(\mathbf{W}^T\bar{\mathbf{W}})^T\}^{-1}\bar{\mathbf{A}}^T(\bar{\mathbf{W}}^T\mathbf{W})$$

$$= (\bar{\mathbf{W}}^T\mathbf{W})^{-1}\bar{\mathbf{A}}^T(\bar{\mathbf{W}}^T\mathbf{W}) = (\mathbf{W}^H\mathbf{W})^{-1}\mathbf{A}^H(\mathbf{W}^H\mathbf{W})$$

13.19 Prove that if an $n \times n$ matrix **A** is self-adjoint, then $\langle \mathbf{AX}, \mathbf{X}\rangle_\mathbf{W}$ is real for all n-dimensional vectors **X**.

Using Property 11.3, we have

$$\langle \mathbf{X}, \mathbf{AX}\rangle_\mathbf{W} = \overline{\langle \mathbf{AX}, \mathbf{X}\rangle_\mathbf{W}} \tag{1}$$

But if **A** is self-adjoint, then also

$$\langle \mathbf{X}, \mathbf{AX}\rangle_\mathbf{W} = \langle \mathbf{AX}, \mathbf{X}\rangle_\mathbf{W} \tag{2}$$

It follows from (*1*) and (*2*) that $\overline{\langle \mathbf{AX}, \mathbf{X}\rangle_\mathbf{W}} = \langle \mathbf{AX}, \mathbf{X}\rangle_\mathbf{W}$, which implies that the inner product is real. For the special case of the Euclidean inner product, this result reduces to Property 13.5 for Hermitian matrices.

Supplementary Problems

13.20 Determine which of the following matrices are Hermitian:

$$\mathbf{A}=\begin{bmatrix}3 & i\\ i & 2\end{bmatrix}\quad \mathbf{B}=\begin{bmatrix}i & 2\\ -2 & -i\end{bmatrix}\quad \mathbf{C}=\begin{bmatrix}2 & 3\\ 3 & 4\end{bmatrix}$$

$$\mathbf{D}=\begin{bmatrix}2 & -3\\ 3 & 4\end{bmatrix}\quad \mathbf{E}=\begin{bmatrix}-1 & 1-i2 & i3\\ 1+i2 & 3 & 2-i5\\ -i3 & 2+i5 & 0\end{bmatrix}\quad \mathbf{F}=\begin{bmatrix}-1 & -1 & 1\\ -1 & -1 & -1\\ 1 & -1 & -1\end{bmatrix}$$

$$\mathbf{G}=\begin{bmatrix}1 & 2\\ 2 & -2\\ 0 & 0\end{bmatrix}\quad \mathbf{H}=\begin{bmatrix}1 & -1 & 1 & 0\\ 1 & 1 & 0 & 1\\ -1 & 0 & 1 & 1\\ 0 & 1 & 1 & -1\end{bmatrix}$$

13.21 Determine which of the matrices in Problem 13.20 are normal.

13.22 Find a canonical basis of orthonormal vectors for matrix **F** in Problem 13.20.

13.23 Find a canonical basis of orthonormal vectors for

$$\mathbf{J}=\begin{bmatrix}3 & 0 & 0 & 0 & 0\\ 0 & 2 & 1 & -1 & 0\\ 0 & 1 & 2 & 1 & 0\\ 0 & -1 & 1 & 2 & 0\\ 0 & 0 & 0 & 0 & 0\end{bmatrix}$$

13.24 Find a canonical basis of orthonormal vectors for

$$\mathbf{K}=\begin{bmatrix}2 & i & 0 & -i\\ -i & 3 & 1 & 0\\ 0 & 1 & 2 & 1\\ i & 0 & 1 & 3\end{bmatrix}$$

13.25 Verify Property 13.4 for (*a*) matrix **F** of Problem 13.20; (*b*) matrix **J** of Problem 13.23; and (*c*) matrix **K** of Problem 13.24.

13.26 Determine the adjoint of matrix **D** of Problem 13.20 for (*a*) the Euclidean inner product and (*b*) the inner product with respect to

$$\mathbf{W}=\begin{bmatrix}1 & 1\\ 0 & 1\end{bmatrix}$$

13.27 Determine the adjoint of matrix **E** of Problem 13.20 for (*a*) the Euclidean inner product and (*b*) the inner product with respect to

$$\mathbf{W}=\begin{bmatrix}1 & 0 & 0\\ 0 & 2 & 1\\ 0 & 0 & 2\end{bmatrix}$$

13.28 Determine the adjoints with respect to the Euclidean inner product for matrices **A**, **B**, and **G** of Problem 13.20.

13.29 Prove that the sum of Hermitian matrices is Hermitian.

13.30 Prove that if **A** and **B** are Hermitian and **AB** = **BA**, then **AB** is Hermitian. What does this imply about powers of Hermitian matrices?

13.31 Prove that the diagonal elements of a Hermitian matrix must be real.

13.32 A matrix **A** is *skew-Hermitian* if $\mathbf{A} = -\mathbf{A}^H$. Show that such a matrix is normal.

13.33 Show that if **A** is skew-Hermitian, then $i\mathbf{A}$ is Hermitian.

13.34 Show that if **A** is an $n \times n$ skew-Hermitian matrix, then $\langle \mathbf{AX}, \mathbf{X} \rangle$ is pure imaginary for every n-dimensional vector **X**.

13.35 Show that if **A** is skew-Hermitian, then every eigenvalue of **A** is pure imaginary.

13.36 A matrix **A** is *skew-symmetric* if $\mathbf{A} = -\mathbf{A}^T$. Show that a real skew-symmetric matrix is skew-Hermitian.

13.37 Show that any real matrix can be written as the sum of a symmetric matrix and a skew-symmetric matrix, and show that any complex-valued matrix can be written as the sum of a Hermitian matrix and a skew-Hermitian matrix.

13.38 Prove that any well-defined function of a Hermitian matrix is Hermitian.

Chapter 14

Positive Definite Matrices

DEFINITE MATRICES

An $n \times n$ Hermitian matrix $\mathbf{A}$ is *positive definite* if

$$\langle \mathbf{AX}, \mathbf{X} \rangle > 0 \tag{14.1}$$

for all nonzero n-dimensional vectors $\mathbf{X}$; and $\mathbf{A}$ is *positive semidefinite* if

$$\langle \mathbf{AX}, \mathbf{X} \rangle \geq 0 \tag{14.2}$$

If the inequalities in (*14.1*) and (*14.2*) are reversed, then $\mathbf{A}$ is *negative definite* and *negative semidefinite*, respectively.

The sum of two definite matrices of the same type is again a definite matrix of that type, as is the Hermitian transpose of such a matrix. Positive (or negative) definite matrices are invertible, and their inverses are also positive (or negative) definite.

TESTS FOR POSITIVE DEFINITENESS

Each of the following three tests stipulates necessary and sufficient conditions for an $n \times n$ Hermitian matrix $\mathbf{A}$ to be positive definite. That is, a Hermitian matrix $\mathbf{A}$ is positive definite if it passes any one of these tests.

Test 14.1: $\mathbf{A}$ is positive definite if and only if it can be reduced to upper triangular form using only elementary row operations E3 and the diagonal elements of the resulting matrix (the pivots) are all positive.

Test 14.2: A *principal minor* of $\mathbf{A}$ is the determinant of any submatrix obtained from $\mathbf{A}$ by deleting its last k rows and k columns ($k = 0, 1, \ldots, n-1$). $\mathbf{A}$ is positive definite if and only if all its principal minors are positive.

Test 14.3: $\mathbf{A}$ is positive definite if and only if all its eigenvalues are positive.

The following tests stipulate necessary conditions for an $n \times n$ matrix $\mathbf{A} = [a_{ij}]$ to be positive definite. A Hermitian matrix that fails any one of these tests is not positive definite, but no conclusions can be drawn about a Hermitian matrix that passes them.

Test 14.4: The diagonal elements of $\mathbf{A}$ must be positive.

Test 14.5: The element of $\mathbf{A}$ having the greatest absolute value must be on the diagonal of $\mathbf{A}$.

Test 14.6: $a_{ii}a_{jj} > |a_{ij}|^2$ $(i \neq j)$.

(See Problems 14.1 through 14.11.) All these tests may be changed to tests for positive semidefiniteness by replacing the word *positive* with *nonnegative* and replacing $>$ with $\geq$. They can also be used to test a matrix $\mathbf{A}$ for negative definiteness or semidefiniteness if they are applied directly to $-\mathbf{A}$. This is equivalent to replacing the word *positive* (or *nonnegative*) with *negative* (or *nonpositive*) in tests 14.1 through 14.4; Tests 14.5 and 14.6 are applied as stated.

SQUARE ROOTS OF MATRICES

The *square root* of a matrix $\mathbf{A}$ is a matrix $\mathbf{A}^{1/2}$ having the property that $\mathbf{A} = \mathbf{A}^{1/2}\mathbf{A}^{1/2}$. If $\mathbf{A}$ and $\mathbf{A}^{1/2}$ are both required to be positive definite or positive semidefinite, then $\mathbf{A}^{1/2}$ is unique, and the square root is a well-defined function. In such cases it may be calculated by the methods given in Chapters 8 and 10. (See Problems 14.13 and 14.14.)

CHOLESKY DECOMPOSITION

Any positive definite matrix **A** may be factored into

$$\mathbf{A} = \mathbf{L}\mathbf{L}^H \tag{14.3}$$

where **L** is a lower triangular matrix having positive values on its diagonal. Equation (*14.3*) defines the *Cholesky decomposition* for **A**, which is unique.

The following algorithm generates the Cholesky decomposition for an $n \times n$ matrix $\mathbf{A} = [a_{ij}]$ by sequentially identifying the columns of **L** on and below the main diagonal. It is a simplification of the **LU** decomposition given in Chapter 3.

STEP 14.1: *Initialization*: Set all elements of **L** above the main diagonal equal to zero, and let $l_{11} = \sqrt{a_{11}}$. The remainder of the first column of **L** is the first column of **A** divided by l_{11}. Set a counter $j = 2$.

STEP 14.2: If $j = n + 1$, stop; the algorithm is complete. Otherwise, define $\mathbf{L}'_i$ $(i = j, j+1, \ldots, n)$ to be a column vector of dimension $j - 1$ whose components are, respectively, the first $j - 1$ elements in the ith row of **L**. These elements have already been computed.

STEP 14.3: Compute

$$l_{jj} = \sqrt{a_{jj} - \langle \mathbf{L}'_j, \mathbf{L}'_j \rangle}\ .$$

STEP 14.4: If $j = n$, skip to Step 14.5; otherwise compute the jth column of **L** below the main diagonal: For each $i = j + 1, j + 2, \ldots, n$, compute

$$l_{ij} = \frac{a_{ij} - \langle \mathbf{L}'_i, \mathbf{L}'_j \rangle}{l_{jj}}$$

STEP 14.5: Increase j by 1, and return to Step 14.2. (See Problems 14.15 and 14.16.)

Solved Problems

14.1 Use Tests 14.1 through 14.3 to verify the positive definiteness of

$$\mathbf{A} = \begin{bmatrix} 6 & 2 & -2 \\ 2 & 6 & -2 \\ -2 & -2 & 10 \end{bmatrix}$$

Test 14.1:

$$\rightarrow \begin{bmatrix} 6 & 2 & -2 \\ 0 & 16/3 & -4/3 \\ -2 & -2 & 10 \end{bmatrix}$$ Adding $-1/3$ times the first row to the second row

$$\rightarrow \begin{bmatrix} 6 & 2 & -2 \\ 0 & 16/3 & -4/3 \\ 0 & -4/3 & 28/3 \end{bmatrix}$$ Adding $1/3$ times the first row to the third row

$$\rightarrow \begin{bmatrix} 6 & 2 & -2 \\ 0 & 16/3 & -4/3 \\ 0 & 0 & 27/3 \end{bmatrix}$$ Adding $1/4$ times the second row to the third row

Since the pivots, 6, 16/3, and 27/3, are all positive, the matrix is positive definite.

Test 14.2: The principal minors of **A** are

$$\det[6] = 6 \qquad \begin{vmatrix} 6 & 2 \\ 2 & 6 \end{vmatrix} = 36 - 4 = 32 \qquad \text{and} \qquad \begin{vmatrix} 6 & 2 & -2 \\ 2 & 6 & -2 \\ -2 & -2 & 10 \end{vmatrix} = 288$$

Since all three principal minors are positive, the matrix is positive definite.

Test 14.3: The eigenvalues of **A** are 4, 6, and 12. Since all three are positive, the matrix is positive definite.

14.2 Use all the tests to determine whether the following matrix is positive definite:

$$\mathbf{A} = \begin{bmatrix} 2 & 10 & -2 \\ 10 & 5 & 8 \\ -2 & 8 & 11 \end{bmatrix}$$

Test 14.1:

$$\rightarrow \begin{bmatrix} 2 & 10 & -2 \\ 0 & -45 & 18 \\ -2 & 8 & 11 \end{bmatrix}$$ Adding -5 times the first row to the second row

Since the second pivot, -45, is negative, **A** is neither positive definite nor positive semidefinite. We can also rule out **A** being either negative definite or negative semidefinite, because the first pivot, 2, is positive.

Test 14.2: $\det[2] = 2$, but

$$\begin{vmatrix} 2 & 10 \\ 10 & 2 \end{vmatrix} = -96$$

so **A** is not positive definite.

Test 14.3: The eigenvalues of **A** are -9, 9, and 18. Since they are not all positive, the matrix is not positive definite.

Test 14.4: The diagonal elements, 2, 5, and 11, are all positive, so no conclusion can be drawn from this test.

Test 14.5: The element of greatest absolute value is 11, which does appear on the main diagonal. No conclusion can be drawn from this test.

Test 14.6: With $i = 1$ and $j = 2$, we have

$$a_{11}a_{22} = 2(5) = 10 < 100 = (10)^2 = |a_{12}|^2$$

so **A** is not positive definite.

14.3 Determine whether the following matrix is positive definite:

$$\mathbf{A} = \begin{bmatrix} 11 & -3 & 5 & -8 \\ -3 & 11 & -5 & -8 \\ 5 & -5 & 19 & 0 \\ -8 & -8 & 0 & 16 \end{bmatrix}$$

To use Test 14.1, we reduce **A** to the following form, using only elementary row operations E3:

$$\begin{bmatrix} 11 & -3 & 5 & -8 \\ 0 & 112/11 & -40/11 & -112/11 \\ 0 & 0 & 108/7 & 0 \\ 0 & 0 & 0 & 0 \end{bmatrix}$$

Since the pivots, 11, 112/11, 108/7, and 0, are not all positive, the matrix is not positive definite. However, these pivots are nonnegative, so **A** is positive semidefinite.

14.4 Determine whether the following matrix is positive definite:

$$\mathbf{A} = \begin{bmatrix} 2 & -17 & 7 \\ -17 & -4 & 1 \\ 7 & 1 & -14 \end{bmatrix}$$

$\mathbf{A}$ is not positive definite because it fails tests 14.4, 14.5, and 14.6: Its diagonal elements are not all positive; the largest element in absolute value, -17, is not on the main diagonal; and $a_{11}a_{22} = -8$ is not greater than $|a_{12}|^2 = 289$.

14.5 Prove that the diagonal elements of a positive definite matrix must be positive.

If $\mathbf{A}$ has order $n \times n$, define $\mathbf{X}$ to be an n-dimensional vector having one of its components, say the kth, equal to unity and all other components equal to zero. For this vector, (*14.1*) becomes

$$0 < \langle \mathbf{AX}, \mathbf{X} \rangle = (\mathbf{AX}) \cdot \bar{\mathbf{X}} = a_{kk}$$

14.6 Prove that if $\mathbf{A} = [a_{ij}]$ is an $n \times n$ positive definite matrix, then for any distinct i and j $(i, j = 1, 2, \ldots, n)$, $a_{ii}a_{jj} > |a_{ij}|^2$.

Define $\mathbf{X}$ to be an n-dimensional vector having all components equal to zero except for the ith and jth components. Denote these as x_i and x_j, respectively. For this vector, (*14.1*) becomes

$$0 < \langle \mathbf{AX}, \mathbf{X} \rangle = (\mathbf{AX}) \cdot \bar{\mathbf{X}} = a_{ii}x_i\bar{x}_i + a_{ij}x_j\bar{x}_i + a_{ji}x_i\bar{x}_j + a_{jj}x_j\bar{x}_j$$

Setting $x_i = -a_{ij}/a_{ii}$ and $x_j = 1$, we find that the first two terms on the right cancel, and we are left with

$$0 < \frac{-a_{ji}a_{ij}}{a_{11}} + a_{jj} = \frac{1}{a_{ii}}(-a_{ji}a_{ij} + a_{jj}a_{ii})$$

The desired inequality follows, since a_{ii} is positive (see Problem 14.5) and, because $\mathbf{A}$ is Hermitian, $a_{ji} = \bar{a}_{ij}$.

14.7 Show that the largest element in absolute value of a positive definite matrix must lie on the main diagonal.

Assume that the largest element in absolute value does not lie on the main diagonal but rather in another location, say the (i, j) position, with $i \neq j$. Then $|a_{ij}| > a_{ii}$ and $|a_{ij}| > a_{jj}$. It follows that

$$|a_{ij}|^2 = |a_{ij}||a_{ij}| > a_{ii}a_{jj}$$

which contradicts the result of Problem 14.6. Thus, the assumption is incorrect.

14.8 Prove that the eigenvalues of a positive definite matrix are positive.

Let $\mathbf{A}$ be positive definite with eigenvalue λ and corresponding eigenvector $\mathbf{X}$. Then for this $\mathbf{X}$, Eq. (14.1) becomes

$$0 < \langle \mathbf{AX}, \mathbf{X} \rangle = \langle \lambda\mathbf{X}, \mathbf{X} \rangle = \lambda \langle \mathbf{X}, \mathbf{X} \rangle \tag{1}$$

Since $\mathbf{X}$ is an eigenvector, it is not zero and $\langle \mathbf{X}, \mathbf{X} \rangle$ is positive (Property 11.1). Dividing (*1*) by $\langle \mathbf{X}, \mathbf{X} \rangle$, we obtain $\lambda > 0$.

14.9 Prove that if all the eigenvalues of a Hermitian matrix are positive, then the matrix is positive definite.

An $n \times n$ Hermitian matrix has a canonical basis of orthonormal eigenvectors (Property 13.2). Denote these basis vectors as $\mathbf{X}_1, \mathbf{X}_2, \ldots, \mathbf{X}_n$, with corresponding eigenvalues $\lambda_1, \lambda_2, \ldots, \lambda_n$. Then $\mathbf{AX}_i = \lambda_i\mathbf{X}_i$ $(i = 1, 2, \ldots, n)$.

If $\mathbf{X}$ is any nonzero n-dimensional vector, then the set $\{\mathbf{X}_1, \mathbf{X}_2, \ldots, \mathbf{X}_n, \mathbf{X}\}$ is linearly dependent

(Property 6.1). But the orthonormal eigenvectors are linearly independent (Problem 11.27), so it follows from Property 6.2 that there exist constants $d_1, d_2, \ldots, d_n$ such that

$$\mathbf{X} = d_1\mathbf{X}_1 + d_2\mathbf{X}_2 + \cdots + d_n\mathbf{X}_n$$

Then

$$\mathbf{AX} = d_1\mathbf{AX}_1 + d_2\mathbf{AX}_2 + \cdots + d_n\mathbf{AX}_n = d_1\lambda_1\mathbf{X}_1 + d_2\lambda_2\mathbf{X}_2 + \cdots + d_n\lambda_n\mathbf{X}_n$$

and

$$\langle \mathbf{AX}, \mathbf{X}\rangle = \langle d_1\lambda_1\mathbf{X}_1 + d_2\lambda_2\mathbf{X}_2 + \cdots + d_n\lambda_n\mathbf{X}_n, d_1\mathbf{X}_1 + d_2\mathbf{X}_2 + \cdots + d_n\mathbf{X}_n\rangle$$
$$= |d_1|^2\lambda_1 + |d_2|^2\lambda_2 + \cdots + |d_n|^2\lambda_n$$

because the eigenvectors are orthonormal. Since the eigenvalues are given to be positive, this last quantity is positive for any nonzero vector $\mathbf{X}$; thus the matrix $\mathbf{A}$ satisfies (*14.1*) and is positive definite.

14.10 Show that the determinant of a positive definite matrix is positive.

The determinant of a matrix is the product of its eigenvalues (Property 7.8), and each eigenvalue of a positive definite matrix is positive (Problem 14.8).

14.11 Show that all principal minors of a positive definite matrix must be positive.

Let $\mathbf{A}$ be an $n \times n$ positive definite matrix, and let $\mathbf{B}$ be a submatrix of $\mathbf{A}$ obtained by deleting from $\mathbf{A}$ its last k rows and k columns ($k = 0, 1, \ldots, n-1$). Then $\mathbf{B}$ has order $(n-k) \times (n-k)$. Let $\mathbf{Y}$ denote an arbitrary nonzero $(n-k)$-dimensional vector, and define $\mathbf{X}$ to be an n-dimensional vector having its first $n-k$ components identical to those of $\mathbf{Y}$ and its last k components equal to zero. It follows from Eq. (14.1) that

$$0 < \langle \mathbf{AX}, \mathbf{X}\rangle = \langle \mathbf{BY}, \mathbf{Y}\rangle$$

Since this is true for any nonzero vector $\mathbf{Y}$, it follows that $\mathbf{B}$ is positive definite and, from Problem 14.10, that $\det \mathbf{B}$ is positive.

14.12 Show that a positive definite matrix is invertible.

The determinant of a positive definite matrix is positive and so nonzero (Problem 14.10), and therefore that matrix must have an inverse as given by (*5.3*).

14.13 Find the square root of matrix $\mathbf{A}$ in Problem 14.1.

The eigenvalues of $\mathbf{A}$ are 4, 6, and 12, with corresponding eigenvectors $[1, -1, 0]^T$, $[1, 1, 1]^T$, and $[1, 1, -2]^T$. Thus, by (*10.2*), $\mathbf{A} = \mathbf{MJM}^{-1}$, where

$$\mathbf{M} = \begin{bmatrix} 1 & 1 & 1 \\ -1 & 1 & 1 \\ 0 & 1 & -2 \end{bmatrix} \quad \text{and} \quad \mathbf{J} = \begin{bmatrix} 4 & 0 & 0 \\ 0 & 6 & 0 \\ 0 & 0 & 12 \end{bmatrix}$$

Also,

$$\mathbf{A}^{1/2} = \mathbf{MJ}^{1/2}\mathbf{M}^{-1} = \begin{bmatrix} 1 & 1 & 1 \\ -1 & 1 & 1 \\ 0 & 1 & -2 \end{bmatrix}\begin{bmatrix} 2 & 0 & 0 \\ 0 & \sqrt{6} & 0 \\ 0 & 0 & \sqrt{12} \end{bmatrix}\begin{bmatrix} 3/6 & -3/6 & 0 \\ 2/6 & 2/6 & 2/6 \\ 1/6 & 1/6 & -2/6 \end{bmatrix}$$

$$= \begin{bmatrix} 2.3938 & 0.39385 & -0.33820 \\ 0.39385 & 2.3938 & -0.33820 \\ -0.33820 & -0.33820 & 3.1259 \end{bmatrix}$$

14.14 Show by example that if the square root of a matrix is not required to be positive definite or positive semidefinite, then it is not unique.

For

$$\mathbf{A} = \mathbf{I} = \begin{bmatrix} 1 & 0 \\ 0 & 1 \end{bmatrix}$$

each of the matrices

$$\mathbf{B}=\begin{bmatrix}1 & -1\\0 & -1\end{bmatrix}\qquad \mathbf{C}=\begin{bmatrix}-1 & 0\\0 & -1\end{bmatrix}\qquad \mathbf{D}=\begin{bmatrix}1 & 0\\0 & 1\end{bmatrix}$$

has the property that its square is **A**. Only **D** is positive definite.

14.15 Determine the Cholesky decomposition for

$$\mathbf{A}=\begin{bmatrix}4 & i2 & -i\\-i2 & 10 & 1\\i & 1 & 9\end{bmatrix}$$

Since **A** is a 3×3 matrix, so too is **L** in (*14.3*).

STEP 14.1: Set $l_{11}=\sqrt{4}=2$; then $l_{21}=-i2/2=-i$ and $l_{22}=i/2$. Set $j=2$. To this point, then, we have

$$\mathbf{L}=\begin{bmatrix}2 & 0 & 0\\-i & - & 0\\i/2 & - & -\end{bmatrix}$$

STEP 14.2: Define $\mathbf{L}_2'=[-i]$ and $\mathbf{L}_3'=[i/2]$.

STEP 14.3: Compute

$$l_{22}=\sqrt{a_{22}-\langle \mathbf{L}_2',\mathbf{L}_2'\rangle}=\sqrt{(10-1)}=3$$

STEP 14.4: Compute

$$l_{32}=\frac{a_{32}-\langle \mathbf{L}_3',\mathbf{L}_2'\rangle}{l_{22}}=\frac{1-(-1/2)}{3}=\frac{1}{2}$$

To this point, we have

$$\mathbf{L}=\begin{bmatrix}2 & 0 & 0\\-i & 3 & 0\\i/2 & 1/2 & -\end{bmatrix}$$

STEP 14.5: Set $j=3$.

STEP 14.2:

$$\mathbf{L}_3'=\begin{bmatrix}i/2\\1/2\end{bmatrix}$$

STEP 14.3: Compute $l_{33}=\sqrt{a_{33}-\langle \mathbf{L}_3',\mathbf{L}_3'\rangle}=\sqrt{9-1/2}=\sqrt{8.5}$

Therefore, the complete decomposition is

$$\begin{bmatrix}4 & i2 & -i\\-i2 & 10 & 1\\i & 1 & 9\end{bmatrix}=\begin{bmatrix}2 & 0 & 0\\-i & 3 & 0\\i/2 & 1/2 & \sqrt{8.5}\end{bmatrix}\begin{bmatrix}2 & i & -i/2\\0 & 3 & 1/2\\0 & 0 & \sqrt{8.5}\end{bmatrix}$$

14.16 Determine the Cholesky decomposition for

$$\mathbf{A}=\begin{bmatrix}16 & -3 & 5 & -8\\-3 & 16 & -5 & -8\\5 & -5 & 24 & 0\\-8 & -8 & 0 & 21\end{bmatrix}$$

Since **A** is a 4×4 matrix, so too is **L** in (*14.3*).

STEP 14.1: Set $l_{11}=\sqrt{16}=4$. Then $l_{21}=-3/4=-0.75$; $l_{31}=5/4=1.25$; $l_{41}=-8/4=-2$. Set $j=2$. To this point we have

$$\mathbf{L}=\begin{bmatrix}4 & 0 & 0 & 0\\-0.75 & - & 0 & 0\\1.25 & - & - & 0\\-2 & - & - & -\end{bmatrix}$$

STEP 14.2: Define $\mathbf{L}_2' = [-0.75]$; $\mathbf{L}_3' = [1.25]$; and $\mathbf{L}_4' = [-2]$.

STEP 14.3: Compute

$$l_{22} = \sqrt{a_{22} - \langle \mathbf{L}_2', \mathbf{L}_2' \rangle} = \sqrt{16 - 0.5625} = 3.92906$$

STEP 14.4: Compute

$$l_{32} = \frac{a_{32} - \langle \mathbf{L}_3', \mathbf{L}_2' \rangle}{l_{22}} = \frac{-5 + 0.9375}{3.92906} = -1.03396 \qquad l_{42} = \frac{a_{42} - \langle \mathbf{L}_4', \mathbf{L}_2' \rangle}{l_{22}} = \frac{-8 - 1.5}{3.92906} = -2.41788$$

To this point, we have

$$\mathbf{L} = \begin{bmatrix} 4 & 0 & 0 & 0 \\ -0.75 & 3.92906 & 0 & 0 \\ 1.25 & -1.03396 & - & 0 \\ -2 & -2.41788 & - & - \end{bmatrix}$$

STEP 14.5: Increase j by 1 to $j = 3$.

STEP 14.2:

$$\mathbf{L}_3' = \begin{bmatrix} 1.25 \\ -1.03396 \end{bmatrix} \qquad \mathbf{L}_4' = \begin{bmatrix} -2 \\ -2.41788 \end{bmatrix}$$

STEP 14.3:

$$l_{33} = \sqrt{a_{33} - \langle \mathbf{L}_3', \mathbf{L}_3' \rangle} = \sqrt{24 - 2.631573} = 4.62260$$

STEP 14.4: Compute

$$l_{43} = \frac{a_{43} - \langle \mathbf{L}_4', \mathbf{L}_3' \rangle}{l_{33}} = \frac{0 - 0}{4.62260} = 0$$

To this point, we have

$$\mathbf{L} = \begin{bmatrix} 4 & 0 & 0 & 0 \\ -0.75 & 3.92906 & 0 & 0 \\ 1.25 & -1.03396 & 4.62260 & 0 \\ -2 & -2.41788 & 0 & - \end{bmatrix}$$

STEP 14.5: Increase j by 1 to $j = 4$.

STEP 14.2:

$$\mathbf{L}_4' = [-2, -2.41788, 0]^T$$

STEP 14.3:

$$l_{44} = \sqrt{a_{44} - \langle \mathbf{L}_4', \mathbf{L}_4' \rangle} = \sqrt{21 - 9.84614} = 3.33974$$

Finally, we have $\mathbf{A} = \mathbf{L}\mathbf{L}^H$, where

$$\mathbf{L} = \begin{bmatrix} 4 & 0 & 0 & 0 \\ -0.75 & 3.92906 & 0 & 0 \\ 1.25 & -1.03396 & 4.62260 & 0 \\ -2 & -2.41788 & 0 & 3.33974 \end{bmatrix}$$

Supplementary Problems

14.17 Determine which of the following matrices are positive definite and which are positive semidefinite:

$$\mathbf{A}=\begin{bmatrix} 3 & 1 & -1 \\ 1 & 3 & -1 \\ -1 & -1 & 5 \end{bmatrix} \quad \mathbf{B}=\begin{bmatrix} 1 & 1 & -1 \\ 1 & 1 & -1 \\ -1 & -1 & 3 \end{bmatrix} \quad \mathbf{C}=\begin{bmatrix} 5 & 1 & 1 \\ 0 & 5 & 2 \\ 0 & 0 & 5 \end{bmatrix}$$

$$\mathbf{D}=\begin{bmatrix} 2 & 2 & -1 \\ 2 & 2 & -1 \\ -1 & -1 & 5 \end{bmatrix} \quad \mathbf{E}=\begin{bmatrix} 4 & i2 & 2 \\ -i2 & 10 & 1-i \\ 2 & 1+i & 9 \end{bmatrix} \quad \mathbf{F}=\begin{bmatrix} 25 & 1+i2 & 3-i2 \\ 1+i2 & 7 & 2+i \\ 3-i2 & 2+i & 9 \end{bmatrix}$$

$$\mathbf{G}=\begin{bmatrix} 9 & -3 & 0 & -3 \\ -3 & 6 & 3 & 0 \\ 0 & 3 & 9 & -3 \\ -3 & 0 & -3 & 6 \end{bmatrix} \quad \mathbf{H}=\begin{bmatrix} 1 & -1 & 2 & -1 \\ -1 & 3 & 4 & 2 \\ 2 & 4 & 3 & 1 \\ -1 & 2 & 1 & 1 \end{bmatrix}$$

14.18 Find the square root of matrix **A** in Problem 14.17, given that its eigenvalues are 2, 3, and 6.

14.19 Find the square root of matrix **B** in Problem 14.17, given that its eigenvalues are 0, 1, and 4.

14.20 Find the square root of

$$\mathbf{K}=\begin{bmatrix} 25 & i24 \\ -i24 & 25 \end{bmatrix}$$

14.21 Find the Cholesky decomposition for matrix **A** in Problem 14.17.

14.22 Find the Cholesky decomposition for matrix **E** in Problem 14.17.

14.23 Find the Cholesky decomposition for matrix **G** in Problem 14.17.

14.24 Prove that the sum of two positive definite matrices is positive definite.

14.25 Prove that if **A** is positive definite, then so too is $\mathbf{A}^H$.

14.26 Prove that if **A** is positive definite, then so too is $\mathbf{A}^{-1}$.

14.27 Prove that if **A** is positive definite and **C** is nonsingular, then $\mathbf{B}=\mathbf{C}^H\mathbf{A}\mathbf{C}$ is also positive definite.

14.28 Show that if **A** is Hermitian, then $e^{\mathbf{A}t}$ is positive definite.

14.29 Show that the requirement $\langle \mathbf{AX}, \mathbf{X} \rangle > 0$ for all complex-valued vectors **X** of suitable dimension is sufficient to guarantee that **A** be Hermitian (as well as positive definite).

14.30 Show that there exist nonsymmetric real matrices that satisfy (*14.1*) for all real-valued vectors of suitable dimension.

Chapter 15

Unitary Transformations

UNITARY MATRICES

A matrix is *unitary* if its inverse equals its Hermitian transpose; that is, $\mathbf{U}$ is unitary if

$$\mathbf{U}^{-1} = \mathbf{U}^H = \bar{\mathbf{U}}^T \tag{15.1}$$

Unitary matrices are normal because $\mathbf{UU}^H = \mathbf{UU}^{-1} = \mathbf{I} = \mathbf{U}^{-1}\mathbf{U} = \mathbf{U}^H\mathbf{U}$. In addition, they have the following properties:

Property 15.1: A matrix is unitary if and only if its columns (or rows) form an orthonormal set of vectors.

Property 15.2: The product of unitary matrices of the same order is a unitary matrix.

Property 15.3: If $\mathbf{U}$ is unitary, then $\langle \mathbf{UX}, \mathbf{UY} \rangle = \langle \mathbf{X}, \mathbf{Y} \rangle$ for all vectors $\mathbf{X}$ and $\mathbf{Y}$ of appropriate dimension.

Property 15.4: All eigenvalues of a unitary matrix have absolute value equal to 1.

Property 15.5: The determinant of a unitary matrix has absolute value equal to 1.

(See Problems 15.2, 15.5 to 15.7, and 15.24.) Unitary matrices are invaluable for constructing similarity transformations (see Chapter 10), because their inverses are so easy to obtain.

An *orthogonal matrix* is a unitary matrix whose elements are all real. If $\mathbf{P}$ is orthogonal, then

$$\mathbf{P}^{-1} = \mathbf{P}^T \tag{15.2}$$

SCHUR DECOMPOSITION

Every square matrix is similar to a matrix in upper triangular form, and a unitary matrix may be chosen to produce the transformation. That is, for any square matrix $\mathbf{A}$, there exists a unitary matrix $\mathbf{U}$ such that

$$\mathbf{U}^H\mathbf{AU} = \mathbf{U}^{-1}\mathbf{AU} = \mathbf{T} \tag{15.3}$$

where $\mathbf{T}$ is a matrix in upper triangular form. Equation (*15.3*) is called a *Schur decomposition* for $\mathbf{A}$. Such a decomposition is not unique, even though the diagonal elements of $\mathbf{T}$ must be the eigenvalues of $\mathbf{A}$.

The following algorithm for producing a Schur decomposition for an $n \times n$ matrix $\mathbf{A}$ is iterative; it sequentially generates, at each stage, matrices $\mathbf{U}_k$ and $\mathbf{T}_k$ $(k = 1, 2, \ldots, n-1)$. Each matrix $\mathbf{U}_k$ is unitary, and each $\mathbf{T}_k$ has only zeros below its main diagonal in its first k columns. $\mathbf{T}_{n-1}$ is in upper triangular form, and $\mathbf{U} = \mathbf{U}_1\mathbf{U}_2 \cdots \mathbf{U}_{n-1}$ is the unitary matrix that transforms $\mathbf{A}$ into $\mathbf{T}_{n-1}$. For notational convenience we set $\mathbf{T}_0 = \mathbf{A}$. The kth iteration of the algorithm is:

STEP 15.1: Denote as $\mathbf{A}_k$ the $(n-k+1) \times (n-k+1)$ submatrix in the lower right portion of $\mathbf{T}_{k-1}$.

STEP 15.2: Determine an eigenvalue and a corresponding unit eigenvector for $\mathbf{A}_k$.

STEP 15.3: Construct a unitary matrix $\mathbf{N}_k$ which has as its first column the eigenvector found in Step 15.2.

STEP 15.4: For $k = 1$, set $\mathbf{U}_1 = \mathbf{N}_1$; for $k > 1$, set

$$\mathbf{U}_k = \left[\begin{array}{c|c} \mathbf{I}_{k-1} & \mathbf{0} \\ \hline \mathbf{0} & \mathbf{N}_k \end{array}\right]$$

where $\boldsymbol{I}_{k-1}$ is the $(k-1) \times (k-1)$ identity matrix.

STEP 15.5: Calculate $\mathbf{T}_k = \mathbf{U}_k^H \mathbf{T}_{k-1} \mathbf{U}_k$.

(See Problems 15.8 and 15.9.)

If $\mathbf{A}$ is normal, then the Schur decomposition implies:

Theorem 15.1: Every normal matrix is similar to a diagonal matrix, and the similarity transformation can be effected with a unitary matrix.

(See Problem 15.11.)

ELEMENTARY REFLECTORS

An *elementary reflector* (or *Householder transformation*) associated with a real n-dimensional column vector $\mathbf{V}$ is the $n \times n$ matrix

$$\mathbf{R} = \mathbf{I} - 2\,\frac{\mathbf{V}\mathbf{V}^T}{\|\mathbf{V}\|_2^2} \tag{15.4}$$

An elementary reflector is both real symmetric and orthogonal, and its square is the identity matrix. (See Problems 15.13, 15.14, and 15.22.)

SUMMARY OF SIMILARITY TRANSFORMATIONS

As indicated by (*10.1*), a similarity transformation requires the computation of an inverse; and inversion is a tedious process for all but unitary matrices, whose inverses are their Hermitian transposes. If the matrix $\mathbf{S}$ in (*10.1*) is unitary, then $\mathbf{A}$ and $\mathbf{B}$ are said to be *unitarily similar*.

Similarity transformations are important because they preserve many basic attributes of a square matrix—in particular, eigenvalues (Problem 10.3)—while yielding matrices that are simpler in form. The simplest form is that of a diagonal matrix, and any matrix possessing a canonical basis of eigenvectors (Chapter 9) is similar to a diagonal matrix. Normal matrices (Chapter 13) have this feature, and they include Hermitian and unitary matrices. The most that can be said of an arbitrary square matrix is it is similar to a block diagonal matrix in Jordan canonical form.

If the matrix $\mathbf{S}$ in (*10.1*) is restricted to be unitary, then the simplest general form that results is no longer Jordan canonical form but upper triangular (via Schur decomposition). Normal matrices are special in that their Schur decompositions are diagonal matrices. Thus, normal matrices are unitarily similar to diagonal matrices.

In practice, to perform a similarity transformation requires knowledge of eigenvalues and eigenvectors, and that information is generally difficult to obtain. Numerical techniques for approximating these quantities are given in Chapters 19 and 20.

Solved Problems

15.1 Determine which of the following matrices are unitary:

$$\mathbf{A} = \begin{bmatrix} 1/2 + i/2 & 1/2 - i/2 \\ 1/2 - i/2 & 1/2 + i/2 \end{bmatrix} \qquad \mathbf{B} = \begin{bmatrix} 6/7 & 2/7 & 3/7 \\ 3/7 & -6/7 & -2/7 \\ 2/7 & 3/7 & -6/7 \end{bmatrix} \qquad \mathbf{C} = 1/\sqrt{3}\begin{bmatrix} 1 & 1 & 1 & 0 \\ 1 & 0 & -1 & -i \\ 1 & -1 & 0 & i \\ 0 & -i & i & 1 \end{bmatrix}$$

All three are unitary, because the product of each with its Hermitian transpose yields an identity matrix. Since the elements of **B** are real, that matrix is also orthogonal.

15.2 Prove that a matrix is unitary if and only if its rows (or columns) form an orthonormal set of vectors.

Designate the rows of **U** as $\mathbf{U}_1, \mathbf{U}_2, \ldots, \mathbf{U}_n$. Then the (i, j) element $(i = 1, 2, \ldots, n;\ j = 1, 2, \ldots, n)$ of $\mathbf{UU}^H$ is

$$(\mathbf{UU}^H)_{i,j} = \mathbf{U}_i \cdot \bar{\mathbf{U}}_j = \langle \mathbf{U}_i, \mathbf{U}_j \rangle$$

If **U** is unitary, then $\mathbf{UU}^H = \mathbf{I}$, and this (i, j) element $\langle \mathbf{U}_i, \mathbf{U}_j \rangle$ must be 1 when $i = j$ and 0 otherwise. This, in turn, implies that the set $\{\mathbf{U}_1, \mathbf{U}_2, \ldots, \mathbf{U}_n\}$ is an orthonormal set of vectors. (The columns of **U** may be shown to be orthonormal by considering the product $\mathbf{U}^H\mathbf{U}$ instead.)

Conversely, if the rows (or columns) of a matrix form an orthonormal set, then the argument given above may be reversed to show that **U** is unitary.

15.3 Show that if λ is an eigenvalue of an $n \times n$ matrix **A**, then there exists an $n \times n$ unitary matrix **U** having as its first column an eigenvector of **A** corresponding to λ.

If λ is an eigenvalue of **A**, then there must exist an eigenvector **X** corresponding to λ. Setting $\mathbf{Y} = \mathbf{X}/\|\mathbf{X}\|$ gives us a unit eigenvector of **A** corresponding to λ.

Consider the set of vectors $\{\mathbf{Y}, \mathbf{E}_1, \mathbf{E}_2, \ldots, \mathbf{E}_n\}$, where $\mathbf{E}_k$ $(k = 1, 2, \ldots, n)$ has a 1 as its kth component and all other components equal to zero. Using the algorithm given in Problem 6.12, we can reduce this set to a maximal set of linearly independent vectors. Such a set must contain n vectors, because the subset $\{\mathbf{E}_1, \mathbf{E}_2, \ldots, \mathbf{E}_n\}$ is linearly independent; and it will contain **Y**, because **Y** is the first nonzero vector in the original set. [The first nonzero component is brought into the (1,1) position of the matrix generated by the algorithm, and it remains nonzero throughout the algorithm. Thus, the first vector, **Y**, remains part of the maximal linearly independent set.]

Now apply the Gram-Schmidt orthogonalization process to this maximal set of linearly independent vectors, with **Y** taken as the first vector; **Y** remains unchanged. Finally, choose the columns of **U** to be the vectors resulting from the Gram-Schmidt process. **U** is unitary as a result of Problem 15.2, and the first column of **U** is the eigenvector **Y**.

15.4 Apply the procedure of Problem 15.3 to construct a unitary matrix having as its first column an eigenvector corresponding to $\lambda = 2$ for the matrix

$$\mathbf{A} = \begin{bmatrix} 3 & 0 & 0 \\ 1 & 3 & 1 \\ 2 & -1 & 1 \end{bmatrix}$$

An eigenvector of **A** corresponding to $\lambda = 2$ is $\mathbf{X} = [0, -1, 1]^T$, which when normalized yields $\mathbf{Y} = [0, -1/\sqrt{2}, 1/\sqrt{2}]^T$. Applying the algorithm given in Problem 6.12 to the set consisting of **Y** and

$$\mathbf{E}_1 = [1, 0, 0]^T \qquad \mathbf{E}_2 = [0, 1, 0]^T \qquad \mathbf{E}_3 = [0, 0, 1]^T\}$$

we construct the matrix

$$\begin{bmatrix} 0 & 1 & 0 & 0 \\ -1/\sqrt{2} & 0 & 1 & 0 \\ 1/\sqrt{2} & 0 & 0 & 1 \end{bmatrix}$$

This has the row-echelon form

$$\begin{bmatrix} 1 & 0 & -\sqrt{2} & 0 \\ 0 & 1 & 0 & 0 \\ 0 & 0 & 1 & 1 \end{bmatrix}$$

which indicates that the first, second, and third vectors of the set form a maximal linearly independent set. Applying the Gram-Schmidt process to the set $\{\mathbf{Y}, \mathbf{E}_1, \mathbf{E}_2\}$, we obtain the orthonormal set $\{\mathbf{Q}_1 = \mathbf{Y}, \mathbf{Q}_2 = \mathbf{E}_1, \mathbf{Q}_3 = [0, 1/\sqrt{2}, 1/\sqrt{2}]^T\}$. Then

$$\mathbf{U} = \begin{bmatrix} 0 & 1 & 0 \\ -1/\sqrt{2} & 0 & 1/\sqrt{2} \\ 1/\sqrt{2} & 0 & 1/\sqrt{2} \end{bmatrix}$$

15.5 Prove that the product of unitary matrices of the same order is also a unitary matrix.

If $\mathbf{A}$ and $\mathbf{B}$ are unitary, then

$$(\mathbf{AB})^{-1} = \mathbf{B}^{-1}\mathbf{A}^{-1} = \mathbf{B}^H\mathbf{A}^H = \bar{\mathbf{B}}^T\bar{\mathbf{A}}^T = (\bar{\mathbf{A}}\bar{\mathbf{B}})^T = (\overline{\mathbf{AB}})^T = (\mathbf{AB})^H$$

15.6 Show that if $\mathbf{U}$ is unitary, then $\langle \mathbf{UX}, \mathbf{UY} \rangle = \langle \mathbf{X}, \mathbf{Y} \rangle$ for all vectors $\mathbf{X}$ and $\mathbf{Y}$ of appropriate dimension.

Under the Euclidean inner product, the adjoint of $\mathbf{U}$ is its Hermitian transpose by *(13.5)*; hence,

$$\langle \mathbf{UX}, \mathbf{UY} \rangle = \langle \mathbf{X}, \mathbf{U}^*\mathbf{UY} \rangle = \langle \mathbf{X}, \mathbf{U}^H\mathbf{UY} \rangle = \langle \mathbf{X}, \mathbf{U}^{-1}\mathbf{UY} \rangle = \langle \mathbf{X}, \mathbf{IY} \rangle = \langle \mathbf{X}, \mathbf{Y} \rangle$$

15.7 Show that if λ is an eigenvalue of a unitary matrix $\mathbf{U}$, then $|\lambda| = 1$.

Let $\mathbf{X}$ be an eigenvector of $\mathbf{U}$ corresponding to λ. Then using Problem 15.6, we have

$$|\lambda|^2\langle \mathbf{X}, \mathbf{X} \rangle = \lambda\bar{\lambda}\langle \mathbf{X}, \mathbf{X} \rangle = \langle \lambda\mathbf{X}, \lambda\mathbf{X} \rangle = \langle \mathbf{UX}, \mathbf{UX} \rangle = \langle \mathbf{X}, \mathbf{X} \rangle \qquad (1)$$

Since $\mathbf{X}$ is an eigenvector, it is nonzero; hence $\langle \mathbf{X}, \mathbf{X} \rangle \neq 0$. Dividing *(1)* by $\langle \mathbf{X}, \mathbf{X} \rangle$, we obtain $|\lambda|^2 = 1$, which implies that $|\lambda| = 1$.

15.8 Find a Schur decomposition for

$$\mathbf{A} = \begin{bmatrix} 4 & 0 & 1 \\ 1 & 3 & -1 \\ -1 & 0 & 2 \end{bmatrix}$$

We follow Steps 15.1 through 15.5, beginning with $k = 1$ and $\mathbf{T}_0 = \mathbf{A}$. For $k = 1$, $\mathbf{A}_1 = \mathbf{T}_0 = \mathbf{A}$. An eigenvalue for $\mathbf{A}$ is $\lambda = 3$, with unit eigenvector $\mathbf{Y} = [0, 1, 0]^T$. Using the procedure given in Problem 15.3 with $n = 3$, we get

$$\mathbf{N}_1 = \begin{bmatrix} 0 & 1 & 0 \\ 1 & 0 & 0 \\ 0 & 0 & 1 \end{bmatrix} = \mathbf{U}_1 \quad \text{and} \quad \mathbf{T}_1 = \mathbf{U}_1^H\mathbf{T}_0\mathbf{U}_1 = \begin{bmatrix} 3 & 1 & -1 \\ 0 & 4 & 1 \\ 0 & -1 & 2 \end{bmatrix}$$

Now we apply Steps 15.1 through 15.5 with $k = 2$. We begin by setting

$$\mathbf{A}_2 = \begin{bmatrix} 4 & 1 \\ -1 & 2 \end{bmatrix}$$

This matrix possesses the eigenvalue $\lambda = 3$ with corresponding unit eigenvector $\mathbf{Y} = [1/\sqrt{2}, -1/\sqrt{2}]^T$. Using the procedure given in Problem 15.3 with $n = 2$, we generate the unitary matrix

$$\mathbf{N}_2 = \begin{bmatrix} 1/\sqrt{2} & 1/\sqrt{2} \\ -1/\sqrt{2} & 1/\sqrt{2} \end{bmatrix}$$

which is expanded into

$$\mathbf{U}_2 = \left[\begin{array}{c|cc} 1 & 0 & 0 \\ \hline 0 & 1/\sqrt{2} & 1/\sqrt{2} \\ 0 & -1/\sqrt{2} & 1/\sqrt{2} \end{array}\right] \qquad \text{so that} \qquad \mathbf{T}_2 = \mathbf{U}_2^H\mathbf{T}_1\mathbf{U}_2 = \begin{bmatrix} 3 & 2/\sqrt{2} & 0 \\ 0 & 3 & 2 \\ 0 & 0 & 3 \end{bmatrix}$$

Setting $\mathbf{U} = \mathbf{U}_1\mathbf{U}_2$, we have $\mathbf{U}^H\mathbf{A}\mathbf{U} = \mathbf{T}_2$, a matrix in upper triangular form. In this case, all the elements of $\mathbf{U}$ are real, so it is orthogonal.

15.9 Find a Schur decomposition for

$$\mathbf{A} = \begin{bmatrix} 3 & 0 & 0 & -1 \\ 1 & 2 & 0 & 1 \\ 2 & 0 & 4 & 2 \\ -1 & 0 & 0 & 3 \end{bmatrix}$$

We follow Steps 15.1 through 15.5, beginning with $k = 1$ and $\mathbf{A}_1 = \mathbf{T}_0 = \mathbf{A}$. An eigenvalue for $\mathbf{A}_1$ is $\lambda = 4$, with corresponding unit eigenvector $\mathbf{Y} = [0, 0, 1, 0]^T$. Using the procedure of Problem 15.3 with $n = 4$, we get

$$\mathbf{N}_1 = \begin{bmatrix} 0 & 1 & 0 & 0 \\ 0 & 0 & 1 & 0 \\ 1 & 0 & 0 & 0 \\ 0 & 0 & 0 & 1 \end{bmatrix} = \mathbf{U}_1 \qquad \text{and} \qquad \mathbf{T}_1 = \mathbf{U}_1^H\mathbf{T}_0\mathbf{U}_1 = \begin{bmatrix} 4 & 2 & 0 & 2 \\ 0 & 3 & 0 & -1 \\ 0 & 1 & 2 & 1 \\ 0 & -1 & 0 & 3 \end{bmatrix}$$

Next we apply Steps 15.1 through 15.5 with $k = 2$. We first determine

$$\mathbf{A}_2 = \begin{bmatrix} 3 & 0 & -1 \\ 1 & 2 & 1 \\ -1 & 0 & 3 \end{bmatrix}$$

This matrix possesses the eigenvalue $\lambda = 2$ with corresponding unit eigenvector $\mathbf{Y} = [0, 1, 0]^T$. Using the procedure of Problem 15.3, now with $n = 3$, we generate the unitary matrix

$$\mathbf{N}_2 = \begin{bmatrix} 0 & 1 & 0 \\ 1 & 0 & 0 \\ 0 & 0 & 1 \end{bmatrix}$$

This is expanded into

$$\mathbf{U}_2 = \left[\begin{array}{c|ccc} 1 & 0 & 0 & 0 \\ \hline 0 & 0 & 1 & 0 \\ 0 & 1 & 0 & 0 \\ 0 & 0 & 0 & 1 \end{array}\right] \qquad \text{so that} \qquad \mathbf{T}_2 = \mathbf{U}_2^H\mathbf{T}_1\mathbf{U}_2 = \begin{bmatrix} 4 & 2 & 0 & 2 \\ 0 & 2 & 1 & 1 \\ 0 & 0 & 3 & -1 \\ 0 & 0 & -1 & 3 \end{bmatrix}$$

We now apply Steps 15.1 through 15.5 with $k = 3$. We start by setting

$$\mathbf{A}_3 = \begin{bmatrix} 3 & -1 \\ -1 & 3 \end{bmatrix}$$

This matrix possesses the eigenvalue $\lambda = 2$, with corresponding unit eigenvector $\mathbf{Y} = [1/\sqrt{2}, 1/\sqrt{2}]^T$. Using the procedure of Problem 15.3 with $n = 2$, we generate the unitary matrix

$$\mathbf{N}_3 = \begin{bmatrix} 1/\sqrt{2} & -1/\sqrt{2} \\ 1/\sqrt{2} & 1/\sqrt{2} \end{bmatrix}$$

which we expand into

$$\mathbf{U}_3 = \left[\begin{array}{cc|cc} 1 & 0 & 0 & 0 \\ 0 & 1 & 0 & 0 \\ \hline 0 & 0 & 1/\sqrt{2} & -1/\sqrt{2} \\ 0 & 0 & 1/\sqrt{2} & 1/\sqrt{2} \end{array}\right] \quad \text{so that} \quad \mathbf{T}_3 = \mathbf{U}_3^H\mathbf{T}_2\mathbf{U}_2 = \begin{bmatrix} 4 & 2 & 2/\sqrt{2} & 2/\sqrt{2} \\ 0 & 2 & 2/\sqrt{2} & 0 \\ 0 & 0 & 2 & 0 \\ 0 & 0 & 0 & 4 \end{bmatrix}$$

Setting $\mathbf{U} = \mathbf{U}_1\mathbf{U}_2\mathbf{U}_3$, we have $\mathbf{U}^H\mathbf{A}\mathbf{U} = \mathbf{T}_3$, a matrix in upper triangular form.

15.10 Show that if $\mathbf{U}$ is unitary and $\mathbf{A} = \mathbf{U}^H\mathbf{B}\mathbf{U}$, then $\mathbf{B}$ is normal if and only if $\mathbf{A}$ is normal.

If $\mathbf{B}$ is normal, then $\mathbf{B}^H\mathbf{B} = \mathbf{B}\mathbf{B}^H$, and

$$\begin{aligned} \mathbf{A}^H\mathbf{A} &= (\mathbf{U}^H\mathbf{B}\mathbf{U})^H(\mathbf{U}^H\mathbf{B}\mathbf{U}) = (\mathbf{U}^H\mathbf{B}^H\mathbf{U})(\mathbf{U}^H\mathbf{B}\mathbf{U}) = (\mathbf{U}^H\mathbf{B}^H)(\mathbf{U}\mathbf{U}^H)(\mathbf{B}\mathbf{U}) \\ &= (\mathbf{U}^H\mathbf{B}^H)(\mathbf{U}\mathbf{U}^{-1})(\mathbf{B}\mathbf{U}) = (\mathbf{U}^H\mathbf{B}^H)(\mathbf{B}\mathbf{U}) = \mathbf{U}^H(\mathbf{B}^H\mathbf{B})\mathbf{U} = \mathbf{U}^H(\mathbf{B}\mathbf{B}^H)\mathbf{U} \\ &= (\mathbf{U}^H\mathbf{B})(\mathbf{B}^H\mathbf{U}) = (\mathbf{U}^H\mathbf{B})(\mathbf{U}\mathbf{U}^{-1})(\mathbf{B}^H\mathbf{U}) = (\mathbf{U}^H\mathbf{B})(\mathbf{U}\mathbf{U}^H)(\mathbf{B}^H\mathbf{U}) \\ &= (\mathbf{U}^H\mathbf{B}\mathbf{U})(\mathbf{U}^H\mathbf{B}^H\mathbf{U}) = (\mathbf{U}^H\mathbf{B}\mathbf{U})(\mathbf{U}^H\mathbf{B}\mathbf{U})^H = \mathbf{A}\mathbf{A}^H \end{aligned}$$

The reverse proposition is proved analogously, using the identity $\mathbf{B} = \mathbf{U}\mathbf{A}\mathbf{U}^H$.

15.11 Prove that every normal matrix is unitarily similar to a diagonal matrix.

Let $\mathbf{A}$ be normal. Using a Schur decomposition, we can write $\mathbf{T} = \mathbf{U}^H\mathbf{A}\mathbf{U}$, where $\mathbf{T}$ is upper triangular and $\mathbf{U}$ is unitary. It follows from Problem 15.10 that $\mathbf{T}$ is normal, and then from Problem 13.14 that $\mathbf{T}$ must be diagonal.

15.12 Prove that $\|\mathbf{A}\|_S$, the matrix norm induced by the Euclidean vector norm, is the square root of the largest eigenvalue of $\mathbf{A}^H\mathbf{A}$.

For any matrix $\mathbf{A}$ (not necessarily square), the product $\mathbf{A}^H\mathbf{A}$ is normal (Problem 13.2) and has nonnegative eigenvalues (Problem 13.3), which we denote as $\lambda_1, \lambda_2, \ldots, \lambda_n$. It follows from Problem 15.11 that there exists a unitary matrix $\mathbf{U}$ such that $\mathbf{U}^H(\mathbf{A}^H\mathbf{A})\mathbf{U} = \mathbf{D}$ or, equivalently, such that $\mathbf{A}^H\mathbf{A} = \mathbf{U}\mathbf{D}\mathbf{U}^H$, where $\mathbf{D}$ is a diagonal matrix whose diagonal elements are the eigenvalues of $\mathbf{A}^H\mathbf{A}$. Then if we set $\lambda = \max(\lambda_1, \lambda_2, \ldots, \lambda_n)$ and $\mathbf{Y} = \mathbf{U}^H\mathbf{X}$, we have, for any nonzero vector $\mathbf{X}$,

$$\begin{aligned} \|\mathbf{A}\mathbf{X}\|_2^2 &= \langle \mathbf{A}\mathbf{X}, \mathbf{A}\mathbf{X}\rangle = \langle \mathbf{A}^*\mathbf{A}\mathbf{X}, \mathbf{X}\rangle = \langle \mathbf{A}^H\mathbf{A}\mathbf{X}, \mathbf{X}\rangle = \langle \mathbf{U}\mathbf{D}\mathbf{U}^H\mathbf{X}, \mathbf{X}\rangle = \langle \mathbf{D}\mathbf{U}^H\mathbf{X}, \mathbf{U}^H\mathbf{X}\rangle = \langle \mathbf{D}\mathbf{Y}, \mathbf{Y}\rangle \\ &= \lambda_1 y_1\bar{y}_1 + \lambda_2 y_2\bar{y}_2 + \cdots + \lambda_n y_n\bar{y}_n = \lambda_1|y_1|^2 + \lambda_2|y_2|^2 + \cdots + \lambda_n|y_n|^2 \\ &\le \lambda|y_1|^2 + \lambda|y_2|^2 + \cdots + \lambda|y_n|^2 = \lambda\langle \mathbf{Y}, \mathbf{Y}\rangle = \lambda\langle \mathbf{U}^H\mathbf{X}, \mathbf{U}^H\mathbf{X}\rangle = \lambda\langle \mathbf{X}, \mathbf{U}\mathbf{U}^H\mathbf{X}\rangle \\ &= \lambda\langle \mathbf{X}, \mathbf{X}\rangle = \lambda\|\mathbf{X}\|_2^2 \end{aligned}$$

Therefore, using the result of Problem 12.14, we have

$$\|\mathbf{A}\|_S = \max_{\mathbf{X}\neq\mathbf{0}}\left(\frac{\|\mathbf{A}\mathbf{X}\|_2}{\|\mathbf{X}\|_2}\right) \le \max_{\mathbf{X}\neq\mathbf{0}}\left(\frac{\sqrt{\lambda}\|\mathbf{X}\|_2}{\|\mathbf{X}\|_2}\right) = \sqrt{\lambda} \tag{1}$$

Denoting as $\mathbf{Z}$ an eigenvector of $\mathbf{A}^H\mathbf{A}$ corresponding to λ, we find that

$$\|\mathbf{A}\mathbf{Z}\|_2^2 = \langle \mathbf{A}\mathbf{Z}, \mathbf{A}\mathbf{Z}\rangle = \langle \mathbf{A}^H\mathbf{A}\mathbf{Z}, \mathbf{Z}\rangle = \langle \lambda\mathbf{Z}, \mathbf{Z}\rangle = \lambda\langle \mathbf{Z}, \mathbf{Z}\rangle = \lambda\|\mathbf{Z}\|_2^2$$

so

$$\|\mathbf{A}\|_S = \max_{\mathbf{X}\neq\mathbf{0}}\left(\frac{\|\mathbf{A}\mathbf{X}\|_2}{\|\mathbf{X}\|_2}\right) \ge \frac{\|\mathbf{A}\mathbf{Z}\|_2}{\|\mathbf{Z}\|_2} = \frac{\sqrt{\lambda}\|\mathbf{Z}\|_2}{\|\mathbf{Z}\|_2} = \sqrt{\lambda} \tag{2}$$

Inequalities (*1*) and (*2*) imply $\|\mathbf{A}\|_S = \sqrt{\lambda}$.

15.13 Find elementary reflectors associated with (*a*) $\mathbf{V}_0 = [1, 2]^T$ and (*b*) $\mathbf{V}_1 = [9, 3, -6]^T$.

(*a*) We compute $\|\mathbf{V}_0\|_2 = \sqrt{5}$, so

$$\mathbf{R}_0 = \begin{bmatrix} 1 & 0 \\ 0 & 1 \end{bmatrix} - \frac{2}{5}\begin{bmatrix} 1 \\ 2 \end{bmatrix}[1, 2] = \begin{bmatrix} 1 & 0 \\ 0 & 1 \end{bmatrix} - \frac{2}{5}\begin{bmatrix} 1 & 2 \\ 2 & 4 \end{bmatrix} = \begin{bmatrix} 3/5 & -4/5 \\ -4/5 & -3/5 \end{bmatrix}$$

(*b*) Similarly, $\|\mathbf{V}_1\|_2 = \sqrt{126}$, so

$$\mathbf{R}_1 = \begin{bmatrix} 1 & 0 & 0 \\ 0 & 1 & 0 \\ 0 & 0 & 1 \end{bmatrix} - \frac{2}{126}\begin{bmatrix} 9 \\ 3 \\ -6 \end{bmatrix}[9, 3, -6] = \begin{bmatrix} 1 & 0 & 0 \\ 0 & 1 & 0 \\ 0 & 0 & 1 \end{bmatrix} - \frac{1}{63}\begin{bmatrix} 81 & 27 & -54 \\ 27 & 9 & -18 \\ -54 & -18 & 36 \end{bmatrix}$$

$$= \begin{bmatrix} -2/7 & -3/7 & 6/7 \\ -3/7 & 6/7 & 2/7 \\ 6/7 & 2/7 & 3/7 \end{bmatrix}$$

15.14 Prove that an elementary reflector $\mathbf{R}$ is both symmetric and orthogonal.

For any constant c,

$$(c\mathbf{V}\mathbf{V}^T)^T = c(\mathbf{V}\mathbf{V}^T)^T = c(\mathbf{V}^T)^T\mathbf{V}^T = c\mathbf{V}\mathbf{V}^T$$

Setting $c = -2/\|\mathbf{V}\|_2^2$, we conclude that $(-2/\|\mathbf{V}\|_2^2)\mathbf{V}\mathbf{V}^T$ is symmetric. Since $\mathbf{I}$ is symmetric and the sum of real symmetric matrices is also real and symmetric, it follows that any elementary reflector is symmetric. In addition,

$$\mathbf{R}^T\mathbf{R} = \mathbf{R}\mathbf{R} = \left(\mathbf{I} - \frac{2}{\|\mathbf{V}\|_2^2}\mathbf{V}\mathbf{V}^T\right)\left(\mathbf{I} - \frac{2}{\|\mathbf{V}\|_2^2}\mathbf{V}\mathbf{V}^T\right)$$

$$= \mathbf{I} - \frac{2}{\|\mathbf{V}\|_2^2}\mathbf{V}\mathbf{V}^T - \frac{2}{\|\mathbf{V}\|_2^2}\mathbf{V}\mathbf{V}^T + \frac{4}{\|\mathbf{V}\|_2^4}\mathbf{V}\mathbf{V}^T\mathbf{V}\mathbf{V}^T$$

$$= \mathbf{I} - \frac{4}{\|\mathbf{V}\|_2^2}\mathbf{V}\mathbf{V}^T + \frac{4}{\|\mathbf{V}\|_2^4}\mathbf{V}(\mathbf{V}^T\mathbf{V})\mathbf{V}^T$$

But if $\mathbf{V}$ is a real column vector, then $\mathbf{V}^T\mathbf{V} = \langle \mathbf{V}, \mathbf{V} \rangle = \|\mathbf{V}\|_2^2$. Thus, the last two terms in the above equality cancel, $\mathbf{R}^T\mathbf{R} = \mathbf{I}$, and $\mathbf{R}$ is orthogonal.

15.15 Let $\mathbf{R}$ be the elementary reflector associated with the vector $\mathbf{V} = \mathbf{X} + \|\mathbf{X}\|_2\mathbf{E}$, where $\mathbf{X}$ is an arbitrary real n-dimensional column vector, and $\mathbf{E}$ is an n-dimensional column vector whose first component is 1 and whose other components are all 0s. Show that $\mathbf{R}\mathbf{X} = -\|\mathbf{X}\|_2\mathbf{E}$.

Denote the first component of $\mathbf{X}$ as x_1. Then

$$\mathbf{V}^T\mathbf{X} = (\mathbf{X} + \|\mathbf{X}\|_2\mathbf{E})^T\mathbf{X} = \mathbf{X}^T\mathbf{X} + \|\mathbf{X}\|_2\mathbf{E}^T\mathbf{X} = \|\mathbf{X}\|_2^2 + \|\mathbf{X}\|_2 x_1$$

and

$$\mathbf{V}^T\mathbf{V} = (\mathbf{X} + \|\mathbf{X}\|_2\mathbf{E})^T(\mathbf{X} + \|\mathbf{X}\|_2\mathbf{E}) = \mathbf{X}^T\mathbf{X} + \|\mathbf{X}\|_2\mathbf{E}^T\mathbf{X} + \|\mathbf{X}\|_2\mathbf{X}^T\mathbf{E} + \|\mathbf{X}\|_2^2\mathbf{E}^T\mathbf{E}$$

$$= \|\mathbf{X}\|_2^2 + \|\mathbf{X}\|_2 x_1 + \|\mathbf{X}\|_2 x_1 + \|\mathbf{X}\|_2^2 = 2(\|\mathbf{X}\|_2^2 + \|\mathbf{X}\|_2 x_1) = 2\mathbf{V}^T\mathbf{X}$$

Then $\mathbf{R}\mathbf{X} = \mathbf{I}\mathbf{X} - 2\dfrac{\mathbf{V}\mathbf{V}^T}{\|\mathbf{V}\|_2^2}\mathbf{X} = \mathbf{X} - 2\dfrac{\mathbf{V}\mathbf{V}^T}{\mathbf{V}^T\mathbf{V}}\mathbf{X} = \mathbf{X} - \dfrac{2\mathbf{V}(\mathbf{V}^T\mathbf{X})}{2\mathbf{V}^T\mathbf{X}} = \mathbf{X} - \mathbf{V} = \mathbf{X} - (\mathbf{X} + \|\mathbf{X}\|_2\mathbf{E}) = -\|\mathbf{X}\|_2\mathbf{E}$

Supplementary Problems

15.16 Determine which of the following matrices are unitary:

$$\mathbf{A}=\begin{bmatrix}1 & 1 & -1\\ 1 & 1 & 1\\ -2 & 1 & 0\end{bmatrix}\qquad \mathbf{B}=\begin{bmatrix}1/\sqrt{3} & 1/\sqrt{2} & 1/\sqrt{2}\\ 1/\sqrt{3} & -1/\sqrt{2} & 0\\ 1/\sqrt{3} & 0 & -1/\sqrt{2}\end{bmatrix}\qquad \mathbf{C}=(1/\sqrt{3})\begin{bmatrix}1 & 1 & 0 & 1\\ 1 & -1 & 1 & 0\\ 0 & 1 & 1 & -1\\ 1 & 0 & -1 & -1\end{bmatrix}$$

$$\mathbf{D}=\begin{bmatrix}i/\sqrt{2} & i/\sqrt{2}\\ 1/\sqrt{2} & 1/\sqrt{2}\end{bmatrix}\qquad \mathbf{E}=\begin{bmatrix}1/\sqrt{2} & i/\sqrt{2}\\ i/\sqrt{2} & 1/\sqrt{2}\end{bmatrix}$$

15.17 Apply the procedure of Problem 15.3 to construct a unitary matrix having, as its first column, an eigenvector corresponding to $\lambda = 3$ for:

$$(a)\ \mathbf{A}=\begin{bmatrix}7 & -9\\ 4 & -6\end{bmatrix}\qquad (b)\ \mathbf{B}=\begin{bmatrix}2 & 0 & 0\\ 1 & 5 & 2\\ 1 & -1 & 2\end{bmatrix}\qquad (c)\ \mathbf{C}=\begin{bmatrix}3 & 1 & 1\\ 1 & 5 & 1\\ 1 & 1 & 3\end{bmatrix}$$

15.18 Find a Schur decomposition for each of the matrices in Problem 15.17.

15.19 Find elementary reflectors associated with

$$(a)\ \begin{bmatrix}2\\ 1\end{bmatrix}\qquad (b)\ \begin{bmatrix}1\\ 0\\ -1\end{bmatrix}\qquad (c)\ \begin{bmatrix}\sqrt{2}\\ 1\end{bmatrix}$$

15.20 Show that if $\mathbf{U}$ is unitary, then $\|\mathbf{UX}\|_2 = \|\mathbf{X}\|_2$ for any vector $\mathbf{X}$ of suitable dimension. Thus, a unitary transformation preserves Euclidean length.

15.21 The *angle* between two real vectors $\mathbf{X}$ and $\mathbf{Y}$ is defined as

$$\theta = \arccos \frac{\langle \mathbf{X}, \mathbf{Y}\rangle}{\|\mathbf{X}\|_2\|\mathbf{Y}\|_2}$$

Show that if $\mathbf{U}$ is unitary, then the angle between $\mathbf{UX}$ and $\mathbf{UY}$ is the same as that between $\mathbf{X}$ and $\mathbf{Y}$. Thus, a unitary transformation preserves angles.

15.22 Prove that $\mathbf{R}^2 = \mathbf{I}$ for any elementary reflector $\mathbf{R}$.

15.23 Determine the eigenvalues of every elementary reflector.

15.24 Prove that the absolute value of the determinant of a unitary matrix is 1.

15.25 A square matrix $\mathbf{R}_{pq}(\theta)$ is a *rotation matrix* if
(1) The (p, p) and (q, q) elements of $\mathbf{R}_{pq}$ are equal to $\cos\theta$ for $p \neq q$, and all other diagonal elements are unity, and
(2) The (p, q) element is equal to $\sin\theta$; the (q, p) element is equal to $-\sin\theta$, and all other off-diagonal elements are zero. Find 5×5 matrices $\mathbf{R}_{23}(\theta)$ and $\mathbf{R}_{42}(\theta)$.

15.26 Show that a rotation matrix is orthogonal.

15.27 Show that θ may be chosen so that $\mathbf{R}_{pq}^T(\theta)\mathbf{A}\mathbf{R}_{pq}(\theta)$ has a zero in the (k, p) position, provided that $\mathbf{A}$ is square and k is different from p and q. (See Problem 15.25.)

Chapter 16

Quadratic Forms and Congruence

QUADRATIC FORM

A *quadratic form* in the real variables $x_1, x_2, \ldots, x_n$ is a polynomial of the type

$$\sum_{i=1}^{n} \sum_{j=1}^{n} a_{ij} x_i x_j \tag{16.1}$$

with real-valued coefficients a_{ij}. This expression has the matrix representation

$$\mathbf{X}^T\mathbf{A}\mathbf{X} \tag{16.2}$$

with $\mathbf{A} = [a_{ij}]$ and $\mathbf{X} = [x_1, x_2, \ldots, x_n]^T$. The quadratic form $\mathbf{X}^T\mathbf{A}\mathbf{X}$ is algebraically equivalent to $\mathbf{X}^T\{(\mathbf{A} + \mathbf{A}^T)/2\}\mathbf{X}$. Since $(\mathbf{A} + \mathbf{A}^T)/2$ is symmetric, it is standard to use it rather than a nonsymmetric matrix in expression *(16.2)*. Thus, in what follows, we shall assume that $\mathbf{A}$ is symmetric. (See Problems 16.1 and 16.2.)

A *complex quadratic form* is one that has the matrix representation

$$\mathbf{X}^H\mathbf{A}\mathbf{X} \tag{16.3}$$

with $\mathbf{A}$ being Hermitian. Expression *(16.3)* reduces to *(16.2)* when $\mathbf{X}$ and $\mathbf{A}$ are real-valued; and both expressions are equivalent to the Euclidean inner product $\langle \mathbf{AX}, \mathbf{X} \rangle$.

The Euclidean inner product $\langle \mathbf{AX}, \mathbf{X} \rangle$ is real whenever $\mathbf{A}$ is Hermitian (Property 13.5). If the inner product is also positive (or nonnegative, negative, or nonpositive) for all nonzero vectors $\mathbf{X}$, then the quadratic form is classified as positive definite (or positive semidefinite, negative definite, or negative semidefinite, respectively). All the tests listed in Chapter 14 may be applied to the matrix representation of a quadratic form to determine definiteness. (See Problems 16.3 and 16.4.)

DIAGONAL FORM

A quadratic form has *diagonal form* if it contains no cross-product terms; that is, if $a_{ij} = 0$ for all $i \neq j$. It follows from Theorem 15.1 that any quadratic form can be transformed into diagonal form with a unitary matrix $\mathbf{U}$ (recall that a quadratic form is Hermitian and therefore normal). If $\mathbf{U}^H\mathbf{A}\mathbf{U} = \mathbf{D}$, then the substitution $\mathbf{X} = \mathbf{UY}$ converts the quadratic form $\langle \mathbf{AX}, \mathbf{X} \rangle$ into the diagonal quadratic form $\langle \mathbf{DY}, \mathbf{Y} \rangle$. This substitution preserves length (Problem 15.20) and angles (Problem 15.21); the diagonal elements of $\mathbf{D}$ are the eigenvalues of $\mathbf{A}$. (See Problems 16.5 and 16.6.)

CONGRUENCE

A square matrix $\mathbf{A}$ is *congruent* to a square matrix $\mathbf{B}$ of the same order if there exists a nonsingular real matrix $\mathbf{P}$ such that

$$\mathbf{A} = \mathbf{P}\mathbf{B}\mathbf{P}^T \tag{16.4}$$

When $\mathbf{P}$ is factored into a product of elementary matrices corresponding to elementary row operations, then $\mathbf{P}^T$ is the product in reverse order of elementary matrices corresponding to identical elementary column operations. Thus, two matrices are congruent if and only if one can be transformed to the other by a sequence of pairs of elementary row and column operations, where each pair consists of one elementary row operation and one elementary column operation of identical type. It follows that congruent matrices have the same rank.

A matrix **A** is *Hermitian congruent* (or *conjunctive*) to a matrix **B** if there exists a nonsingular matrix **P** such that

$$\mathbf{A} = \mathbf{P}\mathbf{B}\mathbf{P}^H \tag{16.5}$$

Hermitian congruence reduces to congruence when **P** is real. Both Hermitian congruence and congruence are reflexive, symmetric, and transitive.

Two quadratic forms $\langle \mathbf{AX}, \mathbf{X} \rangle$ and $\langle \mathbf{BY}, \mathbf{Y} \rangle$ are congruent if and only if **A** and **B** are congruent.

INERTIA

Every $n \times n$ Hermitian matrix of rank r is congruent to a unique matrix in the partitioned form

$$\left[\begin{array}{c|c|c} \mathbf{I}_k & \mathbf{0} & \mathbf{0} \\ \hline \mathbf{0} & -\mathbf{I}_m & \mathbf{0} \\ \hline \mathbf{0} & \mathbf{0} & \mathbf{0} \end{array}\right] \tag{16.6}$$

where $\mathbf{I}_k$ and $\mathbf{I}_m$ are identity matrices of order $k \times k$ and $m \times m$, respectively. An *inertia matrix* is a matrix having form (*16.6*).

Property 16.1: (*Sylvester's law of inertia*) Two Hermitian matrices are congruent if and only if they are congruent to the same inertia matrix, and then they both have k positive eigenvalues, m negative eigenvalues, and $n - k - m$ zero eigenvalues.

The integer k defined by form (16.6) is called the *index* of **A**, and $s = k - m$ is called its *signature*.

An algorithm for obtaining the inertia matrix of a given matrix **A** is the following:

STEP 16.1: Construct the partitioned matrix $[\mathbf{A} \,|\, \mathbf{I}]$, where **I** is an identity matrix having the same order as **A**.

STEP 16.2: Use elementary row operations E1 and E3 to reduce **A** to upper triangular form, applying each operation to the full partitioned matrix of Step 16.1. In addition, whenever two rows are interchanged, also interchange the corresponding columns in the left partition, but make no similar column interchange in the right partition. Denote the result as $[\mathbf{R} \,|\, \mathbf{S}]$, where **R** is upper triangular.

STEP 16.3: Set all the nondiagonal elements of **R** equal to zero. The result is a partitioned matrix of the form $[\mathbf{D} \,|\, \mathbf{S}]$, where **D** is diagonal.

STEP 16.4: If a zero diagonal element of **D** appears in an earlier (higher) row than a nonzero diagonal element of **D**, then interchange the positions of the two diagonal elements; also interchange the order of the corresponding rows of **S**. Continue to perform these interchanges until all nonzero diagonal elements of **D** appear in earlier rows than zero diagonal elements.

STEP 16.5: If a negative diagonal element of **D** appears in an earlier row than a positive diagonal term of **D**, interchange the positions of the two diagonal elements; also interchange the order of the corresponding rows of **S**. Continue to perform these interchanges until all positive diagonal elements of **D** appear in earlier rows than all negative diagonal elements.

STEP 16.6: If any diagonal element of the left partition is not 0, 1, or -1, denote its value by d. Divide that element by $|d|$, and divide the entire corresponding row of the right partition by $\sqrt{|d|}$.

At the completion of the algorithm, the matrix in the left partition is the inertia matrix for **A**; the matrix in the right partition is the matrix **P** that will transform **A** into its inertia matrix. (See Problems 16.7 through 16.9.)

RAYLEIGH QUOTIENT

The *Rayleigh quotient* for a Hermitian matrix **A** is the ratio

$$R(\mathbf{X}) = \frac{\langle \mathbf{AX}, \mathbf{X} \rangle}{\langle \mathbf{X}, \mathbf{X} \rangle} \tag{16.7}$$

Property 16.2: (*Rayleigh's principle*) If the eigenvalues of **A** a Hermitian matrix are ordered so that $\lambda_1 \le \lambda_2 \le \cdots \le \lambda_n$, then

$$\lambda_1 \le R(\mathbf{X}) \le \lambda_n \tag{16.8}$$

$R(\mathbf{X})$ achieves its maximum when **X** is an eigenvector corresponding to λ_n; $R(\mathbf{X})$ achieves its minimum when **X** is an eigenvector corresponding to λ_1.

(See Problem 16.10.)

Solved Problems

16.1 Determine the symmetric-matrix representation for the real quadratic form $2x_1^2 + 5x_2^2 + 11x_3^2 + 20x_1x_2 - 4x_1x_3 + 16x_2x_3$.

We can rewrite this polynomial as

$$2x_1x_1 + 5x_2x_2 + 11x_3x_3 + 10x_1x_2 + 10x_2x_1 - 2x_2x_1 - 2x_3x_1 + 8x_2x_3 + 8x_3x_2$$

which has the symmetric-matrix representation

$$[x_1, x_2, x_3]\begin{bmatrix} 2 & 10 & -2 \\ 10 & 5 & 8 \\ -2 & 8 & 11 \end{bmatrix}\begin{bmatrix} x_1 \\ x_2 \\ x_3 \end{bmatrix}$$

16.2 Determine the symmetric-matrix representation for the real quadratic form $11x_1^2 + 11x_2^2 + 19x_3^2 + 16x_4^2 - 6x_1x_2 + 10x_1x_3 - 16x_1x_4 - 10x_2x_3 - 16x_2x_4$.

We can rewrite this polynomial as

$$11x_1x_1 + 11x_2x_2 + 19x_3x_3 + 16x_4x_4 - 3x_1x_2 - 3x_2x_1 + 5x_1x_3 + 5x_3x_1 - 8x_1x_4 - 8x_4x_1 - 5x_2x_3 - 5x_3x_2$$
$$- 8x_2x_4 - 8x_4x_2 + 0x_3x_4 + 0x_4x_3$$

which has the symmetric-matrix representation

$$[x_1, x_2, x_3, x_4]\begin{bmatrix} 11 & -3 & 5 & -8 \\ -3 & 11 & -5 & -8 \\ 5 & -5 & 19 & 0 \\ -8 & -8 & 0 & 16 \end{bmatrix}\begin{bmatrix} x_1 \\ x_2 \\ x_3 \\ x_4 \end{bmatrix}$$

16.3 Determine whether the quadratic form given in Problem 16.1 is positive definite.

The results of Problems 16.1 and 14.2 indicate that the matrix representation of the quadratic form is not positive definite. Therefore the quadratic form itself is not positive definite.

16.4 Determine whether the quadratic form given in Problem 16.2 is positive definite.

From the results of Problems 16.2 and 14.3, we determine that the quadratic form is not positive definite because its matrix representation is not. The quadratic form is, however, positive semidefinite.

16.5 Transform the quadratic form given in Problem 16.1 into a diagonal quadratic form.

Given the result of Problem 16.1, we set

$$\mathbf{A} = \begin{bmatrix} 2 & 10 & -2 \\ 10 & 5 & 8 \\ -2 & 8 & 11 \end{bmatrix}$$

A has eigenvalues −9, 9, and 18 and corresponding orthonormal eigenvectors $\mathbf{Q}_1 = [2/3, -2/3, 1/3]^T$, $\mathbf{Q}_2 = [2/3, 1/3, -2/3]^T$, and $\mathbf{Q}_2 = [1/3, 2/3, 2/3]^T$, respectively. We take

$$\mathbf{U} = \begin{bmatrix} 2/3 & 2/3 & 1/3 \\ -2/3 & 1/3 & 2/3 \\ 1/3 & -2/3 & 2/3 \end{bmatrix} \quad \text{whereupon} \quad \mathbf{U}^T\mathbf{A}\mathbf{U} = \mathbf{D} = \begin{bmatrix} -9 & 0 & 0 \\ 0 & 9 & 0 \\ 0 & 0 & 18 \end{bmatrix}$$

Set $\mathbf{X} = \mathbf{U}\mathbf{Y}$. Then

$$\mathbf{X}^T\mathbf{A}\mathbf{X} = (\mathbf{U}\mathbf{Y})^T\mathbf{A}(\mathbf{U}\mathbf{Y}) = \mathbf{Y}^T\mathbf{U}^T\mathbf{A}\mathbf{U}\mathbf{Y} = \mathbf{Y}^T\mathbf{D}\mathbf{Y} = -9y_1^2 + 9y_2^2 + 18y_3^2$$

16.6 Transform into diagonal form the complex quadratic form corresponding to the Hermitian matrix

$$\mathbf{A} = \begin{bmatrix} 3 & 1+i3 \\ 1-i3 & 6 \end{bmatrix}$$

The eigenvalues of **A** are 1 and 8, with corresponding orthonormal eigenvectors $\mathbf{U}_1 = [(1+i3)/\sqrt{14}, -2/\sqrt{14}]^T$ and $\mathbf{U}_2 = [(1+i3)/\sqrt{35}, 5/\sqrt{35}]^T$, respectively. We set

$$\mathbf{U} = \begin{bmatrix} (1+i3)/\sqrt{14} & (1+i3)/\sqrt{35} \\ -2/\sqrt{14} & 5/\sqrt{35} \end{bmatrix} \quad \text{whereupon} \quad \mathbf{U}^H\mathbf{A}\mathbf{U} = \mathbf{D} = \begin{bmatrix} 1 & 0 \\ 0 & 8 \end{bmatrix}$$

Set $\mathbf{X} = \mathbf{U}\mathbf{Y}$. Then the original quadratic form

$$\mathbf{X}^H\mathbf{A}\mathbf{X} = \langle \mathbf{A}\mathbf{X}, \mathbf{X} \rangle = 3x_1\bar{x}_1 + 6x_2\bar{x}_2 + (1+i3)\bar{x}_1x_2 + (1-i3)x_1\bar{x}_2$$

is transformed into the diagonal quadratic form $\langle \mathbf{D}\mathbf{Y}, \mathbf{Y} \rangle = y_1\bar{y}_1 + 8y_2\bar{y}_2$.

16.7 Determine the inertia matrix for

$$\mathbf{A} = \begin{bmatrix} 0 & 1 & 2 \\ 1 & 1 & 3 \\ 2 & 3 & 4 \end{bmatrix}$$

We augment the 3 × 3 identity matrix onto **A** and then reduce **A** to upper triangular form. To do so, we must first interchange the first and second rows, and we do this to the entire partitioned matrix. In addition, we interchange the first and second columns of **A** but make no corresponding change to the columns in the right partition. Steps 16.1 through 16.6 are as follows:

$$\left[\begin{array}{ccc|ccc} 0 & 1 & 2 & 1 & 0 & 0 \\ 1 & 1 & 3 & 0 & 1 & 0 \\ 2 & 3 & 4 & 0 & 0 & 1 \end{array}\right]$$

$$\begin{array}{c}\rightarrow\\ \rightarrow\\ \end{array}\left[\begin{array}{ccc|ccc} 1 & 1 & 3 & 0 & 1 & 0 \\ 0 & 1 & 2 & 1 & 0 & 0 \\ 2 & 3 & 4 & 0 & 0 & 1 \end{array}\right]$$ Interchanging the first and second rows

$$\left[\begin{array}{ccc|ccc} 1 & 1 & 3 & 0 & 1 & 0 \\ 1 & 0 & 2 & 1 & 0 & 0 \\ 3 & 2 & 4 & 0 & 0 & 1 \end{array}\right]$$ Interchanging the first and second columns of the left partition only

$$\begin{array}{c}\\ \rightarrow\\ \end{array}\left[\begin{array}{ccc|ccc} 1 & 1 & 3 & 0 & 1 & 0 \\ 0 & -1 & -1 & 1 & -1 & 0 \\ 3 & 2 & 4 & 0 & 0 & 1 \end{array}\right]$$ Adding -1 times the first row to the second row

$$\begin{array}{c}\\ \\ \rightarrow\end{array}\left[\begin{array}{ccc|ccc} 1 & 1 & 3 & 0 & 1 & 0 \\ 0 & -1 & -1 & 1 & -1 & 0 \\ 0 & -1 & -5 & 0 & -3 & 1 \end{array}\right]$$ Adding -3 times the first row to the third row

$$\begin{array}{c}\\ \\ \rightarrow\end{array}\left[\begin{array}{ccc|ccc} 1 & 1 & 3 & 0 & 1 & 0 \\ 0 & -1 & -1 & 1 & -1 & 0 \\ 0 & 0 & -4 & -1 & -2 & 1 \end{array}\right]$$ Adding -1 times the second row to the third row

$$\begin{array}{c}\rightarrow\\ \rightarrow\\ \end{array}\left[\begin{array}{ccc|ccc} 1 & 0 & 0 & 0 & 1 & 0 \\ 0 & -1 & 0 & 1 & -1 & 0 \\ 0 & 0 & -4 & -1 & -2 & 1 \end{array}\right]$$ Setting all the elements above the diagonal equal to zero in the left partition

There are no zero diagonal elements in the left partition, so Step 16.4 is satisfied. Also, the positive diagonal element in the left partition appears in an earlier row than the negative elements, so Step 16.5 is satisfied. However, the third row has a diagonal element in its left partition that is not equal to 0, 1, or -1; following Step 16.6, we divide that diagonal element by $|d| = 4$, and the remainder of the third row by $\sqrt{4} = 2$. This gives us

$$\left[\begin{array}{ccc|ccc} 1 & 0 & 0 & 0 & 1 & 0 \\ 0 & -1 & 0 & 1 & -1 & 0 \\ 0 & 0 & -1 & -1/2 & -1 & 1/2 \end{array}\right]$$

so $$\mathbf{P} = \begin{bmatrix} 0 & 1 & 0 \\ 1 & -1 & 0 \\ -1/2 & -1 & 1/2 \end{bmatrix} \quad \text{and} \quad \mathbf{PAP}^T = \begin{bmatrix} 1 & 0 & 0 \\ 0 & -1 & 0 \\ 0 & 0 & -1 \end{bmatrix}$$

The index of **A** is 1, and its signature is $1 - 2 = -1$. We may conclude from Property 16.1 that **A** has one positive eigenvalue, two negative eigenvalues, and no zero eigenvalues.

16.8 Determine the inertia matrix for

$$\mathbf{A} = \begin{bmatrix} 1 & 2 & 3 & -1 \\ 2 & 2 & 4 & 2 \\ 3 & 4 & 7 & 1 \\ -1 & 2 & 1 & 9 \end{bmatrix}$$

Augmenting onto **A** the 4×4 identity matrix, and then reducing **A** to upper triangular form, one column at a time and without using elementary row operation E2, we finally obtain

$$\left[\begin{array}{cccc|cccc} 1 & 2 & 3 & -1 & 1 & 0 & 0 & 0 \\ 0 & -2 & -2 & 4 & -2 & 1 & 0 & 0 \\ 0 & 0 & 0 & 0 & -1 & -1 & 1 & 0 \\ 0 & 0 & 0 & 16 & -3 & 2 & 0 & 1 \end{array}\right]$$

The left partition is in upper triangular form. Setting all elements above the main diagonal in that partition equal to zero yields

$$\left[\begin{array}{cccc|cccc} 1 & 0 & 0 & 0 & 1 & 0 & 0 & 0 \\ 0 & -2 & 0 & 0 & -2 & 1 & 0 & 0 \\ 0 & 0 & 0 & 0 & -1 & -1 & 1 & 0 \\ 0 & 0 & 0 & 16 & -3 & 2 & 0 & 1 \end{array}\right]$$

Following Step 16.4, we interchange the diagonal elements in the third and fourth rows of the left partition while simultaneously interchanging the entire third and fourth rows of the right partition. The result is

$$\left[\begin{array}{cccc|cccc} 1 & 0 & 0 & 0 & 1 & 0 & 0 & 0 \\ 0 & -2 & 0 & 0 & -2 & 1 & 0 & 0 \\ 0 & 0 & 16 & 0 & -3 & 2 & 0 & 1 \\ 0 & 0 & 0 & 0 & -1 & -1 & 1 & 0 \end{array}\right]$$

Following Step 16.5, we next interchange the (2,2) diagonal element with the (3,3) diagonal element in the left partition and simultaneously interchange the order of the second and third rows in the right partition. That gives us the partitioned matrix

$$\left[\begin{array}{cccc|cccc} 1 & 0 & 0 & 0 & 1 & 0 & 0 & 0 \\ 0 & 16 & 0 & 0 & -3 & 2 & 0 & 1 \\ 0 & 0 & -2 & 0 & -2 & 1 & 0 & 0 \\ 0 & 0 & 0 & 0 & -1 & -1 & 1 & 0 \end{array}\right]$$

Following Step 16.6, we divide the (2,2) element of the left partition by 16, and the entire second row of the right partition by 4. We also divide the (3,3) element of the left partition by 2, and the entire third row of the right partition by $\sqrt{2}$. We get, finally,

$$\left[\begin{array}{cccc|cccc} 1 & 0 & 0 & 0 & 1 & 0 & 0 & 0 \\ 0 & 1 & 0 & 0 & -3/4 & 1/2 & 0 & 1/4 \\ 0 & 0 & -1 & 0 & -2/\sqrt{2} & 1/\sqrt{2} & 0 & 0 \\ 0 & 0 & 0 & 0 & -1 & -1 & 1 & 0 \end{array}\right]$$

so that $\quad \mathbf{P} = \begin{bmatrix} 1 & 0 & 0 & 0 \\ -3/4 & 1/2 & 0 & 1/4 \\ -2/\sqrt{2} & 1/\sqrt{2} & 0 & 0 \\ -1 & -1 & 1 & 0 \end{bmatrix} \quad$ and $\quad \mathbf{PAP}^T = \begin{bmatrix} 1 & 0 & 0 & 0 \\ 0 & 1 & 0 & 0 \\ 0 & 0 & -1 & 0 \\ 0 & 0 & 0 & 0 \end{bmatrix}$

The index of **A** is 2, and its signature is $2 - 1 = 1$. The rank of **A** is 3, and we may conclude from Property 16.1 that **A** has two positive eigenvalues, one negative eigenvalue, and one zero eigenvalue.

16.9 Discuss the rationale of the algorithm given by Steps 16.1 through 16.6 as it pertains to real matrices.

Since **A** is symmetric, any set of elementary row operations of the first and third kind that reduce **A** to upper triangular form yields an analogous set of column operations that will reduce **A** to lower triangular form. Under a congruence transformation, both sets of operations are applied to **A**, resulting in a diagonal matrix. This is the rationale for Steps 16.1 through 16.3.

Interchanging the position of two diagonal elements of a diagonal matrix is equivalent to interchanging both the rows and the columns in which the two diagonal elements appear. We interchange only the designated rows in **P**, since a postmultiplication by $\mathbf{P}^T$ will effect the same type of column interchange automatically. This is the rationale for Steps 16.4 and 16.5.

Finally, a nonzero diagonal element d is made equal to 1 in absolute value by dividing its row and its column by $\sqrt{|d|}$. Since the divisions will be done in tandem, we have Step 16.6.

16.10 Prove Rayleigh's principle.

Let **U** be a unitary matrix that diagonalizes **A**. Then

$$\mathbf{U}^H\mathbf{A}\mathbf{U} = \mathbf{D} = \begin{bmatrix} \lambda_1 & & & 0 \\ & \lambda_2 & & \\ & & \cdots & \\ 0 & & & \lambda_n \end{bmatrix}$$

and we may assume that the columns of **U** have been ordered so that $\lambda_1 \le \lambda_2 \le \cdots \le \lambda_n$. Setting $\mathbf{X} = \mathbf{UY}$ and using Property 15.3, we have

$$R(\mathbf{X}) = \frac{\langle \mathbf{AX}, \mathbf{X}\rangle}{\langle \mathbf{X}, \mathbf{X}\rangle} = \frac{\langle \mathbf{AUY}, \mathbf{UY}\rangle}{\langle \mathbf{UY}, \mathbf{UY}\rangle} = \frac{\langle \mathbf{U}^H\mathbf{AUY}, \mathbf{Y}\rangle}{\langle \mathbf{Y}, \mathbf{Y}\rangle} = \frac{\langle \mathbf{DY}, \mathbf{Y}\rangle}{\langle \mathbf{Y}, \mathbf{Y}\rangle}$$

$$= \frac{\lambda_1|y_1|^2 + \lambda_2|y_2|^2 + \cdots + \lambda_n|y_n|^2}{|y_1|^2 + |y_2|^2 + \cdots + |y_n|^2} \ge \frac{\lambda_1|y_1|^2 + \lambda_1|y_2|^2 + \cdots + \lambda_1|y_n|^2}{|y_1|^2 + |y_2|^2 + \cdots + |y_n|^2} = \lambda_1$$

The other inequality follows from

$$\lambda_1|y_1|^2 + \lambda_2|y_2|^2 + \cdots + \lambda_n|y_n|^2 \le \lambda_n|y_1|^2 + \lambda_n|y_2|^2 + \cdots + \lambda_n|y_n|^2$$

If $\mathbf{X}_1$ is an eigenvector corresponding to λ_1, then

$$R(\mathbf{X}_1) = \frac{\langle \mathbf{AX}_1, \mathbf{X}_1\rangle}{\langle \mathbf{X}_1, \mathbf{X}_1\rangle} = \frac{\langle \lambda_1\mathbf{X}_1, \mathbf{X}_1\rangle}{\langle \mathbf{X}_1, \mathbf{X}_1\rangle} = \frac{\lambda_1\langle \mathbf{X}_1, \mathbf{X}_1\rangle}{\langle \mathbf{X}_1, \mathbf{X}_1\rangle} = \lambda_1$$

so the minimum of $R(\mathbf{X})$ is achieved when $\mathbf{X} = \mathbf{X}_1$. A similar argument shows that the maximum is achieved when **X** is an eigenvector corresponding to λ_n.

Supplementary Problems

16.11 Determine symmetric matrix representations for the following real quadratic forms:

(*a*) $3x_1^2 + 3x_2^2 + 5x_3^2 + 2x_1x_2 - 2x_1x_3 - 2x_2x_3$

(*b*) $2x_1^2 + 2x_2^2 + 5x_3^2 + 4x_1x_2 - 2x_1x_3 - 2x_2x_3$

(*c*) $9x_1^2 + 6x_2^2 + 9x_3^2 + 6x_4^2 - 6x_1x_2 - 6x_1x_4 + 6x_2x_3 - 6x_3x_4$

(*d*) $x_1^2 + 3x_2^2 + 3x_3^2 + x_4^2 - 2x_1x_2 + 4x_1x_3 - 2x_1x_4 + 8x_2x_3 + 4x_2x_4 + 2x_3x_4$

16.12 Determine which of the quadratic forms in Problem 16.11 are positive definite.

16.13 Determine the inertia matrix associated with each of the quadratic forms in Problem 16.11.

16.14 Using the results of Problem 16.13, determine whether quadratic forms (*a*) and (*b*) of Problem 16.11 are congruent.

16.15 Using the results of Problem 16.13, determine how many positive and negative eigenvalues are associated with the symmetric matrix corresponding to each quadratic form in Problem 16.11.

16.16 Characterize the inertia matrix of a positive definite quadratic form

16.17 Find a nonsingular matrix **P** such that $\mathbf{PAP}^T$ is an inertia matrix for

$$(a)\ \begin{bmatrix} 0 & 2 & 4 \\ 2 & 1 & -2 \\ 4 & -2 & 5 \end{bmatrix} \qquad (b)\ \begin{bmatrix} 1 & 2 & 3 \\ 2 & 4 & 6 \\ 3 & 6 & 5 \end{bmatrix} \qquad (c)\ \begin{bmatrix} 1 & -1 & 1 \\ -1 & 1 & 8 \\ 1 & 8 & 0 \end{bmatrix}$$

16.18 Determine the inertia matrix associated with

$$\mathbf{A} = \begin{bmatrix} 1 & 1-i2 & i \\ 1+i2 & 6 & 3+i2 \\ -i & 3-i2 & 2 \end{bmatrix}$$

and find **P** such that $\mathbf{PAP}^H$ is that inertia matrix.

16.19 Show that if **A** is congruent to **B** and **B** is congruent to **C** then **A** is congruent to **C**.

16.20 Show that if **A** is congruent to **B**, then **B** is congruent to **A**.

16.21 Prove that two Hermitian matrices **A** and **B** are congruent to each other if and only if they are congruent to the same inertia matrix.

16.22 Show that a nonsingular Hermitian matrix **A** is congruent to its own inverse.

Chapter 17

Nonnegative Matrices

EIGENVALUES AND EIGENVECTORS

A matrix $\mathbf{A}$ is *nonnegative*, written $\mathbf{A} \geq \mathbf{0}$, if all its elements are real and nonnegative; $\mathbf{A}$ is *positive*, written $\mathbf{A} > \mathbf{0}$, if all its elements are real and positive. A matrix $\mathbf{A}$ is *greater than* a matrix $\mathbf{B}$ of identical order, denoted $\mathbf{A} > \mathbf{B}$, if $\mathbf{A} - \mathbf{B}$ is positive. Similarly, $\mathbf{A} \geq \mathbf{B}$ if $\mathbf{A} - \mathbf{B}$ is nonnegative.

The spectral radii (see Chapter 12) of nonnegative square matrices have the following properties:

Property 17.1: If $\mathbf{0} \leq \mathbf{A} \leq \mathbf{B}$, then $\sigma(\mathbf{A}) \leq \sigma(\mathbf{B})$.

Property 17.2: If $\mathbf{A} \geq \mathbf{0}$ and if the row (or column) sums of $\mathbf{A}$ are a constant k, then $\sigma(\mathbf{A}) = k$.

Property 17.3: If m is the minimum of the row (or column) sums of $\mathbf{A}$, M is the maximum of the row (or column) sums of $\mathbf{A}$, and $\mathbf{A} \geq \mathbf{0}$, then $m \leq \sigma(\mathbf{A}) \leq M$.

Property 17.4: A nonnegative square matrix has an eigenvalue equal to its spectral radius, and there exist a right eigenvector and a left eigenvector corresponding to this eigenvalue that have only nonnegative components.

Property 17.5: **(Perron's theorem)** A positive square matrix has an eigenvalue of multiplicity one equal to its spectral radius, and no other eigenvalue is as large in absolute value. Moreover, there exist a right eigenvector and a left eigenvector corresponding to $\sigma(\mathbf{A})$ that have only positive components.

(See Problems 17.1 to 17.6.)

IRREDUCIBLE MATRICES

A *permutation matrix* is a matrix obtained from an identity matrix by any rearrangement of its rows. Such a matrix is the product of elementary matrices of the first kind and is orthogonal. A nonnegative matrix is *reducible* if there exists a permutation matrix $\mathbf{P}$ such that

$$\mathbf{PAP}^T = \left[\begin{array}{c|c} \mathbf{A}_{11} & \mathbf{A}_{12} \\ \hline \mathbf{0} & \mathbf{A}_{22} \end{array}\right] \tag{17.1}$$

where both $\mathbf{A}_{11}$ and $\mathbf{A}_{22}$ are square matrices having order less than that of $\mathbf{A}$. If no such permutation matrix exists, then A is said to be *irreducible*. (See Problems 17.7 and 17.8.)

Property 17.6: Positive matrices are irreducible.

Property 17.7: An $n \times n$ matrix $\mathbf{A}$ is irreducible if and only if $(\mathbf{I} + \mathbf{A})^{n-1} \geq \mathbf{0}$.

Property 17.8: **(Perron-Frobenius theorem)** A nonnegative, irreducible matrix $\mathbf{A}$ has an eigenvalue of multiplicity one equal to its spectral radius, and corresponding to this eigenvalue is a right (left) eigenvector which has only positive components. If such a matrix has exactly k eigenvalues with absolute value equal to its spectral radius, then they are of the form $w_i\sigma(\mathbf{A})$, where $w_1, w_2, \ldots, w_k$ are the k distinct roots of unity.

(See Problems 17.9 to 17.11.)

PRIMITIVE MATRICES

A nonnegative matrix is *primitive* if it is irreducible and has only one eigenvalue with absolute value equal to its spectral radius. A nonnegative matrix is *regular* if one of its powers is a positive matrix. A nonnegative matrix is primitive if and only if it is regular.

Property 17.9: If $\mathbf{A}$ is a nonnegative primitive matrix, then the limit $\mathbf{L} = \lim_{m\to\infty} (\{1/\sigma(\mathbf{A})\}\mathbf{A})^m$ exists and is positive. Furthermore, if $\mathbf{X}$ and $\mathbf{Y}$ are, respectively, left and right positive eigenvectors of $\mathbf{A}$ corresponding to the eigenvalue equal to $\sigma(\mathbf{A})$ and scaled so that $\mathbf{YX} = 1$, then $\mathbf{L} = \mathbf{XY}$.

Positive matrices are primitive and have the limit described in Property 17.9. (See Problem 17.12.) Reducible matrices may or may not have such a limit. (See Problem 17.13.)

STOCHASTIC MATRICES

A nonnegative matrix is *stochastic* if all its row sums or all its column sums equal 1. It is *doubly stochastic* if all its row sums and all its column sums equal 1. It follows from Property 17.2 that the spectral radius of such a matrix is unity. If the row (column) sums are all 1, then a right (left) eigenvector corresponding to $\lambda = 1$ has all its components equal.

A stochastic matrix is *ergodic* if the only eigenvalue of absolute value 1 is 1 itself, and if the eigenvalue $\lambda = 1$ has multiplicity k, then there exist k linearly independent eigenvectors corresponding to it.

Property 17.10: If $\mathbf{P}$ is ergodic, then $\lim_{m\to\infty} \mathbf{P}^m = \mathbf{L}$ exists.

A primitive stochastic matrix $\mathbf{A}$ is ergodic with $k = 1$ and has a simple form for the limiting matrix $\mathbf{L}$. If the row (column) sums of $\mathbf{A}$ are all 1, then the same is true for the row (column) sums of $\mathbf{L}$, and all the rows (columns) of $\mathbf{L}$ are identical. Each of these rows (columns) is the unique left (right) eigenvector corresponding to $\lambda = 1$ and having the sum of its components equal to unity. (See Problems 17.14 and 17.19.)

The form of the limiting matrix is not as simple for an ergodic matrix that is not primitive. A canonical basis for such a matrix consists solely of eigenvectors. If the multiplicity of $\lambda = 1$ is denoted by k, and if the k linearly independent right eigenvectors corresponding to this eigenvalue are placed into the first k columns of the modal matrix $\mathbf{M}$, then $\mathbf{L} = \mathbf{MDM}^{-1}$, where $\mathbf{D}$ is a diagonal matrix having its first k diagonal elements equal to unity and all others equal to zero. (See Problem 17.15.)

FINITE MARKOV CHAINS

An *N-state Markov chain* consists of a set of objects and a finite set of N different states (where N is a fixed positive integer), such that (1) at any given time each object is in one of the N states, which may be different for different objects, and (2) the probability that an object will move from one state to another state (or remain in the same state) in one time period depends only on the beginning and ending states. The $N \times N$ matrix $\mathbf{P} = [p_{ij}]$, where p_{ij} denotes the probability of an object moving from state i to state j in one time period, is stochastic. The (i, j) element of the mth power of $\mathbf{P}$ represents the probability that an object will move from state i to state j in m time periods.

Denote the proportion of objects in state i at the end of the mth time period as $x_i^{(m)}$, and define

$$\mathbf{X}^{(m)} = [x_1^{(m)}, x_2^{(m)}, \ldots, x_N^{(m)}]$$

to be the *distribution vector* for the end of the mth time period. Then

$$\mathbf{X}^{(0)} = [x_1^{(0)}, x_2^{(0)}, \ldots, x_N^{(0)}]$$

represents the proportion of objects in each state at the beginning of the process. Necessarily, $\mathbf{X}^{(m)} \geq \mathbf{0}$, and the sum of the components of $\mathbf{X}^{(m)}$ is 1 for each $m = 0, 1, 2, \ldots$. Furthermore,

$$\mathbf{X}^{(m)} = \mathbf{X}^{(0)}\mathbf{P}^m \tag{17.2}$$

If $\mathbf{P}$ is primitive, then

$$\mathbf{X}^{(\infty)} = \lim_{m\to\infty} \mathbf{X}^{(m)} = \mathbf{X}^{(0)}\mathbf{L} \tag{17.3}$$

which is the positive left eigenvector of $\mathbf{P}$ corresponding to $\lambda = 1$ and having the sum of its components equal to unity. The ith component of $\mathbf{X}^{(\infty)}$ represents the approximate proportion of objects in state i after a large number of time periods, and this limiting value is independent of the initial distribution defined by $\mathbf{X}^{(0)}$. If $\mathbf{P}$ is ergodic but not primitive, (*17.3*) still may be used to obtain the limiting state distribution, but it will depend on the value of $\mathbf{X}^{(0)}$. (See Problems 17.16 and 17.17.)

Solved Problems

17.1 Estimate the location of the largest eigenvalue of

$$\mathbf{A} = \begin{bmatrix} 0 & 1 & 6 \\ 2 & 8 & 6 \\ 1 & 2 & 2 \end{bmatrix}$$

$\mathbf{A}$ is nonnegative, and its row sums are 7, 16, and 5. It follows, then, from Property 17.3, that $5 \leq \sigma(\mathbf{A}) \leq 16$. However, the column sums of $\mathbf{A}$ are 3, 11, and 14, so $3 \leq \sigma(\mathbf{A}) \leq 14$. Together, the two inequalities imply that the largest eigenvalue is between 5 and 14 in absolute value.

17.2 Verify Property 17.4 for

$$\mathbf{A} = \begin{bmatrix} 1 & 0 & 0 \\ 2 & 0 & 1 \\ 3 & 1 & 0 \end{bmatrix}$$

The eigenvalues of $\mathbf{A}$ are $\lambda_1 = \lambda_2 = 1$ and $\lambda_3 = -1$; and, since $\sigma(\mathbf{A}) = 1$ by definition, there is an eigenvalue equal to the spectral radius. A right eigenvector corresponding to $\lambda = 1$ is $[0, 1, 1]^T$, which is nonnegative. A left eigenvector corresponding to the same eigenvalue is $[1, 0, 0]$, which is also nonnegative.

17.3 Verify Perron's theorem for

$$\mathbf{A} = \begin{bmatrix} 0.1 & 0.5 & 0.4 \\ 0.7 & 0.2 & 0.1 \\ 0.6 & 0.2 & 0.2 \end{bmatrix}$$

The row sums of $\mathbf{A}$ are all equal to 1, so it follows from Property 17.2 that $\sigma(\mathbf{A}) = 1$. The characteristic equation of $\mathbf{A}$ is $d(\lambda) = -\lambda^3 + 0.5\lambda^2 + 0.53\lambda - 0.03$, which has the roots $\lambda_1 = 1$, $\lambda_2 = -0.554138$, and $\lambda_3 = 0.054138$. Thus, the spectral radius is an eigenvalue of multiplicity one, and it is the greatest eigenvalue in absolute value. A right eigenvector corresponding to this eigenvalue is $[1, 1, 1]^T$, while a left eigenvector is $[62, 48, 37]$. Each of these has only positive components.

17.4 Prove that if $\mathbf{0} \leq \mathbf{A} \leq \mathbf{B}$, where $\mathbf{A}$ and $\mathbf{B}$ are square matrices of the same order, then $\sigma(\mathbf{A}) \leq \sigma(\mathbf{B})$.

If $\mathbf{0} \leq \mathbf{A} \leq \mathbf{B}$, then $\mathbf{A}^m \leq \mathbf{B}^m$ for any positive integer m and, therefore, $\|\mathbf{A}^m\|_F \leq \|\mathbf{B}^m\|_F$. It follows from (*12.4*) that

$$\sigma(\mathbf{A}) = \lim_{m\to\infty} \|\mathbf{A}^m\|_F^{1/m} \leq \lim_{m\to\infty} \|\mathbf{B}^m\|_F^{1/m} = \sigma(\mathbf{B})$$

17.5 Prove that if the row (or column) sums of a nonnegative square matrix $\mathbf{A}$ are a constant k, then $\sigma(\mathbf{A}) = k$.

Using (*12.3*), we may write

$$\sigma(\mathbf{A}) \leq \|\mathbf{A}\|_\infty = k \tag{1}$$

If we set $\mathbf{X} = [1, 1, \ldots, 1]^T$, it follows from the row sums being k that $\mathbf{AX} = \mathbf{kX}$, so that k is an eigenvalue of $\mathbf{A}$. Since $\sigma(\mathbf{A})$ must be the largest eigenvalue in absolute value,

$$\sigma(\mathbf{A}) \geq k \tag{2}$$

Together, (*1*) and (*2*) imply $\sigma(\mathbf{A}) = k$. The proof for column sums follows if we consider $\mathbf{A}^T$ in place of $\mathbf{A}$.

17.6 Prove that if m is the minimum row (or column) sum and M is the maximum row (or column) sum of an $n \times n$ nonnegative matrix $\mathbf{A} = [a_{ij}]$, then $m \leq \sigma(\mathbf{A}) \leq M$.

Construct a matrix $\mathbf{B} = [b_{ij}]$ having the same order as $\mathbf{A}$ and such that

$$b_{ij} = \begin{cases} 0 & \text{if } m = 0 \\ \dfrac{m a_{ij}}{\sum_{j=1}^{n} a_{ij}} & \text{if } m \neq 0 \end{cases}$$

The $\mathbf{A} \geq \mathbf{B} \geq \mathbf{0}$, and the row sums of $\mathbf{B}$ are all equal to m. It follows from Problem 17.4 that $\sigma(\mathbf{B}) \leq \sigma(\mathbf{A})$ and from Problem 17.5 that $\sigma(\mathbf{B}) = m$; thus, $m \leq \sigma(\mathbf{A})$. That M is the upper bound on $\sigma(\mathbf{A})$ follows from (*12.3*); that is, $\sigma(\mathbf{A}) \leq \|\mathbf{A}\|_\infty = M$. The analogous result for column sums is obtained through an identical argument applied to $\mathbf{A}^T$.

17.7 Determine whether the matrix in Problem 17.1 is irreducible.

The matrix possesses a single zero element, and it is on the main diagonal. An elementary row operation of the first kind (E1) followed by an elementary column operation of the same kind will change the positions of diagonal elements, but not their values. Since a permutation matrix $\mathbf{P}$ is a product of elementary matrices of the first kind, it follows that $\mathbf{PAP}^T$ also leaves the values of diagonal elements unchanged. Such a transformation cannot move the zero into the (3,1) position and cannot result in additional zero elements. Since there are no other zeros available to move into that position, the matrix is irreducible.

17.8 Determine whether the matrix in Problem 17.2 is irreducible.

An interchange of the first and third rows followed by an interchange of the first and third columns gives us the transformation

$$\mathbf{PAP}^T = \begin{bmatrix} 0 & 0 & 1 \\ 0 & 1 & 0 \\ 1 & 0 & 0 \end{bmatrix} \begin{bmatrix} 1 & 0 & 0 \\ 2 & 0 & 1 \\ 3 & 1 & 0 \end{bmatrix} \begin{bmatrix} 0 & 0 & 1 \\ 0 & 1 & 0 \\ 1 & 0 & 0 \end{bmatrix} = \begin{bmatrix} 0 & 1 & 3 \\ 1 & 0 & 2 \\ 0 & 0 & 1 \end{bmatrix}$$

which has the partitioned form given in (*17.1*) with

$$\mathbf{A}_{11} = \begin{bmatrix} 0 & 1 \\ 1 & 0 \end{bmatrix} \quad \text{and} \quad \mathbf{A}_{22} = [1]$$

Therefore, **A** is reducible.

17.9 Determine whether the following matrix is irreducible:

$$\mathbf{A} = \begin{bmatrix} 0 & 2 & 0 & 0 \\ 0 & 0 & 4 & 0 \\ 0 & 0 & 0 & 2 \\ 1 & 0 & 0 & 0 \end{bmatrix}$$

To use Property 17.7, we calculate

$$(\mathbf{A} + \mathbf{I})^3 = \begin{bmatrix} 1 & 6 & 24 & 16 \\ 8 & 1 & 12 & 24 \\ 6 & 4 & 1 & 6 \\ 3 & 6 & 8 & 1 \end{bmatrix}$$

Since this matrix is positive, **A** is irreducible.

17.10 Verify the Perron-Frobenius theorem for the matrix in Problem 17.1.

The matrix is irreducible (see Problem 17.7), and its eigenvalues may be found, to four decimal places, to be 10.1806, −1.5631, and 1.3825. Its spectral radius is thus 10.1806, which falls within the bounds identified in Problem 17.1. A right eigenvector corresponding to this spectral radius, and with all components rounded to four decimal places, is $[0.2611, 1, 0.2764]^T$. A left eigenvector, rounded similarly, is $[0.2893, 1, 0.9456]$. Both eigenvectors have only positive components.

17.11 Verify the Perron-Frobenius Theorem for the matrix in Problem 17.9.

The characteristic equation of **A** is $d(\lambda) = \lambda^4 - 16$, so its eigenvalues are 2, −2, $i2$, and $-i2$, all of which have absolute value equal to the spectral radius of 2. Each eigenvalue is the product of the spectral radius and one of the four fourth roots of unity, 1, −1, i, and $-i$. A right eigenvector corresponding to $\lambda = 2$ is $[2, 2, 1, 1]^T$, while a left eigenvector is $[1, 1, 2, 2]$; both are positive.

17.12 Determine whether the matrices given in Problems 17.1, 17.2, 17.3, and 17.9 are primitive.

The matrix **A** in Problem 17.1 is irreducible (Problem 17.7) and has only one eigenvalue with absolute value equal to its spectral radius (Problem 17.10), so it is primitive. Alternatively, $\mathbf{A}^2$ is positive, so **A** is regular and, therefore, primitive.

The matrix in Problem 17.2 is reducible (Problem 17.8) and, therefore, cannot be primitive.

The matrix in Problem 17.3 is positive and, therefore, primitive.

The matrix in Problem 17.9 is irreducible, but it has four eigenvalues having absolute value equal to its spectral radius (Problem 17.11), so it is not primitive. Alternatively, one can show that the (1,1) element of every power of **A** is zero, so **A** is not regular and, therefore, not primitive.

17.13 Show that a square matrix **A** need not be primitive to possess the limit $\mathbf{L} = \lim_{m\to\infty}(\{1/\sigma(\mathbf{A})\}\mathbf{A})^m$.

The matrix

$$\mathbf{A} = \begin{bmatrix} 2 & 0 \\ 0 & 2 \end{bmatrix}$$

has spectral radius $\sigma(\mathbf{A}) = 2$. For this matrix, $(\{1/\sigma(\mathbf{A})\}\mathbf{A})^m = \mathbf{I}$ for every positive integer m, so $\mathbf{L} = \mathbf{I}$.

17.14 Find $\mathbf{L} = \lim_{m\to\infty} \mathbf{A}^m$ for the matrix in Problem 17.3.

The matrix is stochastic and primitive (Problem 17.12), and it has a left eigenvector given by [62, 48, 37]. If we divide each component of that eigenvector by the sum of the components, $62+48+37=147$, we obtain a positive left eigenvector whose components sum to unity. Then

$$\mathbf{L} = \begin{bmatrix} 62/147 & 48/147 & 37/147 \\ 62/147 & 48/147 & 37/147 \\ 62/147 & 48/147 & 37/147 \end{bmatrix}$$

17.15 Determine whether the stochastic matrix

$$\mathbf{P} = \begin{bmatrix} 1 & 0 & 0 & 0 \\ 0.4 & 0 & 0.6 & 0 \\ 0.2 & 0 & 0.1 & 0.7 \\ 0 & 0 & 0 & 1 \end{bmatrix}$$

is ergodic, and, if so, calculate $\mathbf{L} = \lim_{m\to\infty} \mathbf{P}^m$.

The eigenvalues of $\mathbf{P}$ are $\lambda_1 = \lambda_2 = 1$, $\lambda_3 = 0.1$, and $\lambda_4 = 0$, so the matrix is not primitive. $\mathbf{P}$ does, however, possess two linearly independent right eigenvectors corresponding to $\lambda = 1$,

$$[45, 24, 10, 0]^T \qquad \text{and} \qquad [-35, -14, 0, 10]^T$$

so it is ergodic and $\mathbf{L}$ exists. As an easy calculation shows, the right eigenvectors

$$[0, 6, 1, 0]^T \qquad \text{and} \qquad [0, 1, 0, 0]^T$$

correspond, respectively, to λ_3 and λ_4. Thus,

$$\mathbf{M} = \begin{bmatrix} 45 & -35 & 0 & 0 \\ 24 & -14 & 6 & 1 \\ 10 & 0 & 1 & 0 \\ 0 & 10 & 0 & 0 \end{bmatrix} \qquad \text{and} \qquad \mathbf{M}^{-1} = \begin{bmatrix} 1/45 & 0 & 0 & 7/90 \\ 0 & 0 & 0 & 1/10 \\ -2/9 & 0 & 1 & -7/9 \\ 4/5 & 1 & -6 & 21/5 \end{bmatrix}$$

and
$$\mathbf{L} = \begin{bmatrix} 45 & -35 & 0 & 0 \\ 24 & -14 & 6 & 1 \\ 10 & 0 & 1 & 0 \\ 0 & 10 & 0 & 0 \end{bmatrix} \begin{bmatrix} 1 & 0 & 0 & 0 \\ 0 & 1 & 0 & 0 \\ 0 & 0 & 0 & 0 \\ 0 & 0 & 0 & 0 \end{bmatrix} \begin{bmatrix} 1/45 & 0 & 0 & 7/90 \\ 0 & 0 & 0 & 1/10 \\ -2/9 & 0 & 1 & -7/9 \\ 4/5 & 1 & -6 & 21/5 \end{bmatrix} = \begin{bmatrix} 1 & 0 & 0 & 0 \\ 8/15 & 0 & 0 & 7/15 \\ 2/9 & 0 & 0 & 7/9 \\ 0 & 0 & 0 & 1 \end{bmatrix}$$

17.16 Formulate the following problem as a Markov chain and solve it: New owners of rental apartments in Atlanta are considering as the operating agent a real estate management firm with a reputation for improving the condition of antiquated housing under its control. Based on initial ratings of *poor*, *average*, and *excellent* for the condition of rental units, it has been documented that 10 percent of all apartments that begin a year in poor condition remain in poor condition at the end of the year, 50 percent are upgraded to average, and the remaining 40 percent are renovated into excellent condition. Of all apartments that begin a year in average condition, 70 percent deteriorate into poor condition by year's end, 20 percent remain in average condition, while 10 percent are upgraded to excellent. Of all apartments that begin a year in excellent condition, 60 percent deteriorate into poor condition by the end of the year, 20 percent are downgraded to average, and 20 percent retain their excellent rating. Assuming that these findings will apply if the firm is hired, determine the condition that the new owners can expect for their apartments over the long run.

We take state 1 to be a rating of poor, state 2 to be a rating of average, and state 3 to be a rating of excellent. With percentages converted into their decimal equivalents, the probabilities of moving from state i to state j $(i, j = 1, 2, 3)$ over a one-year period are given by the elements of the stochastic matrix in Problem 17.3. Using the results of Problem 17.14 and Eq. (*17.3*), we have

$$\mathbf{X}^{(\infty)} = [62/147, 48/147, 37/147] \cong [0.422, 0.327, 0.252]$$

Over the long run, approximately 42 percent of the apartments controlled by this real estate management firm will be in poor condition, 33 percent will be in average condition, and 25 percent will be in excellent condition.

17.17 Formulate the following problem as a Markov chain and solve it: The training program for production supervisors at a particular company consists of two phases. Phase 1, which involves three weeks of classroom work, is followed by phase 2, which is a three-week apprenticeship program under the direction of working supervisors. From past experience, the company expects only 60 percent of those beginning classroom training to be graduated into the apprenticeship phase, with the remaining 40 percent dropped completely from the training program. Of those who make it to the apprenticeship phase, 70 percent are graduated as supervisors, 10 percent are asked to repeat the second phase, and 20 percent are dropped completely from the program. How many supervisors can the company expect from its current training program if it has 45 people in the classroom phase and 21 people in the apprenticeship phase?

We consider one time period to be three weeks, and define states 1 through 4 as the classification of being dropped, a classroom trainee, an apprentice, and a supervisor, respectively. If we assume that discharged individuals never reenter the training program and that supervisors remain supervisors, then the probabilities of moving from one state to the next are given by the stochastic matrix in Problem 17.15. There are $45 + 21 = 66$ people currently in the training program, so the initial probability vector for current trainees is $\mathbf{X}^{(0)} = [0, 45/66, 21/66, 0]$. It follows from Eq. (*17.3*) and the results of Problem 17.15 that

$$\mathbf{X}^{(\infty)} = \mathbf{X}^{(0)}\mathbf{L} = [0, 45/66, 21/66, 0]\begin{bmatrix} 1 & 0 & 0 & 0 \\ 8/15 & 0 & 0 & 7/15 \\ 2/9 & 0 & 0 & 7/9 \\ 0 & 0 & 0 & 1 \end{bmatrix} = [0.4343, 0, 0, 0.5657]$$

Eventually, 43.43 percent of those currently in training (or about 29 people) will be dropped from the program, and the rest (about 37 people) will move into supervisory positions.

17.18 Prove that the product of two stochastic matrices, both of which have their row (or column) sums equal to unity, is itself a stochastic matrix of the same type.

Let $\mathbf{A}$ and $\mathbf{B}$ denote stochastic matrices with row sums equal to unity, and set $\mathbf{C} = \mathbf{AB}$. Then the ith row sum of $\mathbf{C}$ is

$$\sum_{j=1}^{n} c_{ij} = \sum_{j=1}^{n}\left(\sum_{k=1}^{n} a_{ik}b_{kj}\right) = \sum_{k=1}^{n}\left(\sum_{j=1}^{n} b_{kj}\right)a_{ik} = \sum_{k=1}^{n} a_{ik} = 1$$

17.19 Prove that if $\mathbf{P}$ is a primitive stochastic matrix with row sums equal to unity, then all the rows of $\mathbf{L} = \lim_{m\to\infty} \mathbf{P}^m$ are identical.

It follows directly from Problem 17.18 that if $\mathbf{P}$ is an $n \times n$ stochastic matrix, then so too is any positive integral power of $\mathbf{P}$. Therefore, $\mathbf{L}$ is necessarily stochastic too. Now

$$\mathbf{L} = \lim_{m\to\infty} \mathbf{P}^m = \lim_{m\to\infty} (\mathbf{P}^{m-1}\mathbf{P}) = (\lim_{m\to\infty} \mathbf{P}^{m-1})\mathbf{P} = \mathbf{LP}$$

which implies that every row of $\mathbf{L}$ is a left eigenvector of $\mathbf{P}$ corresponding to $\lambda = 1$. Since $\mathbf{P}$ is primitive, $\lambda = 1$ is an eigenvalue of multiplicity one, and all eigenvectors corresponding to it must be scalar multiples of each other. With $\mathbf{L}$ being stochastic, the row sums of $\mathbf{L}$ must all be unity, so it follows that the rows are identical.

Supplementary Problems

In Problems 17.20 through 17.29, determine whether the given matrix is irreducible, primitive, or stochastic, and estimate its spectral radius. For those matrices **P** that are stochastic, determine $\lim_{m\to\infty} \mathbf{P}^m$ if it exists.

17.20 $\begin{bmatrix} 1 & 2 & 0 \\ 2 & 2 & 0 \\ 1 & 1 & 2 \end{bmatrix}$ **17.21** $\begin{bmatrix} 1 & 2 & 1 \\ 1 & 0 & 1 \\ 2 & 2 & 1 \end{bmatrix}$ **17.22** $\begin{bmatrix} 0 & 2 & 0 \\ 0 & 0 & 1 \\ 1 & 0 & 0 \end{bmatrix}$

17.23 $\begin{bmatrix} 1 & 2 & 1 \\ 1 & 1 & 2 \\ 2 & 1 & 1 \end{bmatrix}$ **17.24** $\begin{bmatrix} 1 & 1 & 2 & 3 \\ 3 & 0 & 1 & 1 \\ 1 & 4 & 0 & 2 \\ 2 & 2 & 4 & 1 \end{bmatrix}$ **17.25** $\begin{bmatrix} 0.1 & 0.8 & 0.1 \\ 0.9 & 0 & 0.1 \\ 0.2 & 0.2 & 0.6 \end{bmatrix}$

17.26 $\begin{bmatrix} 1 & 0 & 0 \\ 0.21 & 0.79 & 0 \\ 0.17 & 0.35 & 0.48 \end{bmatrix}$ **17.27** $\begin{bmatrix} 0.5 & 0 & 0.5 \\ 0 & 1 & 0 \\ 0.3 & 0 & 0.7 \end{bmatrix}$ **17.28** $\begin{bmatrix} 1 & 0 & 0 \\ 0 & 0 & 1 \\ 0 & 1 & 0 \end{bmatrix}$

17.29 $\begin{bmatrix} 0.1 & 0.6 & 0.3 \\ 0.6 & 0.2 & 0.2 \\ 0.3 & 0.2 & 0.5 \end{bmatrix}$

17.30 The manufacturer of Hi-Glo toothpaste currently controls 60 percent of the market in a particular city. Data from the previous year show that 88 percent of Hi-Glo's customers remained loyal to Hi-Glo, while 12 percent switched to rival brands. In addition, 85 percent of the competition's customers did not switch to Hi-Glo during the year, while the other 15 percent did. Assuming that these trends continue, determine Hi-Glo's share of the market (*a*) in 5 years and (*b*) over the long run.

17.31 Grape harvests in the Sonoma Valley are classified as either superior, average, or poor. Following a superior harvest, the probabilities of having superior, average, and poor harvests the next year are 0, 0.8, and 0.2, respectively. Following an average harvest, these probabilities are 0.2, 0.6, and 0.2, respectively; following a poor harvest, they are 0.1, 0.8, and 0.1, respectively. Determine the probability of a superior harvest for each of the next 5 years, if the most recent harvest was average.

17.32 The geriatric ward of a hospital lists its patients as bedridden or ambulatory. Historical data indicate that over a one-week period, 30 percent of all ambulatory patients are discharged, 40 percent remain ambulatory, and 30 percent are remanded to complete bed rest. During the same period, 50 percent of all bedridden patients become ambulatory, 20 percent remain bedridden, and 30 percent die. Currently the hospital has 100 patients in its geriatric ward, with 30 bedridden and 70 ambulatory. Determine the expected status of these patients (*a*) after 2 weeks, and (*b*) over the long run. (The status of a discharged patient does not change if that patient later dies away from the hospital.)

Chapter 18

Patterned Matrices

CIRCULANT MATRICES

A *circulant matrix* is a square matrix in which every row beginning with the second can be obtained from the preceding row by moving each of its elements one column to the right, with the last element circling to become the first. Circulant matrices have the general form

$$\mathbf{A} = \begin{bmatrix} a_1 & a_2 & a_3 & \cdots & a_n \\ a_n & a_1 & a_2 & \cdots & a_{n-1} \\ a_{n-1} & a_n & a_1 & \cdots & a_{n-2} \\ \cdots & \cdots & \cdots & \cdots & \cdots \\ a_2 & a_3 & a_4 & \cdots & a_1 \end{bmatrix}$$

Note that each diagonal consists of identical elements. (See Problem 18.1.)

Property 18.1: If a circulant matrix $\mathbf{A}$ has order $n \times n$, then its eigenvalues are

$$\lambda_i = a_1 + a_2 r_i + a_3 r_i^2 + \cdots a_n r^{n-1} \qquad (i = 1, 2, \ldots, n)$$

where $[a_1, a_2, \ldots, a_n]$ is the first row of $\mathbf{A}$ and r_i is one of the n distinct solutions of $r^n = 1$. The corresponding eigenvectors are $\mathbf{X}_i = [1, r_i, r_i^2, \ldots, r_i^{n-1}]^T$. (See Problems 18.2 and 18.3.)

Property 18.2: If $\mathbf{A}$ and $\mathbf{B}$ are circulant matrices of the same order and a and b are any two scalars, then $a\mathbf{A} + b\mathbf{B}$ is also a circulant matrix.

Property 18.3: The product of two circulant matrices of the same order is itself a circulant matrix, and the product is commutative.

Property 18.4: If a circulant matrix is nonsingular, then its inverse is a circulant matrix.

BAND MATRICES

A square matrix $\mathbf{A} = [a_{ij}]$ having order $n \times n$ is a *band matrix* of width $2K + 1$ if $a_{ij} = 0$ when $|i - j| > K$ for some nonnegative integer K between 0 and $n - 1$. In a band matrix, all nonzero elements are positioned on the main diagonal and the first K diagonals directly above and below the main diagonal. The general form of such a matrix is

$\mathbf{A} =$ 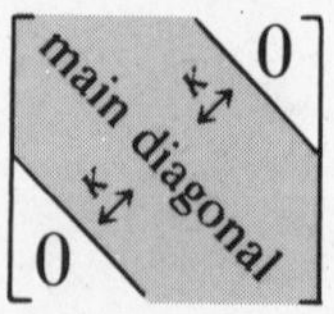

A diagonal matrix is a band matrix with $K = 0$.

Property 18.5: The sums, products, and transposes of $n \times n$ band matrices of width $2K + 1$ are band matrices of the same width.

A *Toeplitz matrix* is a band matrix in which each diagonal consists of identical elements, although different diagonals may contain different elements. Every nonzero circulant matrix is a Toeplitz matrix of full width.

TRIDIAGONAL MATRICES

A *tridiagonal matrix* is a band matrix of width three. Nonzero elements appear only on the main diagonal, the superdiagonal, and the subdiagonal; all other diagonals contain only zero elements.

Property 18.6: The eigenvalues of an $n \times n$ tridiagonal Toeplitz matrix with elements a on the main diagonal, b on the superdiagonal, and c on the subdiagonal are

$$\lambda_k = a + 2\sqrt{bc}\cos\frac{k\pi}{n+1} \qquad (k = 1, 2, \ldots, n)$$

(See Problem 18.4.)

Crout's reduction (see Chapter 3) is an algorithm for obtaining an **LU** factorization of a square matrix such that $\mathbf{L} = [l_{ij}]$ is lower triangular, and $\mathbf{U} = [u_{ij}]$ is upper triangular with unity elements on the main diagonal. For a tridiagonal matrix $\mathbf{A} = [a_{ij}]$ of order $n \times n$, the algorithm simplifies to the following:

STEP 18.1: *Initialization:* If $a_{11} = 0$, stop; factorization is not possible. Otherwise, set $l_{11} = a_{11}$; set the subdiagonal of **L** equal to the subdiagonal of **A**; set each diagonal element of **U** equal to unity; set all other elements of **L** and **U** equal to zero; and set a counter at $i = 2$.

STEP 18.2: Calculate $u_{i-1,i} = a_{i-1,i}/l_{i-1,i-1}$.

STEP 18.3: Calculate $l_{ii} = a_{ii} - l_{i,i-1}u_{i-1,i}$. If $i = n$, stop; the algorithm is complete.

STEP 18.4: If l_{ii} is zero, stop; factorization is not possible. Otherwise, increase i by 1 and return to Step 18.2.

This factorization will produce an **L** matrix having nonzero elements only on its diagonal and subdiagonal, and a **U** matrix having nonzero elements only on its diagonal and superdiagonal. (See Problems 18.5 and 18.6.)

HESSENBERG FORM

A square matrix is in *Hessenberg form* if all elements below the subdiagonal are zero. Every real square matrix **A** is congruent (see Chapter 16) to a matrix in Hessenberg form. An iterative algorithm for effecting this transformation successively generates, at each stage, matrices $\mathbf{A}_k$ and $\mathbf{P}_k$ $(k = 1, 2, \ldots, n-2)$, where $\mathbf{P}_k$ is orthogonal and $\mathbf{A}_k$ has its first k columns in Hessenberg format. Then $\mathbf{P} = \mathbf{P}_1\mathbf{P}_2 \cdots \mathbf{P}_{n-2}$ and $\mathbf{P}^T\mathbf{A}\mathbf{P} = \mathbf{H}$, where **H** is in Hessenberg form. Since **P** is orthogonal, $\mathbf{A} = \mathbf{P}\mathbf{H}\mathbf{P}^T$. For notational convenience, we set $\mathbf{A}_0 = \mathbf{A}$. The kth iteration of the algorithm is:

STEP 18.5: Set $\mathbf{X}_k$ equal to the kth column of $\mathbf{A}_{k-1}$, restricted to those elements that are below the main diagonal. Thus, $\mathbf{X}_k$ contains $n - k$ components.

STEP 18.6: Construct $\mathbf{V}_k = \mathbf{X}_k + \|\mathbf{X}_k\|_2\mathbf{E}_k$, where $\mathbf{E}_k$ is an $(n-k)$-dimensional column vector having its first component equal to 1 and all other components equal to zero.

STEP 18.7: Determine the elementary reflector $\mathbf{R}_k$ associated with $\mathbf{V}_k$, using *(15.4)*.

STEP 18.8: Set

$$\mathbf{P}_k = \left[\begin{array}{c|c} \mathbf{I}_k & \mathbf{0} \\ \hline \mathbf{0} & R_k \end{array}\right]$$

where $\mathbf{I}_k$ denotes the $k \times k$ identity matrix, and calculate $\mathbf{A}_k = \mathbf{P}_k^T\mathbf{A}_{k-1}\mathbf{P}_k$.

If **A** is real symmetric, then the resulting matrix in Hessenberg form is also real symmetric and, therefore, tridiagonal. (See Problems 18.7 and 18.8.)

Solved Problems

18.1 Determine whether the following matrices are circulant, Toeplitz, band matrices, tridiagonal, and/or in Hessenberg form:

$$\mathbf{A} = \begin{bmatrix} 1 & -2 & 3 & -4 \\ -4 & 1 & -2 & 3 \\ 3 & -4 & 1 & -2 \\ -2 & 3 & -4 & 1 \end{bmatrix} \qquad \mathbf{B} = \begin{bmatrix} 2 & -1 & 0 & 0 & 0 & 0 \\ 1 & 2 & -1 & 0 & 0 & 0 \\ 0 & 1 & 2 & -1 & 0 & 0 \\ 0 & 0 & 1 & 2 & -1 & 0 \\ 0 & 0 & 0 & 1 & 2 & -1 \\ 0 & 0 & 0 & 0 & 1 & 2 \end{bmatrix}$$

$$\mathbf{C} = \begin{bmatrix} 1 & 2 & -1 & 0 & 0 & 0 \\ 0 & 1 & 2 & -1 & 0 & 0 \\ 0 & 0 & 1 & 2 & -1 & 0 \\ 0 & 0 & 0 & 1 & 2 & -1 \\ -1 & 0 & 0 & 0 & 1 & 2 \end{bmatrix} \qquad \mathbf{D} = \begin{bmatrix} 1 & 2 & 3 & 0 & 0 & 0 \\ 2 & 1 & 3 & 3 & 0 & 0 \\ 0 & 2 & 2 & 0 & 2 & 0 \\ 0 & 0 & 2 & 3 & 3 & 2 \\ 0 & 0 & 0 & 2 & 1 & 0 \\ 0 & 0 & 0 & 0 & 2 & 2 \end{bmatrix}$$

$$\mathbf{E} = \begin{bmatrix} 1 & 2 & 0 & 0 & 0 \\ -1 & 2 & -2 & 0 & 0 \\ 0 & 2 & 1 & 1 & 0 \\ 0 & 0 & 3 & -1 & 1 \\ 0 & 0 & 0 & -3 & 2 \end{bmatrix} \qquad \mathbf{F} = \begin{bmatrix} 1 & 2 & 3 & 4 & 5 \\ 5 & 1 & 2 & 3 & 4 \\ 0 & 5 & 1 & 2 & 3 \\ 0 & 0 & 5 & 1 & 2 \\ 0 & 0 & 0 & 5 & 1 \end{bmatrix}$$

A is a circulant matrix and a Toeplitz matrix; it is a band matrix of full width which, in this case, is 7.
B is a band matrix of width 3, tridiagonal, and in Hessenberg form; it is also a Toeplitz matrix.
C is a circulant matrix.
D is a band matrix of width 5; it is in Hessenberg form.
E is a band matrix of width 3, tridiagonal, and in Hessenberg form.
F is a Toeplitz matrix in Hessenberg form.

18.2 Determine the eigenvalues of, and a canonical basis for, matrix **A** in Problem 18.1.

A is a circulant matrix having order 4×4. The roots of $r^4 = 1$ are $r_1 = 1$, $r_2 = -1$, $r_3 = i$, $r_4 = -i$, so

$$\lambda_1 = 1 + (-2)(1) + 3(1)^2 + (-4)(1)^3 = -2 \quad \text{with} \quad \mathbf{X}_1 = [1, 1, 1, 1]^T$$
$$\lambda_2 = 1 + (-2)(-1) + 3(-1)^2 + (-4)(-1)^3 = 10 \quad \text{with} \quad \mathbf{X}_2 = [1, -1, 1, -1]^T$$
$$\lambda_3 = 1 + (-2)(i) + 3(i)^2 + (-4)(i)^3 = -2 + i2 \quad \text{with} \quad \mathbf{X}_3 = [1, i, -1, -i]^T$$
$$\lambda_4 = 1 + (-2)(-i) + 3(-i)^2 + (-4)(-i)^3 = -2 - i2 \quad \text{with} \quad \mathbf{X}_4 = [1, -i, -1, i]^T$$

18.3 Derive Property 18.1.

Denote the elements in the first row of a circulant matrix **A** as $a_1, a_2, \ldots, a_n$ from left to right, and let r be any root of

$$r^n = 1 \tag{1}$$

Set

$$y = a_1 + a_2 r + a_3 r^2 + \cdots + a_{n-1} r^{n-2} + a_n r^{n-1} \tag{2}$$

Multiplying (*2*) successively by $r, r^2, \ldots, r^{n-1}$ gives the system of equations

$$\begin{aligned} yr &= a_n + a_1 r + a_2 r^2 + \cdots + a_{n-1} r^{n-1} \\ yr^2 &= a_{n-1} + a_n r + a_1 r^2 + \cdots a_{n-2} r^{n-1} \\ &\cdots\cdots\cdots\cdots\cdots\cdots\cdots\cdots \\ yr^{n-1} &= a_2 + a_3 r + a_4 r^2 + \cdots + a_1 r^{n-1} \end{aligned}$$

This system, with (2), has the matrix form $y\mathbf{X} = \mathbf{AX}$, for $\mathbf{X} = [1, r, r^2, \ldots, r^{n-1}]^T$. Thus, y, as given by (2), is an eigenvalue, and $\mathbf{X}$ is a corresponding eigenvector for every root r.

Given (2) and the fact that $r = 1$ is always a root of (1), it follows that the sum of any row of a circulant matrix is an eigenvalue of that matrix.

18.4 Determine the eigenvalue of matrix **B** in Problem 18.1.

Using Property 18.6 with $a = 2$, $b = -1$, and $c = 1$, we have $\lambda_k = 2 + 2\sqrt{(-1)(1)}\cos(k\pi/7)$, which, for $k = 1, 2, \ldots, 6$, yields

$$\lambda_1 = 2 + i1.801938 \qquad \lambda_2 = 2 + i1.246980 \qquad \lambda_3 = 2 + i0.445042$$

$$\lambda_4 = 2 - i0.445042 \qquad \lambda_5 = 2 - i1.246980 \qquad \lambda_6 = 2 - i1.801938$$

18.5 Determine an **LU** decomposition for matrix **B** in Problem 18.1.

We apply Steps 18.1 through 18.4 to $\mathbf{B} = [b_{ij}]$, initializing to

$$\mathbf{L} = \begin{bmatrix} 2 & 0 & 0 & 0 & 0 & 0 \\ 1 & 0 & 0 & 0 & 0 & 0 \\ 0 & 1 & 0 & 0 & 0 & 0 \\ 0 & 0 & 1 & 0 & 0 & 0 \\ 0 & 0 & 0 & 1 & 0 & 0 \\ 0 & 0 & 0 & 0 & 1 & 0 \end{bmatrix} \qquad \mathbf{U} = \begin{bmatrix} 1 & 0 & 0 & 0 & 0 & 0 \\ 0 & 1 & 0 & 0 & 0 & 0 \\ 0 & 0 & 1 & 0 & 0 & 0 \\ 0 & 0 & 0 & 1 & 0 & 0 \\ 0 & 0 & 0 & 0 & 1 & 0 \\ 0 & 0 & 0 & 0 & 0 & 1 \end{bmatrix}$$

and then calculating as follows:

For $i = 2$:
$$u_{12} = b_{12}/l_{11} = -1/2$$
$$l_{22} = b_{22} - l_{21}u_{12} = 2 - 1(-1/2) = 5/2$$

For $i = 3$:
$$u_{23} = b_{23}/l_{22} = -1/(5/2) = -2/5$$
$$l_{33} = b_{33} - l_{32}u_{23} = 2 - 1(-2/5) = 12/5$$

For $i = 4$:
$$u_{34} = b_{34}/l_{33} = -1/(12/5) = -5/12$$
$$l_{44} = b_{44} - l_{43}u_{34} = 2 - 1(-5/12) = 29/12$$

For $i = 5$:
$$u_{45} = b_{45}/l_{44} = -1/(29/12) = -12/29$$
$$l_{55} = b_{55} - l_{54}u_{45} = 2 - 1(-12/29) = 70/29$$

For $i = 6$:
$$u_{56} = b_{56}/l_{55} = -1(70/29) = -29/70$$
$$l_{66} = b_{66} - l_{65}u_{56} = 2 - 1/(-29/70) = 169/70$$

The factorization is, then,

$$\mathbf{B} = \begin{bmatrix} 2 & 0 & 0 & 0 & 0 & 0 \\ 1 & 5/2 & 0 & 0 & 0 & 0 \\ 0 & 1 & 12/5 & 0 & 0 & 0 \\ 0 & 0 & 1 & 29/12 & 0 & 0 \\ 0 & 0 & 0 & 1 & 70/29 & 0 \\ 0 & 0 & 0 & 0 & 1 & 169/70 \end{bmatrix} \begin{bmatrix} 1 & -1/2 & 0 & 0 & 0 & 0 \\ 0 & 1 & -2/5 & 0 & 0 & 0 \\ 0 & 0 & 1 & -5/12 & 0 & 0 \\ 0 & 0 & 0 & 1 & -12/29 & 0 \\ 0 & 0 & 0 & 0 & 1 & -29/70 \\ 0 & 0 & 0 & 0 & 0 & 1 \end{bmatrix}$$

18.6 Determine an **LU** decomposition for matrix **E** in Problem 18.1.

We apply Steps 18.1 through 18.4 to $\mathbf{E} = [e_{ij}]$, initializing to

$$\mathbf{L} = \begin{bmatrix} 1 & 0 & 0 & 0 & 0 \\ -1 & 0 & 0 & 0 & 0 \\ 0 & 2 & 0 & 0 & 0 \\ 0 & 0 & 3 & 0 & 0 \\ 0 & 0 & 0 & -3 & 0 \end{bmatrix} \qquad \mathbf{U} = \begin{bmatrix} 1 & 0 & 0 & 0 & 0 \\ 0 & 1 & 0 & 0 & 0 \\ 0 & 0 & 1 & 0 & 0 \\ 0 & 0 & 0 & 1 & 0 \\ 0 & 0 & 0 & 0 & 1 \end{bmatrix}$$

and then calculating:

For $i = 2$:
$$u_{12} = e_{12}/l_{11} = 2/1 = 2$$
$$l_{22} = e_{22} - l_{21}u_{12} = 2 - (-1)(2) = 4$$

For $i = 3$:
$$u_{23} = e_{23}/l_{22} = -2/4 = -1/2$$
$$l_{33} = e_{33} - l_{32}u_{23} = 1 - 2(-1/2) = 2$$

For $i = 4$:
$$u_{34} = e_{34}/l_{33} = 1/2$$
$$l_{44} = e_{44} - l_{43}u_{34} = -1 - 3(1/2) = -5/2$$

For $i = 5$:
$$u_{45} = e_{45}/l_{44} = 1/(-5/2) = -2/5$$
$$l_{55} = e_{55} - l_{54}u_{45} = 2 - (-3)(-2/5) = 4/5$$

The factorization is, then,

$$\mathbf{E} = \begin{bmatrix} 1 & 0 & 0 & 0 & 0 \\ -1 & 4 & 0 & 0 & 0 \\ 0 & 2 & 2 & 0 & 0 \\ 0 & 0 & 3 & -5/2 & 0 \\ 0 & 0 & 0 & -3 & 4/5 \end{bmatrix} \begin{bmatrix} 1 & 2 & 0 & 0 & 0 \\ 0 & 1 & -1/2 & 0 & 0 \\ 0 & 0 & 1 & 1/2 & 0 \\ 0 & 0 & 0 & 1 & -2/5 \\ 0 & 0 & 0 & 0 & 1 \end{bmatrix}$$

18.7 Transform to Hessenberg form the matrix

$$\mathbf{A}_0 = \begin{bmatrix} 2 & 1 & -2 \\ -3 & 1 & 0 \\ 4 & 3 & 1 \end{bmatrix}$$

The first iteration ($k = 1$) of Steps 18.5 through 18.8 yields

$$\mathbf{X}_1 = \begin{bmatrix} -3 \\ 4 \end{bmatrix}$$

$$\mathbf{V}_1 = \mathbf{X}_1 + \|\mathbf{X}_1\|_2\mathbf{E}_1 = \begin{bmatrix} -3 \\ 4 \end{bmatrix} + 5\begin{bmatrix} 1 \\ 0 \end{bmatrix} = \begin{bmatrix} 2 \\ 4 \end{bmatrix}$$

$$\mathbf{R}_1 = \mathbf{I} - \frac{2}{\|\mathbf{V}_1\|_2^2}\mathbf{V}_1\mathbf{V}_1^T = \begin{bmatrix} 1 & 0 \\ 0 & 1 \end{bmatrix} - \frac{2}{20}\begin{bmatrix} 2 \\ 4 \end{bmatrix}[2, 4] = \begin{bmatrix} 3/5 & -4/5 \\ -4/5 & -3/5 \end{bmatrix}$$

$$\mathbf{P}_1 = \left[\begin{array}{c:cc} 1 & 0 & 0 \\ \hdashline 0 & 3/5 & -4/5 \\ 0 & -4/5 & -3/5 \end{array}\right]$$

and
$$\mathbf{A}_1 = \mathbf{P}_1^T\mathbf{A}_0\mathbf{P}_1 = \begin{bmatrix} 2 & 11/5 & 2/5 \\ -5 & -11/25 & 48/25 \\ 0 & -27/25 & 61/25 \end{bmatrix}$$

$\mathbf{A}_1$ is in Hessenberg form, so $\mathbf{P} = \mathbf{P}_1$ and the transformation is completed.

18.8 Transform to Hessenberg form the matrix

$$\mathbf{A}_0 = \begin{bmatrix} 1 & 2 & 3 & -6 \\ 2 & 0 & 3 & 4 \\ 3 & 3 & 2 & 1 \\ -6 & 4 & 1 & -1 \end{bmatrix}$$

The first iteration ($k = 1$) of Steps 18.5 through 18.8 yields

$$\mathbf{X}_1 = \begin{bmatrix} 2 \\ 3 \\ -6 \end{bmatrix}$$

and
$$\mathbf{V}_1 = \mathbf{X}_1 + \|\mathbf{X}_1\|_2\mathbf{E}_1 = \begin{bmatrix} 2 \\ 3 \\ -6 \end{bmatrix} + 7\begin{bmatrix} 1 \\ 0 \\ 0 \end{bmatrix} = \begin{bmatrix} 9 \\ 3 \\ -6 \end{bmatrix}$$

for which we have, from Problem 15.13,

$$\mathbf{R}_1 = \begin{bmatrix} -2/7 & -3/7 & 6/7 \\ -3/7 & 6/7 & 2/7 \\ 6/7 & 2/7 & 3/7 \end{bmatrix}$$

so

$$\mathbf{P}_1 = \left[\begin{array}{c|ccc} 1 & 0 & 0 & 0 \\ \hline 0 & -2/7 & -3/7 & 6/7 \\ 0 & -3/7 & 6/7 & 2/7 \\ 0 & 6/7 & 2/7 & 3/7 \end{array}\right]$$

and

$$\mathbf{A}_1 = \mathbf{P}_1^T\mathbf{A}_0\mathbf{P}_1 = \begin{bmatrix} 1 & -7 & 0 & 0 \\ -7 & -114/49 & -115/49 & 27/49 \\ 0 & -115/49 & -64/49 & 142/49 \\ 0 & 27/49 & 142/49 & 227/49 \end{bmatrix}$$

The second iteration ($k = 2$) yields

$$\mathbf{X}_2 = \begin{bmatrix} -115/49 \\ 27/49 \end{bmatrix} \quad \text{for which} \quad \|\mathbf{X}_2\|_2 = \sqrt{5.811745} = 2.410756$$

and

$$\mathbf{V}_2 = \mathbf{X}_2 + \|\mathbf{X}_2\|_2\mathbf{E}_2 = \begin{bmatrix} -115/49 \\ 27/49 \end{bmatrix} + 2.410756\begin{bmatrix} 1 \\ 0 \end{bmatrix} = \begin{bmatrix} 0.0638173 \\ 0.5510204 \end{bmatrix}$$

for which $\|\mathbf{V}_2\|_2 = \sqrt{0.307696}$. Then,

$$\mathbf{R}_2 = \mathbf{I}_2 - \frac{2}{0.307696}\mathbf{V}_2\mathbf{V}_2^T = \begin{bmatrix} 0.973528 & -0.228567 \\ -0.228567 & -0.973528 \end{bmatrix}$$

$$\mathbf{P}_2 = \left[\begin{array}{cc|cc} 1 & 0 & 0 & 0 \\ 0 & 1 & 0 & 0 \\ \hline 0 & 0 & 0.973528 & -0.228567 \\ 0 & 0 & -0.228567 & -0.973528 \end{array}\right]$$

and

$$\mathbf{A}_2 = \mathbf{P}_2^T\mathbf{A}_1\mathbf{P}_2 = \begin{bmatrix} 1 & -7 & 0 & 0 \\ -7 & -2.326531 & -2.410756 & 0 \\ 0 & -2.410756 & -2.285550 & -1.273688 \\ 0 & 0 & -1.273688 & 5.612079 \end{bmatrix}$$

Setting $\mathbf{P} = \mathbf{P}_1\mathbf{P}_2$, we have $\mathbf{P}^T\mathbf{A}_0\mathbf{P} = \mathbf{A}_2$, which is in Hessenberg form. Furthermore, since $\mathbf{A}_0$ is symmetric, $\mathbf{A}_2$ is tridiagonal.

18.9 The (right) *Kronecker product* (or *direct product*) of an $m \times n$ matrix $\mathbf{A} = [a_{ij}]$ and a $p \times q$ matrix $\mathbf{B} = [b_{ij}]$ is the $mp \times nq$ partitioned matrix

$$\mathbf{A} \otimes \mathbf{B} = \begin{bmatrix} a_{11}\mathbf{B} & a_{12}\mathbf{B} & \cdots & a_{1n}\mathbf{B} \\ a_{21}\mathbf{B} & a_{22}\mathbf{B} & \cdots & a_{2n}\mathbf{B} \\ \cdots & \cdots & \cdots & \cdots \\ a_{m1}\mathbf{B} & a_{m2}\mathbf{B} & \cdots & a_{mn}\mathbf{B} \end{bmatrix}$$

Determine $\mathbf{A} \otimes \mathbf{B}$ when

$$\mathbf{A} = \begin{bmatrix} 1 & 2 \\ 4 & 3 \end{bmatrix} \quad \text{and} \quad \mathbf{B} = \begin{bmatrix} 5 & -4 \\ 4 & -5 \end{bmatrix}$$

$$\mathbf{A} \otimes \mathbf{B} = \begin{bmatrix} 1\mathbf{B} & 2\mathbf{B} \\ 4\mathbf{B} & 3\mathbf{B} \end{bmatrix} = \begin{bmatrix} 5 & -4 & 10 & -8 \\ 4 & -5 & 8 & -10 \\ 20 & -16 & 15 & -12 \\ 16 & -20 & 12 & -15 \end{bmatrix}$$

18.10 Any $n \times m$ matrix $\mathbf{X}$ can be converted into an $nm \times 1$ column vector $\mathbf{x}$ (denoted with a lowercase boldface letter) by taking the transpose of all the rows of $\mathbf{X}$ and placing them successively below one another into $\mathbf{x}$. The matrix equation

$$\mathbf{AXB} = \mathbf{C} \tag{18.1}$$

is then equivalent to the *matrix-vector equation*

$$(\mathbf{A} \otimes \mathbf{B}^T)\mathbf{x} = \mathbf{c} \tag{18.2}$$

where $\mathbf{x}$ and $\mathbf{c}$ are the vector representations of the matrices $\mathbf{X}$ and $\mathbf{C}$, respectively. Equation (*18.1*) may be solved for the unknown matrix $\mathbf{X}$ in terms of $\mathbf{A}$, $\mathbf{B}$, and $\mathbf{C}$ by solving (*18.2*) for the vector $\mathbf{x}$ using the methods developed in Chapter 2. Equation (*18.1*) may possess exactly one solution, no solutions, or infinitely many solutions.

Solve the matrix equation $\mathbf{AXB} = \mathbf{C}$ for $\mathbf{X}$ when

$$\mathbf{A} = \begin{bmatrix} 1 & 2 \\ 3 & 4 \\ 5 & -1 \end{bmatrix} \qquad \mathbf{B} = \begin{bmatrix} 2 & -1 & 1 \\ 1 & 0 & 3 \end{bmatrix} \qquad \mathbf{C} = \begin{bmatrix} 7 & 1 & 26 \\ 21 & 1 & 68 \\ 35 & -6 & 75 \end{bmatrix}$$

Let

$$\mathbf{X} = \begin{bmatrix} x_{11} & x_{12} \\ x_{21} & x_{22} \end{bmatrix} \qquad \text{and} \qquad \mathbf{B}^T = \begin{bmatrix} 2 & 1 \\ -1 & 0 \\ 1 & 3 \end{bmatrix}$$

Then (*18.2*) becomes

$$\begin{bmatrix} 2 & 1 & 4 & 2 \\ -1 & 0 & -2 & 0 \\ 1 & 3 & 2 & 6 \\ 6 & 3 & 8 & 4 \\ -3 & 0 & -4 & 0 \\ 3 & 9 & 4 & 12 \\ 10 & 5 & -2 & -1 \\ -5 & 0 & 1 & 0 \\ 5 & 15 & -1 & -3 \end{bmatrix} \begin{bmatrix} x_{11} \\ x_{12} \\ x_{21} \\ x_{22} \end{bmatrix} = \begin{bmatrix} 7 \\ 1 \\ 26 \\ 21 \\ 1 \\ 68 \\ 35 \\ -6 \\ 75 \end{bmatrix}$$

Solving by Gaussian elimination, we find that $x_{11} = 1$, $x_{12} = 5$, $x_{21} = -1$, and $x_{22} = 2$; hence,

$$\mathbf{X} = \begin{bmatrix} 1 & 5 \\ -1 & 2 \end{bmatrix}$$

18.11 Solve the matrix equation $\mathbf{AXB} = \mathbf{C}$ for $\mathbf{X}$ when

$$\mathbf{A} = [1, 2] \qquad \mathbf{B} = \begin{bmatrix} 0 & 1 & 2 \\ 1 & -1 & 1 \end{bmatrix} \qquad \mathbf{C} = [1, 2, 7]$$

Let

$$\mathbf{X} = \begin{bmatrix} x_{11} & x_{12} \\ x_{21} & x_{22} \end{bmatrix} \qquad \text{and} \qquad \mathbf{B}^T = \begin{bmatrix} 0 & 1 \\ 1 & -1 \\ 2 & 1 \end{bmatrix}$$

Then (*18.2*) (Problem 18.10) becomes

$$\begin{bmatrix} 0 & 1 & 0 & 2 \\ 1 & -1 & 2 & -2 \\ 2 & 1 & 4 & 2 \end{bmatrix} \begin{bmatrix} x_{11} \\ x_{12} \\ x_{21} \\ x_{22} \end{bmatrix} = \begin{bmatrix} 1 \\ 2 \\ 7 \end{bmatrix}$$

Solving by Gaussian elimination, we find that this system has infinitely many solutions given by $x_{11} = 3 - 2x_{21}$ and $x_{12} = 1 - 2x_{22}$, with x_{21} and x_{22} arbitrary. Therefore,

$$\mathbf{X} = \begin{bmatrix} 3 - 2x_{21} & 1 - 2x_{22} \\ x_{21} & x_{22} \end{bmatrix}$$

Supplementary Problems

18.12 Determine whether the following matrices are circulant, Toeplitz, band matrices, tridiagonal, and/or in Hessenberg form:

$$(a)\ \begin{bmatrix} 3 & 1 & 0 & 0 & 0 \\ -1 & 3 & 1 & 0 & 0 \\ 0 & -1 & 3 & 1 & 0 \\ 0 & 0 & -1 & 3 & 1 \\ 0 & 0 & 0 & -1 & 3 \end{bmatrix} \qquad (b)\ \begin{bmatrix} 1 & 0 & 1 & 0 \\ 0 & 1 & 0 & 1 \\ 1 & 0 & 1 & 0 \\ 0 & 1 & 0 & 1 \end{bmatrix} \qquad (c)\ \begin{bmatrix} 1 & 0 & 1 & 0 \\ 1 & 0 & 1 & 0 \\ 0 & 1 & 0 & 1 \\ 0 & 0 & 1 & 1 \end{bmatrix}$$

$$(d)\ \begin{bmatrix} 1 & 2 & 3 \\ 3 & 1 & 2 \\ 2 & 3 & 1 \end{bmatrix} \qquad (e)\ \begin{bmatrix} 1 & 2 & 0 \\ 3 & 1 & 2 \\ 0 & 3 & 1 \end{bmatrix} \qquad (f)\ \begin{bmatrix} 1 & 2 & 3 \\ 3 & 2 & 1 \\ 2 & 1 & 3 \end{bmatrix}$$

$$(g)\ \begin{bmatrix} -1 & 2 & 0 & 0 \\ 2 & 1 & 1 & 0 \\ 0 & 2 & -2 & 1 \\ 0 & 0 & 1 & -1 \end{bmatrix}$$

18.13 Find the eigenvalues of, and a canonical basis for, the matrix in Problem 18.12(*b*).

18.14 Find the eigenvalues of, and a canonical basis for, the matrix in Problem 18.12(*d*).

18.15 Determine the eigenvalues of the matrix in Problem 18.12(*e*).

18.16 Construct an **LU** factorization for the matrix in Problem 18.12(*e*).

18.17 Construct an **LU** factorization for the matrix in Problem 18.12(*g*).

In Problems 18.18 to 18.20, transform $\mathbf{A}_0$ into Hessenberg form.

18.18 $\mathbf{A}_0 = \begin{bmatrix} 2 & 0 & 0 \\ 1 & 5 & 2 \\ 1 & -1 & 2 \end{bmatrix}$ **18.19** $\mathbf{A}_0 = \begin{bmatrix} 3 & 1 & 1 \\ 1 & 5 & 1 \\ 1 & 1 & 3 \end{bmatrix}$ **18.20** $\mathbf{A}_0 = \begin{bmatrix} 1 & 1 & 0 & -1 \\ 1 & 1 & 1 & 0 \\ 0 & 1 & -1 & 1 \\ -1 & 0 & 1 & 0 \end{bmatrix}$

18.21 An alternative procedure for transforming a matrix into Hessenberg form is *Given's method*, which utilizes rotation matrices (see Problems 15.25 through 15.27). The method is iterative, transforming to zero one element below the subdiagonal at a time. Zeros are introduced from left to right and from the last row up to the third row. For an $n \times n$ matrix **A**, with $n \geq 3$, Given's method is as follows:

STEP 1: Initialize counters k and i to $k = n$ and $i = 1$.

STEP 2: Set $j = i + 1$.

STEP 3: Determine θ so that $\tan\theta = a_{ki}/a_{kj}$.

STEP 4: Construct the rotation matrix $\mathbf{R}_{ij}(\theta)$.

STEP 5: Calculate $\mathbf{R}_{ij}^T(\theta)\mathbf{A}\mathbf{R}_{ij}(\theta)$ and designate the product as the new matrix **A**. It will have a zero in the (k, i) position.

STEP 6: If $i < k - 2$, then increase i by 1 and return to Step 2; if $i = k - 2$, go to Step 7.

STEP 7: If $k = 3$, stop; the algorithm is completed. If $k > 3$, then decrease k by 1, set $i = 1$, and return to Step 2.

Use this algorithm to reduce the matrix in Problem 18.7 to Hessenberg form.

18.22 Use the algorithm given in Problem 18.21 to reduce the matrix in Problem 18.8 to Hessenberg form.

18.23 Construct $\mathbf{C} \otimes \mathbf{D}$ when

$$\mathbf{C} = \begin{bmatrix} 1 & 3 \\ 1 & 3 \end{bmatrix} \quad \text{and} \quad \mathbf{D} = \begin{bmatrix} 5 & -4 \\ 4 & -3 \end{bmatrix}$$

18.24 Rework Problem 18.11 with $\mathbf{C} = [1, 2, 3]$.

18.25 Solve the matrix equation $\mathbf{AXB} = \mathbf{C}$ for $\mathbf{X}$ when

$$\mathbf{A} = \begin{bmatrix} 1 & 2 \\ 3 & 4 \end{bmatrix} \quad \mathbf{B} = \begin{bmatrix} 2 & 0 \\ 1 & -1 \end{bmatrix} \quad \mathbf{C} = \begin{bmatrix} 1 & 1 \\ 1 & 1 \end{bmatrix}$$

18.26 Solve the matrix equation $\mathbf{AXB} = \mathbf{C}$ for $\mathbf{X}$ when

$$\mathbf{A} = \begin{bmatrix} 1 & 0 & 1 \\ 2 & 1 & 1 \\ 1 & 3 & -3 \end{bmatrix} \quad \mathbf{B} = [1, 1, 1] \quad \mathbf{C} = \begin{bmatrix} 1 & 1 & 1 \\ 1 & 1 & 1 \\ -2 & -2 & -2 \end{bmatrix}$$

Chapter 19

Power Methods for Locating Real Eigenvalues

NUMERICAL METHODS

Algebraic procedures for determining eigenvalues and eigenvectors, as described in Chapter 7, are impractical for most matrices of large order. Instead, numerical methods that are efficient and stable when programmed on high-speed computers have been developed for this purpose. Such methods are iterative, and, in the ideal case, converge to the eigenvalues and eigenvectors of interest. Included with each method are termination criteria, generally a test to determine when a specified precision has been achieved (if the results are converging) and an upper bound on the number of iterations to be performed (in case convergence does not occur).

This chapter describes algorithms for locating a single real eigenvalue and its associated eigenvector. The first method presented is the simplest; the last is the most powerful. Chapter 20 describes a procedure for obtaining all eigenvalues of a matrix; it is usually packaged with the shifted inverse power method as an excellent general-purpose algorithm.

THE POWER METHOD

Applied to a matrix $\mathbf{A}$, the *power method* consists in choosing a vector $\mathbf{X}$ and forming the sequence

$$c_0\mathbf{X},\ c_1\mathbf{AX},\ c_2\mathbf{A}^2\mathbf{X},\ c_3\mathbf{A}^3\mathbf{X},\ \ldots$$

where $c_0, c_1, c_2, \ldots$ are scaling constants selected to avoid computer overflow due to extremely large vector components. The sequence will generally converge to an eigenvector of $\mathbf{A}$, and if the scaling constants are wisely chosen, the eigenvalue will be obvious too. This eigenvalue is usually the *dominant eigenvalue* of $\mathbf{A}$, the one having greatest absolute value, provided such an eigenvalue is real. The usual implementation of the power method is as follows:

STEP 19.1: Initialize $\mathbf{X}_0$ so that its largest component in absolute value is 1, and initialize $\lambda_0 = 0$ as the first approximation to the eigenvalue. Specify a desired precision PRE for the eigenvalue, and the maximum number of iterations to be performed; set an iteration counter $k = 1$.

STEP 19.2: Calculate $\mathbf{Y}_k = \mathbf{AX}_{k-1}$.

STEP 19.3: Determine the component of $\mathbf{Y}_k$ that is largest in absolute value. Denote it as λ_k.

STEP 19.4: Set $\mathbf{X}_k = (1/\lambda_k)\mathbf{Y}_k$.

STEP 19.5: If $|\lambda_k - \lambda_{k-1}| < \text{PRE}$, stop; the eigenvalue and associated eigenvector are λ_k and $\mathbf{X}_k$. Otherwise, continue.

STEP 19.6: Increase k by 1. If k is greater than the maximum number of iterations to be performed, stop. Otherwise, return to Step 19.2.

(See Problems 19.1 through 19.4.)

The power method will not converge if the dominant eigenvalue is complex. (See Problem 19.5.) Once convergence occurs, however, the procedure may be attempted again on $\mathbf{A} - \lambda\mathbf{I}$ to determine a second eigenvalue-eigenvector pair. (See Problem 19.3.)

Convergence toward an eigenvalue may occur without an accompanying convergence of the associated eigenvector. A component that oscillates in sign with decreasing magnitude is converging to zero.

THE INVERSE POWER METHOD

The *inverse power method* is the power method applied to $\mathbf{A}^{-1}$, provided the matrix is nonsingular. The procedure will converge to the dominant eigenvalue of $\mathbf{A}^{-1}$, the reciprocal of which is the eigenvalue of $\mathbf{A}$ having the smallest absolute value. The associated eigenvector is the same for both (Property 7.4). The steps are identical to those of the power method with the exception of the following:

STEP 19.2′: Calculate $\mathbf{Y}_k = \mathbf{A}^{-1}\mathbf{X}_{k-1}$ by solving the system $\mathbf{AY}_k = \mathbf{X}_{k-1}$ using **LU** decomposition. If this system does not have a unique solution, stop; zero is an eigenvalue of $\mathbf{A}$.

(See Problems 19.8 and 19.9.)

THE SHIFTED INVERSE POWER METHOD

The inverse power method may be used to find all real eigenvalues of a matrix if estimates of their locations are available. If u is an estimate of λ, then $\mathbf{A} - u\mathbf{I}$ will have an eigenvalue near zero, and its reciprocal will be the dominant eigenvalue of $(\mathbf{A} - u\mathbf{I})^{-1}$. Therefore, if λ and $\mathbf{X}$ are the eigenvalue and eigenvector obtained by applying the inverse power method to $\mathbf{A} - u\mathbf{I}$, then $u + 1/\lambda$ and $\mathbf{X}$ are approximations to an eigenvalue and eigenvector of $\mathbf{A}$. (See Problem 19.11.)

GERSCHGORIN'S THEOREM

Each row of a square matrix generates a *Gerschgorin disk*, which is bounded by a circle, whose center is the diagonal element in the row and whose radius is the sum of the absolute values of all other elements in that row.

Example 19.1 The Gerschgorin disks for

$$\mathbf{A} = \begin{bmatrix} 15 & 3+i4 & 0 \\ -3 & i32 & 1-i3 \\ 0.01 & -0.25 & 2-i3 \end{bmatrix}$$

are

$$|z-15| \le |3+i4| + |0| = 5$$

$$|z-i32| \le |-3| + |1-i3| = 6.162$$

and

$$|z-(2-i3)| \le |0.01| + |-0.25| = 0.26$$

Property 19.1: (*Gerschgorin's theorem*) Every eigenvalue of a matrix (real or complex) must lie in one of its Gerschgorin disks. Furthermore, if the union of N of these disks is disjoint from all the rest, then there are exactly N eigenvalues in the union of those N disks.

Gerschgorin's theorem is used to estimate the locations of the eigenvalues of a matrix. (See Problem 19.10.) Moreover, since the eigenvalues of a matrix are preserved under transposition, a second set of estimates may be developed by applying Gerschgorin's theorem to the matrix transpose. Still other estimates are provided by (*12.3*).

Solved Problems

19.1 Use the power method to locate an eigenvalue and eigenvector for

$$\mathbf{A} = \begin{bmatrix} 5 & -1 & 7 \\ -1 & -1 & 1 \\ 7 & 1 & 5 \end{bmatrix}$$

We choose $\mathbf{X}_0 = [1, 1, 1]^T$. Then we have:

First iteration:

$$\mathbf{Y}_1 = \mathbf{AX}_0 = [11, -1, 13]^T$$
$$\lambda_1 = 13$$
$$\mathbf{X}_1 = \frac{1}{\lambda_1}\mathbf{Y}_1 = [0.846154, -0.076923, 1.000000]^T$$

Second iteration:

$$\mathbf{Y}_2 = \mathbf{AX}_1 = [11.307692, 0.230767, 10.846157]^T$$
$$\lambda_2 = 11.307692$$
$$\mathbf{X}_2 = \frac{1}{\lambda_2}\mathbf{Y}_2 = [1.000000, 0.020408, 0.959184]^T$$

Third iteration:

$$\mathbf{Y}_3 = \mathbf{AX}_2 = [11.693874, -0.061220, 11.816237]^T$$
$$\lambda_3 = 11.816327$$
$$\mathbf{X}_3 = \frac{1}{\lambda_3}\mathbf{Y}_3 = [0.989637, -0.005181, 1.000000]^T$$

Continuing in this manner, we generate Table 19.1, where all entries are rounded to four decimal places. The algorithm is converging to the eigenvector $[1, 0, 1]^T$ with corresponding eigenvalue $\lambda = 12$. Note how the second component of the eigenvector oscillates in sign as it converges to zero.

Table 19.1

Iteration	Eigenvector components			Eigenvalue
0	1.0000	1.0000	1.0000	
1	0.8462	−0.0769	1.0000	13.0000
2	1.0000	0.0204	0.9592	11.3077
3	0.9896	−0.0052	1.0000	11.8163
4	1.0000	0.0013	0.9974	11.9534
5	0.9993	−0.0003	1.0000	11.9883
6	1.0000	0.0001	0.9998	11.9971
7	1.0000	−0.0000	1.0000	11.9993
8	1.0000	0.0000	1.0000	11.9998

19.2 Use the power method to determine an eigenvalue and eigenvector for

$$\mathbf{A} = \begin{bmatrix} 5 & 2 & 2 \\ 3 & 6 & 3 \\ 6 & 6 & 9 \end{bmatrix}$$

We initialize by choosing $\mathbf{X}_0 = [1, 1, 1]^T$. Then we have:

***First iteration*:**

$$\mathbf{Y}_1 = \mathbf{AX}_0 = [9, 12, 21]^T$$
$$\lambda_1 = 21$$
$$\mathbf{X}_1 = \frac{1}{\lambda_1}\mathbf{Y}_1 = [0.428571, 0.571429, 1.000000]^T$$

***Second iteration*:**

$$\mathbf{Y}_2 = \mathbf{AX}_1 = [5.285714, 7.714286, 15.000000]^T$$
$$\lambda_2 = 15$$
$$\mathbf{X}_2 = \frac{1}{\lambda_2}\mathbf{Y}_2 = [0.352381, 0.514286, 1.000000]^T$$

***Third iteration*:**

$$\mathbf{Y}_3 = \mathbf{AX}_2 = [4.790476, 7.142857, 14.200000]^T$$
$$\lambda_3 = 14.2$$
$$\mathbf{X}_3 = \frac{1}{\lambda_3}\mathbf{Y}_3 = [0.337357, 0.503018, 1.000000]^T$$

Continuing in this manner, we generate Table 19.2, where all entries are rounded to four decimal places. The algorithm is converging to the eigenvector $[1/3, 1/2, 1]^T$ with corresponding eigenvalue $\lambda = 14$.

Table 19.2

Iteration	Eigenvector components			Eigenvalue
0	1.0000	1.0000	1.0000	
1	0.4286	0.5714	1.0000	21.0000
2	0.3524	0.5143	1.0000	15.0000
3	0.3374	0.5030	1.0000	14.2000
4	0.3342	0.5006	1.0000	14.0423
5	0.3335	0.5001	1.0000	14.0090
6	0.3334	0.5000	1.0000	14.0019
7	0.3333	0.5000	1.0000	14.0004
8	0.3333	0.5000	1.0000	14.0001

19.3 Use the power method to determine a second eigenvalue for the matrix given in Problem 19.1.

Instead of the matrix $\mathbf{A}$ given in that problem, we consider the matrix $\mathbf{A} - 12\mathbf{I}$, which has the same eigenvectors as $\mathbf{A}$ but whose eigenvalues are those of $\mathbf{A}$ reduced by 12. One eigenvalue of $\mathbf{A} - 12\mathbf{I}$ is, therefore, zero, corresponding to the eigenvalue found in Problem 19.1. Since zero most likely is not the dominant eigenvalue of $\mathbf{A} - 12\mathbf{I}$, we can use the power method on this new matrix to locate another eigenvalue and eigenvector. We set

$$\mathbf{B} = \mathbf{A} - 12\mathbf{I} = \begin{bmatrix} -7 & -1 & 7 \\ -1 & -13 & 1 \\ 7 & 1 & -7 \end{bmatrix}$$

and apply the power method to $\mathbf{B}$. The results are summarized in Table 19.3. The algorithm is converging to the eigenvector $[1, 1, -1]^T$ with associated eigenvalue $\lambda = -15$. The corresponding eigenvalue for the matrix $\mathbf{A}$ is $\lambda = -15 + 12 = -3$ with the same associated eigenvector.

Having two eigenvalues of a 3×3 matrix, we can produce the third easily. The trace of $\mathbf{A}$ equals the sum of the eigenvalues, so $5 + (-1) + 5 = 12 + (-3) + \lambda$. The last eigenvalue of $\mathbf{A}$ is thus $\lambda = 0$.

Table 19.3

Iteration	Eigenvector components			Eigenvalue
0	1.0000	1.0000	1.0000	
1	0.0769	1.0000	−0.0769	−13.0000
2	0.1579	1.0000	−0.1579	−13.1538
3	0.2411	1.0000	−0.2411	−13.3158
4	0.3245	1.0000	−0.3245	−13.4822
5	0.4061	1.0001	−0.4061	−13.6491
10	0.7348	1.0000	−0.7348	−14.3651
20	0.9662	1.0000	−0.9662	−14.9160
30	0.9963	1.0000	−0.9963	−14.9907
40	0.9996	1.0000	−0.9996	−14.9990
45	0.9999	1.0000	−0.9999	−14.9997

19.4 Derive the power method.

Assume that the matrix **A** has order $n \times n$ and possesses n real eigenvalues $\lambda_1, \lambda_2, \ldots, \lambda_n$ such that

$$|\lambda_1| > |\lambda_2| \geq \cdots \geq |\lambda_n|$$

Furthermore, assume that the eigenvectors $\mathbf{V}_1, \mathbf{V}_2, \ldots, \mathbf{V}_n$ corresponding to each of these eigenvalues form a linearly independent set. Then for any n-dimensional vector **X** there exist constants $d_1, d_2, \ldots, d_n$, not all zero, such that

$$\mathbf{X} = d_1\mathbf{V}_1 + d_2\mathbf{V}_2 + \cdots + d_n\mathbf{V}_n$$

Multiplying on the left by **A** repeatedly, we get

$$\mathbf{A}^k\mathbf{X} = d_1\lambda_1^k\mathbf{V}_1 + d_2\lambda_2^k\mathbf{V}_2 + \cdots + d_n\lambda_n^k\mathbf{V}_n$$

or

$$\frac{\mathbf{A}^k\mathbf{X}}{\lambda_1^k} = d_1\mathbf{V}_1 + d_2\left(\frac{\lambda_2}{\lambda_1}\right)^k\mathbf{V}_2 + \cdots + d_n\left(\frac{\lambda_n}{\lambda_1}\right)^k\mathbf{V}_n$$

Since λ_1 is the dominant eigenvalue, the sequence converges to $d_1\mathbf{V}_1$. Therefore, the power method will converge to the eigenvector associated with the dominant eigenvalue provided (1) the dominant eigenvalue is real and of multiplicity one, and (2) the initial guess **X** is not a linear combination of the remaining $n-1$ eigenvectors ($d_1 \neq 0$). The rate of convergence is a function of the ratio $|\lambda_2/\lambda_1|$.

19.5 Apply the power method to

$$\mathbf{A} = \begin{bmatrix} 4 & -8 & 0 \\ 9 & -8 & 0 \\ 4 & 5 & 1 \end{bmatrix}$$

and explain the result.

Applying the power method to this matrix, we generate Table 19.4, from which we conclude that the algorithm is not converging. The reason is that the eigenvalues of **A** are 1, $-2+i6$, and $-2-i6$, with the dominant ones being complex.

Table 19.4

Iteration	Eigenvector components			Eigenvalue
0	1.0000	1.0000	1.0000	
1	−0.4000	0.1000	1.0000	10.0000
2	0.5455	1.0000	0.0227	−4.4000
3	−0.8076	−0.4290	1.0000	7.2045
50	0.7229	1.0000	−0.2567	−6.0533
51	−0.6691	−0.1957	1.0000	7.6347
52	0.2492	1.0000	0.5958	−4.4563
53	1.0000	0.8220	−0.9414	−7.0030

19.6 A modification of the power method particularly suited to real symmetric matrices is initialized with a unit vector (in the Euclidean norm) having all its components equal.

At each iteration $\mathbf{Y}_k$ is determined as before, but the eigenvalue is approximated as $\lambda_k = \mathbf{X}_{k-1} \cdot \mathbf{Y}_k$, an approximation to the Rayleigh quotient. Then $\mathbf{X}_k = \mathbf{Y}_k / \|\mathbf{Y}_k\|_2$, unless $\|\mathbf{Y}_k\|_2 = 0$, in which case zero is an eigenvalue and the algorithm is terminated. Use this modified power method to determine an eigenvalue and eigenvector for

$$\mathbf{A} = \begin{bmatrix} 10 & 7 & 8 & 7 \\ 7 & 5 & 6 & 5 \\ 8 & 6 & 10 & 9 \\ 7 & 5 & 9 & 10 \end{bmatrix}$$

We initialize with $\mathbf{X}_0 = [0.5, 0.5, 0.5, 0.5]^T$, which is the vector $[1, 1, 1, 1]^T$ normalized. Then, with all calculations rounded to four decimal places, we have:

First iteration:

$$\mathbf{Y}_1 = \mathbf{AX}_0 = [16, 11.5, 16.5, 15.5]^T$$
$$\lambda_1 = \mathbf{X}_0 \cdot \mathbf{Y}_1 = 29.75$$
$$\|\mathbf{Y}_1\|_2 = 30.0125$$
$$\mathbf{X}_1 = \frac{\mathbf{Y}_1}{\|\mathbf{Y}_1\|_2} = [0.5331, 0.3832, 0.5498, 0.5165]^T$$

Second iteration:

$$\mathbf{Y}_2 = \mathbf{AX}_1 = [16.0267, 11.5285, 16.7097, 15.7601]^T$$
$$\lambda_2 = \mathbf{X}_1 \cdot \mathbf{Y}_2 = 30.2873$$
$$\|\mathbf{Y}_2\|_2 = 30.2879$$
$$\mathbf{X}_2 = \frac{\mathbf{Y}_2}{\|\mathbf{Y}_2\|_2} = [0.5291, 0.3806, 0.5517, 0.5203]^T$$

Third iteration:

$$\mathbf{Y}_3 = \mathbf{AX}_2 = [16.0118, 11.5191, 16.7170, 15.7759]^T$$
$$\lambda_3 = \mathbf{X}_2 \cdot \mathbf{Y}_1 = 30.2887$$
$$\|\mathbf{Y}_3\|_2 = 30.2887$$
$$\mathbf{X}_3 = \frac{\mathbf{Y}_3}{\|\mathbf{Y}_3\|_2} = [0.5287, 0.3803, 0.5519, 0.5209]^T$$

At this point, the eigenvector is already accurate to four decimal places. The standard power method would take six iterations to achieve similar accuracy.

19.7 Use the modified power method described in Problem 19.6 to determine a second eigenvalue and associated eigenvector for the matrix in Problem 19.6.

Having determined that 30.2887 is an eigenvalue of **A**, we can apply the modified power method to

$$\mathbf{B} = \mathbf{A} - 30.2887\mathbf{I} = \begin{bmatrix} -20.2887 & 7 & 8 & 7 \\ 7 & -25.2887 & 6 & 5 \\ 8 & 6 & -20.2887 & 9 \\ 7 & 5 & 9 & -20.2887 \end{bmatrix}$$

We initialize with $\mathbf{X}_0 = [0.5, 0.5, 0.5, 0.5]^T$. Then with all calculations rounded to four decimal places, we have:

First iteration:

$$\mathbf{Y}_1 = \mathbf{B}\mathbf{X}_0 = [0.8557, -3.6444, 1.3557, 0.3557]^T$$
$$\lambda_1 = \mathbf{X}_0 \cdot \mathbf{Y}_1 = -0.5387$$
$$\|\mathbf{Y}_1\|_2 = 3.9972$$
$$\mathbf{X}_1 = \frac{\mathbf{Y}_1}{\|\mathbf{Y}_1\|_2} = [0.2141, -0.9117, 0.3391, 0.0890]^T$$

Second iteration:

$$\mathbf{Y}_2 = \mathbf{B}\mathbf{X}_1 = [-7.3891, 27.0345, -9.8380, -1.8130]^T$$
$$\lambda_2 = \mathbf{X}_1 \cdot \mathbf{Y}_2 = -29.7275$$
$$\|\mathbf{Y}_2\|_2 = 29.7579$$
$$\mathbf{X}_2 = \frac{\mathbf{Y}_2}{\|\mathbf{Y}_2\|_2} = [-0.2483, 0.9085, -0.3306, -0.0609]^T$$

Continuing in this manner, we generate Table 19.5. Four-place precision is attained by iteration number 65, although it takes quite a few additional iterations before confidence in the result is established. The algorithm is converging to -30.2785, so a second eigenvalue of the original matrix is $-30.2785 + 30.2887 = 0.0102$.

At this point, however, there is no convergence to an eigenvector, because the sign of each component is changing at each iteration and none of the components has stabilized. This suggests a third eigenvalue very close to the second one. An alternative approach to finding the desired eigenvector is given in Problem 19.9.

Table 19.5

Iteration	Eigenvector components				Eigenvalue
0	0.5000	0.5000	0.5000	0.5000	
1	0.2141	−0.9117	0.3391	0.0890	−0.5387
2	−0.2483	0.9085	−0.3306	−0.0609	−29.7275
3	0.2788	−0.9040	0.3222	0.0357	−29.8433
63	0.4981	−0.8294	0.2170	−0.1299	−30.2784
64	−0.4982	0.8294	−0.2168	0.1298	−30.2784
65	0.4983	−0.8294	0.2165	−0.1296	−30.2785
66	−0.4984	0.8295	−0.2163	0.1294	−30.2785

19.8 Use the inverse power method to determine a second eigenvalue and second eigenvector for the matrix in Problem 19.2.

For this matrix, **LU** decomposition yields

$$\mathbf{L} = \begin{bmatrix} 5 & 0 & 0 \\ 3 & 4.8 & 0 \\ 6 & 3.6 & 5.25 \end{bmatrix} \quad \text{and} \quad \mathbf{U} = \begin{bmatrix} 1 & 0.4 & 0.4 \\ 0 & 1 & 0.375 \\ 0 & 0 & 1 \end{bmatrix}.$$

With $\mathbf{X}_0 = [1, 1, 1]^T$, the algorithm yields the following:

First iteration: Solve $\mathbf{LZ}_1 = \mathbf{X}_0$ to obtain

$$\mathbf{Z}_1 = [0.200000, 0.083333, -0.095238]^T$$

Solve $\mathbf{UY}_1 = \mathbf{Z}_1$ to obtain

$$\mathbf{Y}_1 = [0.190476, 0.119048, -0.095238]^T$$
$$\lambda_1 = 0.190476$$
$$\mathbf{X}_1 = \frac{1}{\lambda_1}\mathbf{Y}_1 = [1.000000, 0.652500, -0.500000]^T$$

Second iteration: Solve $\mathbf{LZ}_2 = \mathbf{X}_1$ to obtain

$$\mathbf{Z}_2 = [0.200000, 0.005208, -0.327381]^T$$

Solve $\mathbf{UY}_2 = \mathbf{Z}_2$ to obtain

$$\mathbf{Y}_2 = [0.279762, 0.127976, -0.327381]^T$$
$$\lambda_2 = -0.327381$$
$$\mathbf{X}_2 = \frac{1}{\lambda_2}\mathbf{Y}_2 = [-0.854545, -0.390909, 1.000000]^T$$

Third iteration: Solve $\mathbf{LZ}_3 = \mathbf{X}_2$ to obtain

$$\mathbf{Z}_3 = [-0.170909, 0.025379, 0.368398]^T$$

Solve $\mathbf{UY}_3 = \mathbf{Z}_3$ to obtain

$$\mathbf{Y}_3 = [-0.273160, -0.112771, 0.368398]^T$$
$$\lambda_3 = 0.368398$$
$$\mathbf{X}_3 = \frac{1}{\lambda_3}\mathbf{Y}_3 = [-0.741481, -0.306110, 1.000000]^T$$

Continuing in this manner, we generate Table 19.6, where all entries are rounded to four decimal places. The algorithm is converging to an eigenvalue of $1/3$ for $\mathbf{A}^{-1}$, or its reciprocal 3 for $\mathbf{A}$. The associated eigenvector is the same for both $\mathbf{A}^{-1}$ and $\mathbf{A}$; it is converging toward $[-0.7143, -0.2857, 1]^T$.

Having produced two eigenvalues for a 3×3 matrix, we can obtain the third one easily. Using Property 7.1 and the results of Problem 19.2, we have $5 + 6 + 9 = 14 + 3 + \lambda$, so the last eigenvalue is also 3.

Table 19.6

Iteration	Eigenvector components			Eigenvalue
0	1.0000	1.0000	1.0000	
1	1.0000	0.6250	−0.5000	0.1905
2	−0.8545	−0.3909	1.0000	−0.3274
3	−0.7415	−0.3061	1.0000	0.3684
4	−0.7200	−0.2900	1.0000	0.3401
5	−0.7155	−0.2866	1.0000	0.3348
6	−0.7145	−0.2859	1.0000	0.3336
7	−0.7143	−0.2858	1.0000	0.3334
8	−0.7143	−0.2857	1.0000	0.3333

19.9 Use the inverse power method to obtain an eigenvalue and eigenvector for the matrix in Problem 19.6.

For this matrix **LU** decomposition yields

$$\begin{bmatrix} 10 & 0 & 0 & 0 \\ 7 & 0.1 & 0 & 0 \\ 8 & 0.4 & 2 & 0 \\ 7 & 0.1 & 3 & 0.5 \end{bmatrix} \quad \text{and} \quad \mathbf{U} = \begin{bmatrix} 1 & 0.7 & 0.8 & 0.7 \\ 0 & 1 & 4 & 1 \\ 0 & 0 & 1 & 1.5 \\ 0 & 0 & 0 & 1 \end{bmatrix}$$

With $\mathbf{X}_0 = [1, 1, 1, 1]^T$, the algorithm yields the following:

***First iteration*:** Solve $\mathbf{LZ}_1 = \mathbf{X}_0$ to obtain

$$\mathbf{Z}_1 = [0.100000, 3.000000, -0.500000, 3.000000]^T$$

Solve $\mathbf{UY}_1 = \mathbf{Z}_1$ to obtain

$$\mathbf{Y}_1 = [-12.000000, 20.000000, -5.000000, 3.000000]^T$$
$$\lambda_1 = 20.000000$$
$$\mathbf{X}_1 = \frac{1}{\lambda_1}\mathbf{Y}_1 = [-0.600000, 1.000000, -0.250000, 0.150000]^T$$

***Second iteration*:** Solve $\mathbf{LZ}_2 = \mathbf{X}_1$ to obtain

$$\mathbf{Z}_2 = [-0.060000, 14.200000, -2.725000, 14.650000]^T$$

Solve $\mathbf{UY}_2 = \mathbf{Z}_2$ to obtain

$$\mathbf{Y}_2 = [-59.400000, 98.350000, -24.700000, 14.650000]^T$$
$$\lambda_2 = 98.350000$$
$$\mathbf{X}_2 = \frac{1}{\lambda_2}\mathbf{Y}_2 = [-0.603965, 1.000000, -0.251144, 0.148958]^T$$

Two more iterations yield

$$\mathbf{X}_3 = [-0.603972, 1.000000, -0.251135, 0.148954]^T$$

and

$$\mathbf{X}_4 = [-0.603972, 1.000000, -0.251135, 0.148953]^T$$

with $\lambda_3 = 98.521606$ and $\lambda_4 = 98.521698$. The fifth iteration is identical to the fourth, so $\mathbf{X}_4$ approximates an eigenvector of **A** corresponding to the eigenvalue $1/98.521698 = 0.010150$.

19.10 Use Gerschgorin's theorem to estimate the eigenvalues of

$$\mathbf{A} = \begin{bmatrix} 29 & -1 & 4 \\ -1 & -18 & 2 \\ 4 & 2 & 1 \end{bmatrix}$$

The Gerschgorin disks are

$$|z - 29| \le |-1| + |4| = 5$$
$$|z + 18| \le |-1| + |2| = 3$$
$$|z - 1| \le |4| + |2| = 6$$

Since **A** is a real symmetric matrix, its eigenvalues must be real (Property 13.3), and the Gerschgorin disks reduce to the intervals $24 \le z \le 34$, $-21 \le z \le -15$, and $-5 \le z \le 7$. Furthermore, these intervals (disks) are disjoint, so there must be one eigenvalue in each.

19.11 Find the eigenvalues and a corresponding set of eigenvectors for the matrix in Problem 19.10.

From Problem 19.10, we know that one real eigenvalue is located in the interval $24 \le z \le 34$. We take $u = 28$ as an estimate of this eigenvalue and apply the inverse power method to $\mathbf{A} - 28\mathbf{I}$. A better estimate for the eigenvalue might be the center of the interval, $u = 29$, but an **LU** decomposition for $\mathbf{A} - 29\mathbf{I}$ is not possible because that matrix has a zero in the (1,1) position. For $\mathbf{A} - 28\mathbf{I}$, we have

$$\mathbf{L} = \begin{bmatrix} 1 & 0 & 0 \\ -1 & -47 & 0 \\ 4 & 6 & -42.234043 \end{bmatrix} \quad \text{and} \quad \mathbf{U} = \begin{bmatrix} 1 & -1 & 4 \\ 0 & 1 & -0.127660 \\ 0 & 0 & 1 \end{bmatrix}$$

Applying the inverse power method with these matrices, we obtain, after five iterations, $\mathbf{X}_5 = [1.0, -0.015180, 0.138939]^T$ with $\lambda_5 = 0.636563$. The corresponding eigenvalue for $\mathbf{A}$ is $\lambda = 28 + 1/0.636563 = 29.5709$.

From Problem 19.10 we know that a second real eigenvalue lies between -15 and -21. We estimate this eigenvalue as $u = -19$. The **LU** decomposition for $\mathbf{A} + 19\mathbf{I}$ has

$$\mathbf{L} = \begin{bmatrix} 48 & 0 & 0 \\ -1 & 0.979167 & 0 \\ 4 & 2.083333 & 15.234043 \end{bmatrix} \quad \text{and} \quad \mathbf{U} = \begin{bmatrix} 1 & -0.020833 & 0.083333 \\ 0 & 1 & 2.127660 \\ 0 & 0 & 1 \end{bmatrix}$$

With these matrices the inverse power method yields, after five iterations, $\mathbf{X}_5 = [0.030495, 1, -0.110227]^T$ with $\lambda_5 = 1.335023$. The corresponding eigenvalue for $\mathbf{A}$ is $-19 + 1/1.335023 = -18.2509$.

The last real eigenvalue is between -5 and 7. We estimate it as $u = 0$, and apply the inverse power method directly to $\mathbf{A}$. After five iterations we find $\mathbf{X}_5 = [-0.137203, 0.114411, 1]^T$ with $\lambda_5 = 1.470566$. The corresponding eigenvalue for $\mathbf{A}$ is $1/1.470566 = 0.6800$. As a check we note that the sum of the three eigenvalues is $29.5709 + (-18.2509) + 0.6800 = 12$, which is the trace of $\mathbf{A}$.

19.12 Prove that each eigenvalue of a square matrix $\mathbf{A}$ lies in at least one Gerschgorin disk generated by $\mathbf{A}$.

Let λ be an eigenvalue of $\mathbf{A} = [a_{ij}]$ corresponding to the eigenvector $\mathbf{X} = [x_1, x_2, \ldots, x_n]^T$. Denote the largest component of $\mathbf{X}$ in absolute value as x_M. Then we have

$$\mathbf{AX} = \lambda\mathbf{X} \tag{1}$$

and

$$|x_M| = \max(|x_1|, |x_2|, \ldots, |x_n|) \tag{2}$$

Equating the Mth components of both sides of (*1*) gives

$$\sum_{j=1}^{n} a_{Mj}x_j = \lambda x_M$$

so

$$(\lambda - a_{MM})x_M = \sum_{j \ne M} a_{Mj}x_j$$

It follows from this last equation and (*2*) that

$$|\lambda - a_{MM}| = \left|\sum_{j \ne M} a_{Mj}x_j\right| \Big/ |x_M| \le \sum_{j \ne M} |a_{Mj}|(|x_j|/|x_M|) \le \sum_{j \ne M} |a_{Mj}|$$

Thus λ is in the Mth Gerschgorin disk, so it is certainly in one of them.

Supplementary Problems

19.13 Apply five iterations of the power method to

$$\mathbf{A}=\begin{bmatrix}4 & 0 & 5\\ 1 & 4 & 2\\ 3 & 0 & 4\end{bmatrix}$$

19.14 Use the power method to locate a second eigenvector and eigenvalue for the matrix in Problem 19.2. Observe that convergence occurs even though that eigenvalue has multiplicity two.

19.15 Apply the power method to the matrix in Problem 19.11 and stop after four iterations.

19.16 Apply the power method to

$$\mathbf{A}=\begin{bmatrix}-3 & 0 & 1\\ 0 & 1 & 0\\ 1 & 0 & -3\end{bmatrix}$$

and show that convergence to the eigenvalue is rapid even though there is no convergence to the second component of the eigenvector. Deduce the value to which this second component is converging.

19.17 Determine why the power method did not converge to the dominant eigenvalue in Problem 19.16.

19.18 Apply the power method to

$$\mathbf{A}=\begin{bmatrix}2 & 1 & -2\\ -3 & 1 & 0\\ 4 & 3 & 1\end{bmatrix}$$

19.19 Apply the power method to the matrix in Problem 19.6.

19.20 Apply the modified power method to

$$\mathbf{A}=\begin{bmatrix}5 & -1 & 7\\ -1 & -1 & 1\\ 7 & 1 & 5\end{bmatrix}$$

19.21 Apply three iterations of the modified power method to find a second eigenvector and eigenvalue for the matrix in Problem 19.20.

19.22 Apply the inverse power method to

$$\mathbf{A}=\begin{bmatrix}4 & 0 & 5\\ 1 & 4 & 2\\ 3 & 0 & 4\end{bmatrix}$$

19.23 Apply the inverse power method to the matrix in Problem 19.1.

19.24 Apply three iterations of the inverse power method to

$$\mathbf{A}=\begin{bmatrix}-1 & 0 & 0 & 8\\ 2 & -3 & 0 & 3\\ 1 & 0 & 5 & 3\\ 3 & 0 & 0 & 7\end{bmatrix}$$

19.25 The matrix in Problem 19.24 is known to have an eigenvalue near 9. Use the shifted inverse power method to find it.

19.26 The matrix in Problem 19.18 is known to have an eigenvalue near 2.5. Use the shifted inverse power method to find it.

19.27 A modification of the shifted inverse power method uses the Rayleigh quotient as an estimate for the eigenvalue and then shifts by that amount. At the kth iteration, the shift is $\lambda_k = \mathbf{X}_k^T\mathbf{A}\mathbf{X}_k / \mathbf{X}_k^T\mathbf{X}_k$. Thus, the shift is different for each iteration. Termination of the algorithm occurs when two successive λ iterates are within the prescribed tolerance of each other. Use this variable shift method on the matrix in Problem 19.20.

Chapter 20

The QR Algorithm

THE MODIFIED GRAM-SCHMIDT PROCESS

The Gram-Schmidt orthogonalization process (as presented in Chapter 11) may yield grossly inaccurate results due to roundoff error under finite-digit arithmetic (see Problems 20.10 and 20.11). A modification of that algorithm exists which is more stable and which generates the same vectors in the absence of rounding (see Problem 20.12). This modification also transforms a set of linearly independent vectors $\{\mathbf{X}_1, \mathbf{X}_2, \ldots, \mathbf{X}_n\}$ into a set of orthonormal vectors $\{\mathbf{Q}_1, \mathbf{Q}_2, \ldots, \mathbf{Q}_n\}$ such that each vector $\mathbf{Q}_k$ $(k = 1, 2, \ldots, n)$ is a linear combination of $\mathbf{X}_1$ through $\mathbf{X}_{k-1}$. The modified algorithm is iterative, with the kth iteration given by the following steps:

STEP 20.1: Set $r_{kk} = \|\mathbf{X}_k\|_2$ and $\mathbf{Q}_k = (1/r_{kk})\mathbf{X}_k$.

STEP 20.2: For $j = k+1, k+2, \ldots, n$, set $r_{kj} = \langle \mathbf{X}_j, \mathbf{Q}_k \rangle$

STEP 20.3: For $j = k+1, k+2, \ldots, n$, replace $\mathbf{X}_j$ by $\mathbf{X}_j - r_{kj}\mathbf{Q}_k$.

(See Problems 20.1 and 20.3.)

QR DECOMPOSITION

Every $m \times n$ matrix $\mathbf{A}$ $(m \geq n)$ can be factored into the product of a matrix $\mathbf{Q}$, having orthonormal vectors for its columns, and an upper (right) triangular matrix $\mathbf{R}$. The product

$$\mathbf{A} = \mathbf{QR} \tag{20.1}$$

is the **QR** *decomposition* of $\mathbf{A}$. If $\mathbf{A}$ is square, then $\mathbf{Q}$ is unitary.

The **QR** decomposition follows immediately from the modified Gram-Schmidt process applied to the columns of $\mathbf{A}$, provided those columns are linearly independent. If they are, then the columns of $\mathbf{Q}$ and the elements r_{ij} $(i \leq j)$ of $\mathbf{R}$ are the quantities generated by the modified Gram-Schmidt process. (See Problems 20.2 and 20.4.)

If the columns of $\mathbf{A}$ are not linearly independent, then one or more of the r_{kk} values determined in Step 20.1 will be zero. That step must be modified to:

STEP 20.1′: Calculate $r_{kk} = \|\mathbf{X}_k\|_2$. If $r_{kk} \neq 0$, then $\mathbf{Q}_k = \mathbf{X}_k / r_{kk}$. If $r_{kk} = 0$, then choose $\mathbf{Q}_k$ to be any normalized vector which is orthogonal to $\mathbf{Q}_1, \mathbf{Q}_2, \ldots, \mathbf{Q}_{k-1}$.

In practice, r_{kk} is rarely zero. Even if the columns of $\mathbf{A}$ are linearly dependent, roundoff will produce an r_{kk} value close to but not equal to zero, permitting $\mathbf{Q}_k$ to be calculated in the usual manner. The result is however, an incorrect vector. (See Problem 20.13.) Thus, whenever an r_{kk} value is sufficiently small, $\mathbf{Q}_k$ must be checked to guarantee it is orthogonal to the previously calculated $\mathbf{Q}$ vectors; if it is not, the modification given as Step 20.1′ must be implemented.

THE QR ALGORITHM

The **QR** *algorithm* is a procedure for determining all eigenvalues of a real matrix $\mathbf{A}_0$. The algorithm sequentially constructs matrices $\mathbf{A}_k$ $(k = 1, 2, 3, \ldots)$ by forming **QR** decompositions

$$\mathbf{A}_{k-1} = \mathbf{Q}_{k-1}\mathbf{R}_{k-1} \tag{20.2}$$

for $\mathbf{A}_{k-1}$, and then reversing the order of the products to define

$$\mathbf{A}_k = \mathbf{R}_{k-1}\mathbf{Q}_{k-1} \tag{20.3}$$

Each $\mathbf{A}_k$ is similar to its predecessor and has the same eigenvalues (see Problem 20.9). In general, the sequence $\{\mathbf{A}_k\}$ converges to a partitioned matrix having either of two forms:

$$\left[\begin{array}{c|c} \mathbf{E} & \mathbf{F} \\ \hline 0\,0\cdots 0 & a \end{array}\right] \tag{20.4}$$

and

$$\left[\begin{array}{c|c} \mathbf{G} & \mathbf{H} \\ \hline 0\,0\cdots 0 & b\;c \\ 0\,0\cdots 0 & d\;e \end{array}\right] \tag{20.5}$$

If form (*20.4*) occurs, then the element a is an eigenvalue, and the remaining eigenvalues are obtained by applying the **QR** algorithm anew to the matrix **E**. If form (*20.5*) arises, then two eigenvalues can be determined from the characteristic equation of the 2×2 submatrix in the lower right partition, and the remaining eigenvalues are obtained by applying the **QR** algorithm to the matrix **G**. If **E** or **G** is already a 2×2 matrix, its eigenvalues are determined from its characteristic equation.

ACCELERATING CONVERGENCE

Convergence of the **QR** algorithm is markedly accelerated by a shift at each iteration: If the matrices have order $n \times n$, then the element in the (n, n) position of $\mathbf{A}_{k-1}$ is denoted as s_{k-1}, and a **QR** decomposition is constructed for the shifted matrix $\mathbf{A}_{k-1} - s_{k-1}\mathbf{I}$. Equation (*20.2*) is modified to

$$\mathbf{A}_{k-1} - s_{k-1}\mathbf{I} = \mathbf{Q}_{k-1}\mathbf{R}_{k-1} \tag{20.6}$$

and (*20.3*) is replaced with

$$\mathbf{A}_k = \mathbf{R}_{k-1}\mathbf{Q}_{k-1} + s_{k-1}\mathbf{I} \tag{20.7}$$

Equations (*20.6*) and (*20.7*) constitute the *shifted* **QR** *algorithm*. (See Problems 20.5 through 20.8.)

For matrices of large order, significant computation time is also saved by first reducing the given matrix to Hessenberg form and then applying the shifted **QR** algorithm to it.

At each stage the **QR** (or shifted **QR**) algorithm is halted once the zeros in form (*20.4*) or (*20.5*) are obtained to whatever degree of precision is specified by the user. If the eigenvalues are real, only estimates are needed for them. These estimates are then incorporated into the shifted inverse power method (Chapter 19) to quickly obtain better values for the eigenvalue and corresponding eigenvector.

Solved Problems

20.1 Use the modified Gram-Schmidt process to construct an orthogonal set of vectors from the linearly independent set $\{\mathbf{X}_1, \mathbf{X}_2, \mathbf{X}_3\}$ when

$$\mathbf{X}_1 = \begin{bmatrix} -4 \\ 3 \\ 6 \end{bmatrix} \quad \mathbf{X}_2 = \begin{bmatrix} 2 \\ -3 \\ 6 \end{bmatrix} \quad \mathbf{X}_3 = \begin{bmatrix} 2 \\ 3 \\ 0 \end{bmatrix}$$

First iteration:

$$r_{11} = \|\mathbf{X}_1\|_2 = \sqrt{61} = 7.810250$$

$$\mathbf{Q}_1 = \frac{1}{r_{11}} \mathbf{X}_1 = \left[-\frac{4}{\sqrt{61}}, \frac{3}{\sqrt{61}}, \frac{6}{\sqrt{61}} \right]^T = [-0.512148, 0.384111, 0.768221]^T$$

$$r_{12} = \langle \mathbf{X}_2, \mathbf{Q}_1 \rangle = \frac{19}{\sqrt{61}} = 2.432701$$

$$r_{13} = \langle \mathbf{X}_3, \mathbf{Q}_1 \rangle = \frac{1}{\sqrt{61}} = 0.128037$$

$$\mathbf{X}_2 \leftarrow \mathbf{X}_2 - r_{12}\mathbf{Q}_1 = [198/61, -240/61, 252/61]^T$$

$$\mathbf{X}_3 \leftarrow \mathbf{X}_3 - r_{13}\mathbf{Q}_1 = [126/61, 180/61, -6/61]^T$$

Second iteration (using vectors from the first iteration):

$$r_{22} = \|\mathbf{X}_2\|_2 = \sqrt{(198/61)^2 + (-240/61)^2 + (252/61)^2} = \frac{\sqrt{160{,}308}}{61} = 6.563686$$

$$\mathbf{Q}_2 = \frac{1}{r_{22}} \mathbf{X}_2 = \left[\frac{198}{\sqrt{160{,}308}}, -\frac{240}{\sqrt{160{,}308}}, \frac{252}{\sqrt{160{,}308}} \right]^T = [0.494524, -0.599423, 0.629395]^T$$

$$r_{23} = \langle \mathbf{X}_3, \mathbf{Q}_2 \rangle = -\frac{324}{\sqrt{160{,}308}} = -0.809222$$

$$\mathbf{X}_3 \leftarrow \mathbf{X}_3 - r_{23}\mathbf{Q}_2 = \begin{bmatrix} 126/61 \\ 180/61 \\ -6/61 \end{bmatrix} - (-0.809222) \begin{bmatrix} 0.494524 \\ -0.599423 \\ 0.629395 \end{bmatrix} = \begin{bmatrix} 2.465753 \\ 2.465753 \\ 0.410959 \end{bmatrix}$$

Third iteration (using the vector from the second iteration):

$$r_{33} = \sqrt{(2.465753)^2 + (2.465753)^2 + (0.410959)^2} = 3.511234$$

$$\mathbf{Q}_3 = \frac{1}{r_{33}} \mathbf{X}_3 = [0.702247, 0.702247, 0.117041]^T$$

An orthonormal set is $\{\mathbf{Q}_1, \mathbf{Q}_2, \mathbf{Q}_3\}$.

20.2 Construct a **QR** decomposition for the matrix

$$\mathbf{A} = \begin{bmatrix} -4 & 2 & 2 \\ 3 & -3 & 3 \\ 6 & 6 & 0 \end{bmatrix}$$

From Problem 20.1, we have

$$\mathbf{Q} = \begin{bmatrix} -0.512148 & 0.494524 & 0.702247 \\ 0.384111 & -0.599423 & 0.702247 \\ 0.768221 & 0.629395 & 0.117041 \end{bmatrix} \quad \text{and} \quad \mathbf{R} = \begin{bmatrix} 7.810250 & 2.432701 & 0.128037 \\ 0 & 6.563686 & -0.809222 \\ 0 & 0 & 3.511234 \end{bmatrix}$$

A direct calculation shows that $\mathbf{A} = \mathbf{QR}$.

20.3 Use the modified Gram-Schmidt process to construct an orthogonal set of vectors from the linearly independent set $\{\mathbf{X}_1, \mathbf{X}_2, \mathbf{X}_3, \mathbf{X}_4\}$ when

$$\mathbf{X}_1 = \begin{bmatrix} 0 \\ 1 \\ 1 \\ 1 \end{bmatrix} \quad \mathbf{X}_2 = \begin{bmatrix} 1 \\ 0 \\ 1 \\ 1 \end{bmatrix} \quad \mathbf{X}_3 = \begin{bmatrix} 1 \\ 1 \\ 0 \\ 1 \end{bmatrix} \quad \mathbf{X}_4 = \begin{bmatrix} 1 \\ 1 \\ 1 \\ 0 \end{bmatrix}$$

First iteration:

$$r_{11} = \|\mathbf{X}_1\|_2 = \sqrt{3}$$

$$\mathbf{Q}_1 = \frac{1}{r_{11}} \mathbf{X}_1 = \left[0, \frac{1}{\sqrt{3}}, \frac{1}{\sqrt{3}}, \frac{1}{\sqrt{3}}\right]^T$$

$$r_{12} = \langle \mathbf{X}_2, \mathbf{Q}_1 \rangle = \frac{2}{\sqrt{3}}$$

$$r_{13} = \langle \mathbf{X}_3, \mathbf{Q}_1 \rangle = \frac{2}{\sqrt{3}}$$

$$r_{14} = \langle \mathbf{X}_4, \mathbf{Q}_1 \rangle = \frac{2}{\sqrt{3}}$$

$$\mathbf{X}_2 \leftarrow \mathbf{X}_2 - r_{12}\mathbf{Q}_1 = [1, -2/3, 1/3, 1/3]^T$$

$$\mathbf{X}_3 \leftarrow \mathbf{X}_3 - r_{13}\mathbf{Q}_1 = [1, 1/3, -2/3, 1/3]^T$$

$$\mathbf{X}_4 \leftarrow \mathbf{X}_4 - r_{14}\mathbf{Q}_1 = [1, 1/3, 1/3, -2/3]^T$$

Second iteration (using vectors from the first iteration):

$$r_{22} = \|\mathbf{X}_2\|_2 = \sqrt{(1)^2 + (-2/3)^2 + (1/3)^2 + (1/3)^2} = \frac{\sqrt{15}}{3}$$

$$\mathbf{Q}_2 = \frac{1}{r_{22}} \mathbf{X}_2 = \left[\frac{3}{\sqrt{15}}, \frac{-2}{\sqrt{15}}, \frac{1}{\sqrt{15}}, \frac{1}{\sqrt{15}}\right]^T$$

$$r_{23} = \langle \mathbf{X}_3, \mathbf{Q}_2 \rangle = \frac{2}{\sqrt{15}}$$

$$r_{24} = \langle \mathbf{X}_4, \mathbf{Q}_2 \rangle = \frac{2}{\sqrt{15}}$$

$$\mathbf{X}_3 \leftarrow \mathbf{X}_3 - r_{23}\mathbf{Q}_2 = \begin{bmatrix} 1 \\ 1/3 \\ -2/3 \\ 1/3 \end{bmatrix} - \frac{2}{\sqrt{15}} \begin{bmatrix} 3/\sqrt{15} \\ -2/\sqrt{15} \\ 1/\sqrt{15} \\ 1/\sqrt{15} \end{bmatrix} = \begin{bmatrix} 3/5 \\ 3/5 \\ -4/5 \\ 1/5 \end{bmatrix}$$

$$\mathbf{X}_4 \leftarrow \mathbf{X}_4 - r_{24}\mathbf{Q}_2 = \begin{bmatrix} 1 \\ 1/3 \\ 1/3 \\ -2/3 \end{bmatrix} - \frac{2}{\sqrt{15}} \begin{bmatrix} 3/\sqrt{15} \\ -2/\sqrt{15} \\ 1/\sqrt{15} \\ 1/\sqrt{15} \end{bmatrix} = \begin{bmatrix} 3/5 \\ 3/5 \\ 1/5 \\ -4/5 \end{bmatrix}$$

Third iteration (using vectors from the second iteration):

$$r_{33} = \|\mathbf{X}_3\|_2 = \sqrt{(3/5)^2 + (3/5)^2 + (-4/5)^2 + (1/5)^2} = \frac{\sqrt{35}}{5}$$

$$\mathbf{Q}_3 = \frac{1}{r_{33}}\mathbf{X}_3 = \left[\frac{3}{\sqrt{35}}, \frac{3}{\sqrt{35}}, \frac{-4}{\sqrt{35}}, \frac{1}{\sqrt{35}}\right]^T$$

$$r_{34} = \langle \mathbf{X}_4, \mathbf{Q}_3 \rangle = \frac{2}{\sqrt{35}}$$

$$\mathbf{X}_4 \leftarrow \mathbf{X}_4 - r_{34}\mathbf{Q}_3 = \begin{bmatrix} 3/5 \\ 3/5 \\ 1/5 \\ -4/5 \end{bmatrix} - \frac{2}{\sqrt{35}} \begin{bmatrix} 3/\sqrt{35} \\ 3/\sqrt{35} \\ -4/\sqrt{35} \\ 1/\sqrt{35} \end{bmatrix} = \begin{bmatrix} 3/7 \\ 3/7 \\ 3/7 \\ -6/7 \end{bmatrix}$$

Fourth iteration (using the vector from the third iteration):

$$r_{44} = \|\mathbf{X}_4\|_2 = \sqrt{(3/7)^2 + (3/7)^2 + (3/7)^2 + (-6/7)^2} = \frac{\sqrt{63}}{7}$$

$$\mathbf{Q}_4 = \frac{1}{r_{44}}\mathbf{X}_4 = \left[\frac{3}{\sqrt{63}}, \frac{3}{\sqrt{63}}, \frac{3}{\sqrt{63}}, \frac{-6}{\sqrt{63}}\right]^T = \left[\frac{1}{\sqrt{7}}, \frac{1}{\sqrt{7}}, \frac{1}{\sqrt{7}}, \frac{-2}{\sqrt{7}}\right]^T$$

An orthonormal set is $\{\mathbf{Q}_1, \mathbf{Q}_2, \mathbf{Q}_3, \mathbf{Q}_4\}$. (Compare with Problem 11.7.)

20.4 Construct a **QR** decomposition for the matrix

$$\mathbf{A} = \begin{bmatrix} 0 & 1 & 1 & 1 \\ 1 & 0 & 1 & 1 \\ 1 & 1 & 0 & 1 \\ 1 & 1 & 1 & 0 \end{bmatrix}$$

Using the results of Problem 20.3, we have

$$\mathbf{Q} = \begin{bmatrix} 0 & 3/\sqrt{15} & 3/\sqrt{35} & 1/\sqrt{7} \\ 1/\sqrt{3} & -2/\sqrt{15} & 3/\sqrt{35} & 1/\sqrt{7} \\ 1/\sqrt{3} & 1/\sqrt{15} & -4/\sqrt{35} & 1/\sqrt{7} \\ 1/\sqrt{3} & 1/\sqrt{15} & 1/\sqrt{35} & -2/\sqrt{7} \end{bmatrix} \quad \text{and} \quad \mathbf{R} = \begin{bmatrix} \sqrt{3} & 2/\sqrt{3} & 2/\sqrt{3} & 2/\sqrt{3} \\ 0 & \sqrt{15}/3 & 2/\sqrt{15} & 2/\sqrt{15} \\ 0 & 0 & \sqrt{35}/5 & 2/\sqrt{35} \\ 0 & 0 & 0 & \sqrt{63}/7 \end{bmatrix}$$

A direct calculation shows that $\mathbf{A} = \mathbf{QR}$.

20.5 Apply the shifted **QR** algorithm to

$$\mathbf{A} = \begin{bmatrix} 3 & 1 \\ 1 & 5 \end{bmatrix}$$

Using (*20.6*) and (*20.7*), we calculate the following:

First iteration:

$$\mathbf{A}_0 - 5\mathbf{I} = \begin{bmatrix} -2 & 1 \\ 1 & 0 \end{bmatrix} = \mathbf{Q}_0\mathbf{R}_0 = \begin{bmatrix} -0.894427 & 0.447214 \\ 0.447214 & 0.894427 \end{bmatrix} \begin{bmatrix} 2.236068 & -0.894427 \\ 0 & 0.447124 \end{bmatrix}$$

$$\mathbf{A}_1 = \mathbf{R}_0\mathbf{Q}_0 = 5\mathbf{I} = \begin{bmatrix} 2.236068 & -0.894427 \\ 0 & 0.447214 \end{bmatrix} \begin{bmatrix} -0.894427 & 0.447214 \\ 0.447214 & 0.894427 \end{bmatrix} + 5\begin{bmatrix} 1 & 0 \\ 0 & 1 \end{bmatrix} = \begin{bmatrix} 2.6 & 0.2 \\ 0.2 & 5.4 \end{bmatrix}$$

Second iteration:

$$\mathbf{A}_1 - 5.4\mathbf{I} = \begin{bmatrix} -2.8 & 0.2 \\ 0.2 & 0 \end{bmatrix} = \mathbf{Q}_1\mathbf{R}_1 = \begin{bmatrix} -0.997459 & 0.071247 \\ 0.071247 & 0.997459 \end{bmatrix}\begin{bmatrix} 2.807134 & -0.199492 \\ 0 & 0.014249 \end{bmatrix}$$

$$\mathbf{A}_2 = \mathbf{R}_1\mathbf{Q}_1 + 5.4\mathbf{I} = \begin{bmatrix} 2.807134 & -0.199492 \\ 0 & 0.014249 \end{bmatrix}\begin{bmatrix} -0.997459 & 0.071247 \\ 0.071247 & 0.997459 \end{bmatrix} + 5.4\begin{bmatrix} 1 & 0 \\ 0 & 1 \end{bmatrix}$$

$$= \begin{bmatrix} 2.585787 & 0.001015 \\ 0.001015 & 5.414213 \end{bmatrix}$$

Third iteration:

$$\mathbf{A}_2 - 5.414213\mathbf{I} = \begin{bmatrix} -2.828426 & 0.001015 \\ 0.001015 & 0 \end{bmatrix} = \mathbf{Q}_2\mathbf{R}_2 = \begin{bmatrix} -1.000000 & 0.000359 \\ 0.000359 & 1.000000 \end{bmatrix}\begin{bmatrix} 2.828427 & -0.001015 \\ 0 & 0.000000 \end{bmatrix}$$

$$\mathbf{A}_3 = \mathbf{R}_2\mathbf{Q}_2 + 5.414214\mathbf{I} = \begin{bmatrix} 2.828427 & -0.001015 \\ 0 & 0.000000 \end{bmatrix}\begin{bmatrix} -1.000000 & 0.000359 \\ 0.000359 & 1.000000 \end{bmatrix} + 5.414213\begin{bmatrix} 1 & 0 \\ 0 & 1 \end{bmatrix}$$

$$= \begin{bmatrix} 2.585786 & -0.000000 \\ 0.000000 & 5.414213 \end{bmatrix}$$

At this point we have generated form (*20.4*). It follows that one eigenvalue is 5.414213 and the second is 2.585786. (Observe the roundoff error in $\mathbf{Q}_2$, which results in columns that are only approximately unit vectors.)

20.6 Apply the shifted **QR** algorithm to

$$\mathbf{A}_0 = \begin{bmatrix} 5 & 2 & 2 \\ 3 & 6 & 3 \\ 6 & 6 & 9 \end{bmatrix}$$

Using (*20.6*) and (*20.7*), we obtain the following:

First iteration:

$$\mathbf{A}_0 - 9\mathbf{I} = \begin{bmatrix} -4 & 2 & 2 \\ 3 & -3 & 3 \\ 6 & 6 & 0 \end{bmatrix} = \mathbf{Q}_0\mathbf{R}_0$$

$$= \begin{bmatrix} -0.512148 & 0.494524 & 0.702247 \\ 0.384111 & -0.599423 & 0.702247 \\ 0.768221 & 0.629395 & 0.117041 \end{bmatrix}\begin{bmatrix} 7.810250 & 2.432701 & 0.128037 \\ 0 & 6.563686 & -0.809222 \\ 0 & 0 & 3.511234 \end{bmatrix}$$

as already shown in Problem 20.2. Then we have

$$\mathbf{A}_1 = \mathbf{R}_0\mathbf{Q}_0 + 9\mathbf{I} = \begin{bmatrix} 6.032787 & 2.484726 & 7.208066 \\ 1.899520 & 4.556254 & 4.514615 \\ 2.697405 & 2.209952 & 9.410959 \end{bmatrix}$$

Second iteration:

$$\mathbf{A}_1 - 9.410959\mathbf{I} = \begin{bmatrix} -3.378172 & 2.484726 & 7.208066 \\ 1.899520 & -4.854705 & 4.514615 \\ 2.697405 & 2.209952 & 0 \end{bmatrix} = \mathbf{Q}_1\mathbf{R}_1$$

$$= \begin{bmatrix} -0.715428 & 0.134590 & 0.685600 \\ 0.402280 & -0.722952 & 0.561704 \\ 0.571256 & 0.677662 & 0.463077 \end{bmatrix}\begin{bmatrix} 4.721887 & -2.468146 & -3.340716 \\ 0 & 5.341738 & -2.293716 \\ 0 & 0 & 7.477730 \end{bmatrix}$$

$$\mathbf{A}_2 = \mathbf{R}_1\mathbf{Q}_1 + 9.410959\mathbf{I} = \begin{bmatrix} 3.131498 & 0.155992 & 0.303949 \\ 0.838575 & 3.994775 & 1.983808 \\ 4.271696 & 5.067376 & 12.873727 \end{bmatrix}$$

Continuing in this manner, we generate

$$\mathbf{A}_3 = \begin{bmatrix} 3.174780 & 0.197407 & 4.602285 \\ 0.127357 & 3.143844 & 3.353551 \\ 0.405644 & 0.458158 & 13.681376 \end{bmatrix}$$

$$\mathbf{A}_4 = \begin{bmatrix} 2.995229 & -0.005512 & -4.152641 \\ -0.003362 & 2.996115 & -2.926538 \\ 0.012647 & 0.014613 & 14.008656 \end{bmatrix}$$

$$\mathbf{A}_5 = \begin{bmatrix} 2.999996 & -0.000004 & -4.166180 \\ -0.000003 & 2.999997 & -2.939863 \\ 0.000010 & 0.000011 & 14.000007 \end{bmatrix}$$

$$\mathbf{A}_6 = \begin{bmatrix} 3.000000 & -0.000000 & -4.166189 \\ -0.000000 & 3.000000 & -2.939868 \\ 0.000000 & 0.000000 & 14.000000 \end{bmatrix}$$

Convergence is established to the number of decimal places shown. One eigenvalue, 14, appears in the (3,3) position; it is obvious that the other two eigenvalues are 3 and 3. (Compare with Problem 19.2.)

20.7 Apply the shifted **QR** algorithm to

$$\mathbf{A}_0 = \begin{bmatrix} 1 & 3 & 2 & -1 \\ 1 & 1 & 2 & -3 \\ 3 & 1 & 1 & -1 \\ 2 & -2 & 1 & 2 \end{bmatrix}$$

After fifteen iterations of the algorithm, we obtain

$$\mathbf{A}_{15} = \begin{bmatrix} -1.942221 & 0.528619 & 1.156169 & 0.023944 \\ -1.278276 & -1.310739 & 0.878963 & 1.339487 \\ 0.000000 & 0.000000 & 4.639537 & -2.951749 \\ 0.000000 & -0.000000 & 0.630294 & 3.613422 \end{bmatrix}$$

which has form (*20.5*). The characteristic equation of the lower right 2×2 matrix

$$\begin{bmatrix} 4.639537 & -2.951749 \\ 0.630294 & 3.613422 \end{bmatrix}$$

is $\lambda^2 - 8.252959\lambda + 18.625075 = 0$, which has as its roots the eigenvalue estimates $4.126480 \pm i1.263820$. The upper left 2×2 submatrix has the characteristic equation $\lambda^2 + 3.252960\lambda + 3.221466 = 0$, which has as its roots the eigenvalue estimates $-1.626480 \pm i0.758965$.

20.8 Apply the shifted **QR** algorithm to

$$\mathbf{A}_0 = \begin{bmatrix} 10 & 7 & 8 & 7 \\ 7 & 5 & 6 & 5 \\ 8 & 6 & 10 & 9 \\ 7 & 5 & 9 & 10 \end{bmatrix}$$

After four iterations of the algorithm, we obtain

$$\mathbf{A}_4 = \begin{bmatrix} 30.288604 & 0.049230 & 0.006981 & -0.000000 \\ 0.049230 & 0.021553 & 0.096466 & 0.000000 \\ 0.006981 & 0.096466 & 0.831786 & -0.000000 \\ 0.000000 & -0.000000 & 0.000000 & 3.858057 \end{bmatrix}$$

This matrix has form (*20.4*), so one eigenvalue estimate is 3.858057. To determine the others, we apply the shifted **QR** algorithm anew to

$$\mathbf{E}_0 = \begin{bmatrix} 30.288604 & 0.049230 & 0.006981 \\ 0.049230 & 0.021553 & 0.096466 \\ 0.006981 & 0.096466 & 0.831786 \end{bmatrix}$$

After two iterations, we obtain

$$\mathbf{E}_2 = \begin{bmatrix} 30.288686 & 0.000038 & 0.000000 \\ 0.000038 & 0.010150 & 0.000000 \\ 0.000000 & 0.000000 & 0.843107 \end{bmatrix}$$

so a second eigenvalue estimate is 0.843107. The characteristic equation of the upper left 2×2 submatrix in $\mathbf{E}_2$ is $\lambda^2 - 30.298836\lambda + 0.307430 = 0$, which can be solved explicitly to obtain 30.288686 and 0.010150 as the two remaining eigenvalues. (Compare these values with those obtained in Problems 19.6 and 19.7.)

20.9 Show that the shifted **QR** algorithm is a series of similarity transformations that leave the eigenvalues invariant.

Since the **Q** matrix in any **QR** decomposition is unitary, it has an inverse. Therefore, (*20.6*) may be rewritten as

$$\mathbf{R}_{k-1} = \mathbf{Q}_{k-1}^{-1}(\mathbf{A}_{k-1} - s_{k-1}\mathbf{I})$$

Substituting this equation into (*20.7*), we obtain

$$\begin{aligned}\mathbf{A}_k = \mathbf{Q}_{k-1}^{-1}(\mathbf{A}_{k-1} - s_{k-1}\mathbf{I})\mathbf{Q}_{k-1} + s_{k-1}\mathbf{I} &= \mathbf{Q}_{k-1}^{-1}\mathbf{A}_{k-1}\mathbf{Q}_{k-1} - s_{k-1}\mathbf{Q}_{k-1}^{-1}\mathbf{I}\mathbf{Q}_{k-1} + s_{k-1}\mathbf{I} \\ &= \mathbf{Q}_{k-1}^{-1}\mathbf{A}_{k-1}\mathbf{Q}_{k-1}\end{aligned}$$

Therefore, $\mathbf{A}_k$ is similar to $\mathbf{A}_{k-1}$, and the invariance of their eigenvalues follows from Property 10.1.

20.10 Working to four significant figures, show that the unmodified Gram-Schmidt orthogonalization process does not generate an orthogonal set of vectors when applied to

$$\mathbf{X}_1 = \begin{bmatrix} 1 \\ 1.01 \\ 1 \\ 1 \end{bmatrix} \qquad \mathbf{X}_2 = \begin{bmatrix} 1 \\ 1 \\ 1.01 \\ 1 \end{bmatrix} \qquad \mathbf{X}_3 = \begin{bmatrix} 1 \\ 1 \\ 1 \\ 1.01 \end{bmatrix}$$

Using the algorithm given in Chapter 11 and rounding all stored numerical values to four significant digits, we obtain

$$\sqrt{\langle \mathbf{X}_1, \mathbf{X}_1 \rangle} = 2.005$$

so $$\mathbf{Q}_1 = \frac{1}{2.005}\mathbf{X}_1 = [0.4988, 0.5037, 0.4988, 0.4988]^T$$

Then $$\begin{aligned}\mathbf{Y}_2 &= \mathbf{X}_2 - \langle \mathbf{X}_2, \mathbf{Q}_1 \rangle \mathbf{Q}_1 = \mathbf{X}_2 - 2.005\mathbf{Q}_1 \\ &= [-0.9400 \times 10^{-4}, -0.9919 \times 10^{-2}, 0.9906 \times 10^{-2}, -0.9400 \times 10^{-4}]^T\end{aligned}$$

and $$\sqrt{\langle \mathbf{Y}_2, \mathbf{Y}_2 \rangle} = 0.1402 \times 10^{-1}$$

so $$\mathbf{Q}_2 = \frac{1}{0.1402 \times 10^{-1}}\mathbf{Y}_2 = [-0.6705 \times 10^{-2}, -0.7075, 0.7066, -0.6705 \times 10^{-2}]^T$$

Lastly, $$\begin{aligned}\mathbf{Y}_3 &= \mathbf{X}_3 - \langle \mathbf{X}_3, \mathbf{Q}_1 \rangle \mathbf{Q}_1 - \langle \mathbf{X}_3, \mathbf{Q}_2 \rangle \mathbf{Q}_2 = \mathbf{X}_3 - 2.005\mathbf{Q}_1 - (-0.1438 \times 10^{-1})\mathbf{Q}_2 \\ &= [-0.1904 \times 10^{-3}, -0.2009 \times 10^{-1}, 0.1007 \times 10^{-1}, 0.9810 \times 10^{-2}]^T\end{aligned}$$

and $\sqrt{\langle \mathbf{Y}_3, \mathbf{Y}_3 \rangle} = 0.2452 \times 10^{-1}$

so $$\mathbf{Q}_3 = \frac{1}{0.2452 \times 10^{-1}} \mathbf{Y}_3 = [-0.7765 \times 10^{-2}, -0.8193, 0.4107, 0.4001]^T$$

For these vectors, $\langle \mathbf{Q}_2, \mathbf{Q}_3 \rangle = 0.8672$, which is not near zero as it should be. Similar results are obtained wherever the components 1.01 are of the form $1 + 10^{-k}$ and all numerical values are rounded to $2k$ significant digits.

20.11 Redo Problem 20.10 using the modified Gram-Schmidt process and show that the results are better.

First iteration:

$$r_1 = \|\mathbf{X}_1\|_2 = 2.005$$

$$\mathbf{Q}_1 = \frac{1}{2.005} \mathbf{X}_1 = [0.4988, 0.5037, 0.4988, 0.4988]^T$$

$$r_{12} = \langle \mathbf{X}_2, \mathbf{Q}_1 \rangle = 2.005$$

$$r_{13} = \langle \mathbf{X}_3, \mathbf{Q}_1 \rangle = 2.005$$

$$\mathbf{X}_2 \leftarrow \mathbf{X}_2 - 2.005\mathbf{Q}_1 = [-0.9400 \times 10^{-4}, -0.9919 \times 10^{-2}, 0.9906 \times 10^{-2}, -0.9400 \times 10^{-4}]^T$$

$$\mathbf{X}_3 \leftarrow \mathbf{X}_3 - 2.005\mathbf{Q}_1 = [-0.9400 \times 10^{-4}, -0.9919 \times 10^{-2}, -0.9400 \times 10^{-4}, 0.9906 \times 10^{-2}]^T$$

Second iteration:

$$r_{22} = \|\mathbf{X}_2\|_2 = 0.1402 \times 10^{-1}$$

$$\mathbf{Q}_2 = \frac{1}{0.1402 \times 10^{-1}} \mathbf{X}_2 = [-0.6705 \times 10^{-2}, -0.7075, 0.7066, -0.6705 \times 10^{-2}]^T$$

$$r_{23} = \langle \mathbf{X}_3, \mathbf{Q}_2 \rangle = 0.6885 \times 10^{-2}$$

$$\mathbf{X}_3 \leftarrow \mathbf{X}_3 - 0.6885 \times 10^{-2}\mathbf{Q}_2 = [0.4784 \times 10^{-4}, -0.5048 \times 10^{-2}, -0.4959 \times 10^{-2}, 0.9952 \times 10^{-2}]^T$$

Third iteration:

$$r_{33} = \|\mathbf{X}_3\|_2 = 0.1221 \times 10^{-1}$$

$$\mathbf{Q}_3 = \frac{1}{0.1221 \times 10^{-1}} \mathbf{X}_3$$

$$= [0.3918 \times 10^{-2}, -0.4134, -0.4061, 0.8151]$$

For these vectors, $\langle \mathbf{Q}_2, \mathbf{Q}_3 \rangle = 0.00003872$, which is much better than the result obtained in Problem 20.10. All other inner products formed from the vectors obtained here are at least accurate as those formed from vectors found in Problem 20.10.

20.12 Show that the modified Gram-Schmidt process yields the same vectors as the unmodified process in the absence of rounding

The proof is by induction. $\mathbf{Q}_1$ is the same for both methods because it is computed in exactly the same way in both algorithms. Assume that the two methods provide identical results for $\mathbf{Q}_1, \mathbf{Q}_2, \ldots, \mathbf{Q}_k$. We need only show that they do so for $\mathbf{Q}_{k+1}$ as well.

It follows from the induction hypothesis that $\mathbf{Q}_1, \mathbf{Q}_2, \ldots, \mathbf{Q}_k$ are mutually orthogonal because they are the vectors obtained from the unmodified Gram-Schmidt process. Let $\mathbf{X}_i^{(j)}$ $(i > j)$ denote the value of $\mathbf{X}_i$ after it has been altered by j iterations of the modified process. Then, for the modified process, we have

$$\mathbf{X}_{k+1}^{(1)} = \mathbf{X}_{k+1} - \langle \mathbf{X}_{k+1}, \mathbf{Q}_1 \rangle \mathbf{Q}_1 \qquad (1)$$
$$\mathbf{X}_{k+1}^{(2)} = \mathbf{X}_{k+1}^{(1)} - \langle \mathbf{X}_{k+1}^{(1)}, \mathbf{Q}_2 \rangle \mathbf{Q}_2 \qquad (2)$$
$$\mathbf{X}_{k+1}^{(3)} = \mathbf{X}_{k+1}^{(2)} - \langle \mathbf{X}_{k+1}^{(2)}, \mathbf{Q}_3 \rangle \mathbf{Q}_3 \qquad (3)$$
$$\cdots\cdots\cdots\cdots\cdots\cdots$$
$$\mathbf{X}_{k+1}^{(k)} = \mathbf{X}_{k+1}^{(k-1)} - \langle \mathbf{X}_{k+1}^{(k-1)}, \mathbf{Q}_k \rangle \mathbf{Q}_k$$

Substituting (*1*) into (*2*) and noting that $\mathbf{Q}_1$ and $\mathbf{Q}_2$ are orthogonal, we obtain

$$\begin{aligned}\mathbf{X}_{k+1}^{(2)} &= \mathbf{X}_{k+1} - \langle \mathbf{X}_{k+1}, \mathbf{Q}_1 \rangle \mathbf{Q}_1 - \langle (\mathbf{X}_{k+1} - \langle \mathbf{X}_{k+1}, \mathbf{Q}_1 \rangle \mathbf{Q}_1), \mathbf{Q}_2 \rangle \mathbf{Q}_2 \\ &= \mathbf{X}_{k+1} - \langle \mathbf{X}_{k+1}, \mathbf{Q}_1 \rangle \mathbf{Q}_1 - \langle \mathbf{X}_{k+1}, \mathbf{Q}_2 \rangle \mathbf{Q}_2 + \langle \mathbf{X}_{k+1}, \mathbf{Q}_1 \rangle \langle \mathbf{Q}_1, \mathbf{Q}_2 \rangle \mathbf{Q}_2 \\ &= \mathbf{X}_{k+1} - \langle \mathbf{X}_{k+1}, \mathbf{Q}_1 \rangle \mathbf{Q}_1 - \langle \mathbf{X}_{k+1}, \mathbf{Q}_2 \rangle \mathbf{Q}_2\end{aligned}$$

Substituting this result into (*3*) and noting that $\mathbf{Q}_3$ is orthogonal to both $\mathbf{Q}_1$ and $\mathbf{Q}_2$, we obtain

$$\begin{aligned}\mathbf{X}_{k+1}^{(3)} &= \mathbf{X}_{k+1} - \langle \mathbf{X}_{k+1}, \mathbf{Q}_1 \rangle \mathbf{Q}_1 - \langle \mathbf{X}_{k+1}, \mathbf{Q}_2 \rangle \mathbf{Q}_2 - \langle (\mathbf{X}_{k+1} - \langle \mathbf{X}_{k+1}, \mathbf{Q}_1 \rangle \mathbf{Q}_1 - \langle \mathbf{X}_{k+1}, \mathbf{Q}_2 \rangle \mathbf{Q}_2), \mathbf{Q}_3 \rangle \mathbf{Q}_3 \\ &= \mathbf{X}_{k+1} - \langle \mathbf{X}_{k+1}, \mathbf{Q}_1 \rangle \mathbf{Q}_1 - \langle \mathbf{X}_{k+1}, \mathbf{Q}_2 \rangle \mathbf{Q}_2 - \langle \mathbf{X}_{k+1}, \mathbf{Q}_3 \rangle \mathbf{Q}_3 + \langle \mathbf{X}_{k+1}, \mathbf{Q}_1 \rangle \langle \mathbf{Q}_1, \mathbf{Q}_3 \rangle \mathbf{Q}_3 \\ &\quad + \langle \mathbf{X}_{k+1}, \mathbf{Q}_2 \rangle \langle \mathbf{Q}_2, \mathbf{Q}_3 \rangle \mathbf{Q}_3 \\ &= \mathbf{X}_{k+1} - \langle \mathbf{X}_{k+1}, \mathbf{Q}_1 \rangle \mathbf{Q}_1 - \langle \mathbf{X}_{k+1}, \mathbf{Q}_2 \rangle \mathbf{Q}_2 - \langle \mathbf{X}_{k+1}, \mathbf{Q}_3 \rangle \mathbf{Q}_3\end{aligned}$$

Continuing in this manner, we find that $\mathbf{X}_{k+1}^{(k)}$ is identical to $\mathbf{Y}_{k+1}$ in the unmodified Gram-Schmidt process, and, since $\mathbf{Q}_{k+1}$ is obtained in one method by normalizing $\mathbf{X}_{k+1}^{(k)}$ and in the other by normalizing $\mathbf{Y}_{k+1}$, it follows that $\mathbf{Q}_{k+1}$ is the same in both methods.

20.13 Construct a **QR** decomposition for

$$\mathbf{A} = \begin{bmatrix} 1 & 0 & 1 \\ 0 & 1 & 1 \\ 1 & 1 & 2 \end{bmatrix}$$

working to six significant digits, and show how roundoff error can generate an incorrect **Q** matrix when the columns of **A** are linearly dependent.

Designate the columns of **A** as $\mathbf{X}_1, \mathbf{X}_2$, and $\mathbf{X}_3$ from left to right. Then:

***First iteration*:**

$$r_{11} = \|\mathbf{X}_1\|_2 = \sqrt{2} = 0.141421 \times 10^1$$

$$\mathbf{Q}_1 = \frac{1}{r_{11}} \mathbf{X}_1 = [0.707109,\ 0,\ 0.707109]^T$$

$$r_{12} = \langle \mathbf{X}_2, \mathbf{Q}_1 \rangle = 0.707109$$
$$r_{13} = \langle \mathbf{X}_3, \mathbf{Q}_1 \rangle = 0.212133 \times 10^1$$
$$\mathbf{X}_2 \leftarrow \mathbf{X}_2 - r_{12}\mathbf{Q}_1 = [-0.500003,\ 1,\ 0.499997]^T$$
$$\mathbf{X}_3 \leftarrow \mathbf{X}_3 - r_{13}\mathbf{Q}_1 = [-0.500012,\ 1,\ 0.499988]^T$$

***Second iteration*:**

$$r_{22} = \|\mathbf{X}_2\|_2 = 0.122474 \times 10^1$$

$$\mathbf{Q}_2 = \frac{1}{r_{22}} \mathbf{X}_2 = [-0.408252,\ 0.816500,\ 0.408247]^T$$

$$r_{23} = \langle \mathbf{X}_3, \mathbf{Q}_2 \rangle = 0.122475 \times 10^1$$
$$\mathbf{X}_3 \leftarrow \mathbf{X}_3 - r_{23}\mathbf{Q}_2 = [-0.536300 \times 10^{-5},\ -0.837500 \times 10^{-5},\ -0.125133 \times 10^{-4}]^T$$

***Third iteration*:**

$$r_{33} = \|\mathbf{X}_3\|_2 = 0.159839 \times 10^{-4}$$

$$\mathbf{Q}_3 = \frac{1}{r_{33}} \mathbf{X}_3 = [-0.335525, -0.523965, -0.782869]^T$$

Observe that r_{33} is very close to zero, and the last $\mathbf{X}_3$ vector is very close to the zero vector; if we were not rounding intermediate results, they would not exist. However, because of the rounding neither is zero, and $\mathbf{Q}_3$ can be calculated with what are, in effect, error terms. The result is a vector which is not orthogonal to either $\mathbf{Q}_1$ or $\mathbf{Q}_2$.

Supplementary Problems

20.14 Construct **QR** decompositions for the following matrices:

(*a*) $\begin{bmatrix} -4 & 4 & 2 \\ 4 & -4 & 1 \\ 2 & 1 & 0 \end{bmatrix}$ (*b*) $\begin{bmatrix} 4 & 3 & -2 \\ 3 & 1 & -1 \\ -2 & -1 & 0 \end{bmatrix}$ (*c*) $\begin{bmatrix} 0 & -1 & 0 \\ -1 & 0 & -1 \\ 0 & -1 & 0 \end{bmatrix}$

(*d*) $\begin{bmatrix} 25 & -131 & -86 \\ 11 & -41 & -18 \\ -4 & 28 & 0 \end{bmatrix}$ (*e*) $\begin{bmatrix} 0 & -1 & 0 & 0 \\ -1 & 0 & -1 & 0 \\ 0 & -1 & 0 & -1 \\ 0 & 0 & -1 & 0 \end{bmatrix}$ (*f*) $\begin{bmatrix} 880 & 0 & 990 & 0 \\ 990 & 660 & 330 & 330 \\ 0 & -25 & 440 & -25 \\ 0 & -880 & -8558 & 0 \end{bmatrix}$

In Problems 20.15 through 20.24, use the shifted **QR** algorithm to determine estimates for all eigenvalues of the given matrix.

20.15 $\begin{bmatrix} 4 & 4 & 2 \\ 4 & 4 & 1 \\ 2 & 1 & 8 \end{bmatrix}$ **20.16** $\begin{bmatrix} 8 & 3 & -2 \\ 3 & 5 & -1 \\ -2 & -1 & 4 \end{bmatrix}$ **20.17** $\begin{bmatrix} 2 & -1 & 0 \\ -1 & 2 & -1 \\ 0 & -1 & 2 \end{bmatrix}$

20.18 $\begin{bmatrix} 42 & -131 & -86 \\ 11 & -24 & -18 \\ -4 & 28 & 17 \end{bmatrix}$ **20.19** $\begin{bmatrix} 2 & -1 & 0 & 0 \\ -1 & 2 & -1 & 0 \\ 0 & -1 & 2 & -1 \\ 0 & 0 & -1 & 2 \end{bmatrix}$ **20.20** $\begin{bmatrix} 990 & 0 & 990 & 0 \\ 990 & 770 & 330 & 330 \\ 0 & -25 & 550 & -25 \\ 0 & -880 & -8{,}558 & 110 \end{bmatrix}$

20.21 $\begin{bmatrix} 9 & -2 & 2 & 1 \\ 2 & 8 & -1 & 0 \\ 1 & 0 & 8 & 2 \\ -1 & 1 & 0 & 11 \end{bmatrix}$ **20.22** $\begin{bmatrix} 0 & 1 & 0 & 0 \\ 0 & 0 & 1 & 0 \\ 0 & 0 & 0 & 1 \\ -4 & -2 & -5 & -2 \end{bmatrix}$ **20.23** $\begin{bmatrix} 1 & 0 & -1 & 0 \\ 0 & 2 & 0 & 4 \\ 5 & 0 & 3 & 0 \\ 0 & -7 & 0 & -4 \end{bmatrix}$

20.24 $\begin{bmatrix} 100 & 42 & 54 & 66 \\ 42 & 100 & 32 & 44 \\ 54 & 32 & 100 & 22 \\ 66 & 44 & 22 & 100 \end{bmatrix}$

Chapter 21

Generalized Inverses

PROPERTIES

The (Moore-Penrose) *generalized inverse* (or *pseudoinverse*) of a matrix $\mathbf{A}$, not necessarily square, is a matrix $\mathbf{A}^+$ that satisfies the conditions:

(I1): $\mathbf{AA}^+$ and $\mathbf{A}^+\mathbf{A}$ are Hermitian.

(I2): $\mathbf{AA}^+\mathbf{A} = \mathbf{A}$.

(I3): $\mathbf{A}^+\mathbf{AA}^+ = \mathbf{A}^+$.

A generalized inverse exists for every matrix. If $\mathbf{A}$ has order $n \times m$, then $\mathbf{A}^+$ has order $m \times n$ and has the following properties:

Property 21.1: $\mathbf{A}^+$ is unique.

Property 21.2: $\mathbf{A}^+ = \mathbf{A}^{-1}$ for nonsingular $\mathbf{A}$.

Property 21.3: $(\mathbf{A}^+)^+ = \mathbf{A}$.

Property 21.4: $(k\mathbf{A})^+ = (1/k)\mathbf{A}^+$ for $k \neq 0$.

Property 21.5: $(\mathbf{A}^H)^+ = (\mathbf{A}^+)^H$.

Property 21.6: $\mathbf{0}^+ = \mathbf{0}$.

Property 21.7: The rank of $\mathbf{A}^+$ equals the rank of $\mathbf{A}$.

Property 21.8: If $\mathbf{P}$ and $\mathbf{Q}$ are unitary matrices of appropriate orders so that the product $\mathbf{PAQ}$ is defined, then $(\mathbf{PAQ})^+ = \mathbf{Q}^H\mathbf{A}^+\mathbf{P}^H$.

Property 21.9: If $\mathbf{A}$ has order $m \times k$, $\mathbf{B}$ has order $k \times n$, and both matrices have rank k, then $(\mathbf{AB})^+ = \mathbf{B}^+\mathbf{A}^+$.

Property 21.10: For square matrix $\mathbf{A}$, $\mathbf{AA}^+ = \mathbf{A}^+\mathbf{A}$ if and only if $\mathbf{A}^+$ can be expressed as a polynomial in $\mathbf{A}$.

(See Problems 21.13, 21.14, 21.17, and 21.36 to 21.39.)

A FORMULA FOR GENERALIZED INVERSES

The following procedure will provide the generalized inverse for any matrix $\mathbf{A}$:

STEP 21.1: Determine the rank of $\mathbf{A}$, and denote it as K.

STEP 21.2: Locate a $K \times K$ submatrix of $\mathbf{A}$ having rank K.

STEP 21.3: Through a sequence of elementary row and column operations of the first kind (E1), move the submatrix identified in Step 21.2 into the upper left portion of $\mathbf{A}$. That is, determine

$$\mathbf{PAQ} = \left[\begin{array}{c:c} \mathbf{A}_{11} & \mathbf{A}_{12} \\ \hdashline \mathbf{A}_{21} & \mathbf{A}_{22} \end{array}\right]$$

where $\mathbf{P}$ and $\mathbf{Q}$ are each the product of elementary matrices of the first kind, and $\mathbf{A}_{11}$ is a submatrix of $\mathbf{A}$ that is nonsingular and of rank K. If no elementary operations were necessary, then $\mathbf{P}$ and $\mathbf{Q}$ are identity matrices. $\mathbf{A}_{12}$, $\mathbf{A}_{21}$, or $\mathbf{A}_{22}$ may be empty.

STEP 21.4: Set $\mathbf{B} = \left[\begin{array}{c} \mathbf{A}_{11} \\ \hdashline \mathbf{A}_{21} \end{array}\right]$, $\mathbf{F} = \mathbf{A}_{11}^{-1}\mathbf{A}_{12}$, and $\mathbf{C} = [\mathbf{I}_K | \mathbf{F}]$, where $\mathbf{I}_K$ is the $K \times K$ identity matrix.

STEP 21.5: $\mathbf{A}^+ = \mathbf{Q}\mathbf{C}^H(\mathbf{C}\mathbf{C}^H)^{-1}(\mathbf{B}^H\mathbf{B})^{-1}\mathbf{B}^H\mathbf{P}$ (*21.1*)

(See Problems 21.1, 21.2, and 21.16.) When the columns of **A** form a linearly independent set of vectors, (*21.1*) reduces to

$$\mathbf{A}^+ = (\mathbf{A}^H\mathbf{A})^{-1}\mathbf{A}^H \qquad (21.2)$$

(See Problem 21.3.)

SINGULAR-VALUE DECOMPOSITION

Equations (*21.1*) and (*21.2*) are useful formulas for calculating generalized inverses. However, they are not stable when roundoff error is involved, because small errors in the elements of a matrix **A** can result in large errors in the computed elements of $\mathbf{A}^+$. (See Problem 21.12.) In such situtations a better algorithm exists.

For any matrix **A**, not necessarily square, the product $\mathbf{A}^H\mathbf{A}$ is normal and has nonnegative eigenvalues (see Problems 13.2 and 13.3). The positive square roots of these eigenvalues are the *singular values* of **A**. Moreover, there exist unitary matrices **U** and **V** such that

$$\mathbf{A} = \mathbf{U}\left[\begin{array}{c|c} \mathbf{D} & \mathbf{0} \\ \hline \mathbf{0} & \mathbf{0} \end{array}\right]\mathbf{V}^H \qquad (21.3)$$

where **D** is a diagonal matrix having as its main diagonal all the positive singular values of **A**. The block diagonal matrix

$$\mathbf{\Sigma} = \left[\begin{array}{c|c} \mathbf{D} & \mathbf{0} \\ \hline \mathbf{0} & \mathbf{0} \end{array}\right]$$

has the same order as **A** and, therefore, is square only when **A** is square.

Equation (*21.3*) is a *singular-value decomposition* for **A**. An algorithm for constructing such a decomposition is the following:

STEP 21.6: Determine the eigenvalues of $\mathbf{A}^H\mathbf{A}$ and a canonical basis of orthonormal eigenvectors for $\mathbf{A}^H\mathbf{A}$.

STEP 21.7: Construct **D** as a square diagonal matrix whose diagonal elements are the positive singular values of **A**.

STEP 21.8: Set $\mathbf{V} = [\mathbf{V}_1 \mid \mathbf{V}_2]$, where the columns of $\mathbf{V}_1$ are the eigenvectors identified in Step 21.6 that correspond to positive eigenvalues, and the columns of $\mathbf{V}_2$ are the remaining eigenvectors.

STEP 21.9: Calculate $\mathbf{U}_1 = \mathbf{A}\mathbf{V}_1\mathbf{D}^{-1}$.

STEP 21.10: Augment onto $\mathbf{U}_1$ the identity matrix having the same number of rows as $\mathbf{U}_1$.

STEP 21.11: Identify those columns of the augmented matrix that form a maximal set of linearly independent column vectors, and delete the others. Orthonormalize the columns that remain, and denote the resulting matrix as **U**.

If **A** is real, then both **U** and **V** may be chosen to be orthogonal. (See Problems 21.4 and 21.5.)

A STABLE FORMULA FOR THE GENERALIZED INVERSE

Decomposition (*21.3*) generates the numerically stable formula

$$\mathbf{A}^+ = \mathbf{V}\left[\begin{array}{c|c} \mathbf{D}^{-1} & \mathbf{0} \\ \hline \mathbf{0} & \mathbf{0} \end{array}\right]\mathbf{U}^H$$

which can be simplified to

$$\mathbf{A}^{+} = \mathbf{V}_1\mathbf{D}^{-1}\mathbf{U}_1^H \tag{21.4}$$

where $\mathbf{V}_1$ and $\mathbf{U}_1$ are defined by Steps 21.8 and 21.9, respectively. For the purpose of calculating a generalized inverse, Steps 21.10 and 21.11 can be ignored. (See Problems 21.6 and 21.7.)

LEAST-SQUARES SOLUTIONS

A *least-squares solution* to a set of simultaneous linear equations $\mathbf{AX} = \mathbf{B}$ is the vector of smallest Euclidean norm that minimizes $\|\mathbf{AX} - \mathbf{B}\|_2$. That vector is

$$\mathbf{X} = \mathbf{A}^{+}\mathbf{B} \tag{21.5}$$

When $\mathbf{A}$ has an inverse, (*21.5*) reduces to $\mathbf{X} = \mathbf{A}^{-1}\mathbf{B}$, which is the unique solution. For consistent systems (see Chapter 2) that admit infinitely many solutions, (*21.5*) identifies the solution having minimum Euclidean norm. Equation (*21.5*) also identifies a solution for inconsistent systems, the one that is best in the least-squares sense. (See Problems 21.8 through 21.11.)

Solved Problems

21.1 Find the generalized inverse of

$$\mathbf{A} = \begin{bmatrix} 2 & 2 & -2 \\ 2 & 2 & -2 \\ -2 & -2 & 6 \end{bmatrix}$$

Using Steps 21.1 through 21.5, we first determine that $\mathbf{A}$ has rank 2. A 2×2 submatrix of $\mathbf{A}$ having rank 2 is obtained by deleting the second row and second column of $\mathbf{A}$. This submatrix can be moved into the upper left position by interchanging the order of the second and third rows and then the second and third columns. Then, setting

$$\mathbf{P} = \mathbf{Q} = \begin{bmatrix} 1 & 0 & 0 \\ 0 & 0 & 1 \\ 0 & 1 & 0 \end{bmatrix} \qquad \text{gives us} \qquad \mathbf{PAQ} = \left[\begin{array}{cc:c} 2 & -2 & 2 \\ -2 & 6 & -2 \\ \hdashline 2 & -2 & 2 \end{array}\right]$$

where $\qquad \mathbf{A}_{11} = \begin{bmatrix} 2 & -2 \\ -2 & 6 \end{bmatrix} \qquad \mathbf{A}_{12} = \begin{bmatrix} 2 \\ -2 \end{bmatrix} \qquad$ and $\qquad \mathbf{A}_{21} = [2, -2]$

and $\mathbf{A}_{11}$ has rank 2. Then

$$\mathbf{B} = \left[\begin{array}{cc} 2 & -2 \\ -2 & 6 \\ \hdashline 2 & -2 \end{array}\right] \qquad \mathbf{F} = \mathbf{A}_{11}^{-1}\mathbf{A}_{12} = \begin{bmatrix} 6/8 & 2/8 \\ 2/8 & 2/8 \end{bmatrix}\begin{bmatrix} 2 \\ -2 \end{bmatrix} = \begin{bmatrix} 1 \\ 0 \end{bmatrix} \qquad \text{and} \qquad \mathbf{C} = \left[\begin{array}{cc:c} 1 & 0 & 1 \\ 0 & 1 & 0 \end{array}\right]$$

so $\qquad (\mathbf{CC}^H)^{-1} = \begin{bmatrix} 2 & 0 \\ 0 & 1 \end{bmatrix}^{-1} = \begin{bmatrix} 1/2 & 0 \\ 0 & 1 \end{bmatrix} \qquad$ and $\qquad (\mathbf{B}^H\mathbf{B})^{-1} = \begin{bmatrix} 12 & -20 \\ -20 & 44 \end{bmatrix}^{-1} = \begin{bmatrix} 11/32 & 5/32 \\ 5/32 & 3/32 \end{bmatrix}$

and

$$\begin{aligned} \mathbf{A}^{+} &= \mathbf{QC}^H(\mathbf{CC}^H)^{-1}(\mathbf{B}^H\mathbf{B})^{-1}\mathbf{B}^H\mathbf{P} \\ &= \begin{bmatrix} 1 & 0 & 0 \\ 0 & 0 & 1 \\ 0 & 1 & 0 \end{bmatrix}\begin{bmatrix} 1 & 0 \\ 0 & 1 \\ 1 & 0 \end{bmatrix}\begin{bmatrix} 1/2 & 0 \\ 0 & 1 \end{bmatrix}\begin{bmatrix} 11/32 & 5/32 \\ 5/32 & 3/32 \end{bmatrix}\begin{bmatrix} 2 & -2 & 2 \\ -2 & 6 & -2 \end{bmatrix}\begin{bmatrix} 1 & 0 & 0 \\ 0 & 0 & 1 \\ 0 & 1 & 0 \end{bmatrix} \\ &= \begin{bmatrix} 3/16 & 3/16 & 1/8 \\ 3/16 & 3/16 & 1/8 \\ 1/8 & 1/8 & 1/4 \end{bmatrix} \end{aligned}$$

21.2 Find the generalized inverse of

$$\mathbf{A} = \begin{bmatrix} 0 & 0 & 1 & 2 \\ 1 & 2 & 2 & 3 \end{bmatrix}$$

The matrix has rank 2. A 2 × 2 submatrix of **A** having rank 2 (but not the only one) is obtained by deleting the second and fourth columns of **A**. This submatrix can be moved into the upper left position by interchanging the order of the second and third columns. Then, setting

$$\mathbf{P} = \begin{bmatrix} 1 & 0 \\ 0 & 1 \end{bmatrix} \quad \text{and} \quad \mathbf{Q} = \begin{bmatrix} 1 & 0 & 0 & 0 \\ 0 & 0 & 1 & 0 \\ 0 & 1 & 0 & 0 \\ 0 & 0 & 0 & 1 \end{bmatrix}$$

we get
$$\mathbf{PAQ} = \left[\begin{array}{cc:cc} 0 & 1 & 0 & 2 \\ 1 & 2 & 2 & 3 \end{array}\right] \quad \text{with} \quad \mathbf{A}_{11} = \begin{bmatrix} 0 & 1 \\ 1 & 2 \end{bmatrix}$$

and with both $\mathbf{A}_{21}$ and $\mathbf{A}_{22}$ empty. Then

$$\mathbf{B} = \mathbf{A}_{11} = \begin{bmatrix} 0 & 1 \\ 1 & 2 \end{bmatrix} \quad \mathbf{F} = \mathbf{A}_{11}^{-1}\mathbf{A}_{12} = \begin{bmatrix} -2 & 1 \\ 1 & 0 \end{bmatrix}\begin{bmatrix} 0 & 2 \\ 2 & 3 \end{bmatrix} = \begin{bmatrix} 2 & -1 \\ 0 & 2 \end{bmatrix} \quad \mathbf{C} = \left[\begin{array}{cc|cc} 1 & 0 & 2 & -1 \\ 0 & 1 & 0 & 2 \end{array}\right]$$

so
$$(\mathbf{B}^H\mathbf{B})^{-1} = \begin{bmatrix} 1 & 2 \\ 2 & 5 \end{bmatrix}^{-1} = \begin{bmatrix} 5 & -2 \\ -2 & 1 \end{bmatrix} \quad \text{and} \quad (\mathbf{CC}^H)^{-1} = \begin{bmatrix} 6 & -2 \\ -2 & 5 \end{bmatrix}^{-1} = \begin{bmatrix} 5/26 & 2/26 \\ 2/26 & 6/26 \end{bmatrix}$$

and
$$\mathbf{A}^+ = \mathbf{QC}^H(\mathbf{CC}^H)^{-1}(\mathbf{B}^H\mathbf{B})^{-1}\mathbf{B}^H\mathbf{P}$$

$$= \begin{bmatrix} 1 & 0 & 0 & 0 \\ 0 & 0 & 1 & 0 \\ 0 & 1 & 0 & 0 \\ 0 & 0 & 0 & 1 \end{bmatrix}\begin{bmatrix} 1 & 0 \\ 0 & 1 \\ 2 & 0 \\ -1 & 2 \end{bmatrix}\begin{bmatrix} 5/26 & 2/26 \\ 2/26 & 6/26 \end{bmatrix}\begin{bmatrix} 5 & -2 \\ -2 & 1 \end{bmatrix}\begin{bmatrix} 0 & 1 \\ 1 & 2 \end{bmatrix}\begin{bmatrix} 1 & 0 \\ 0 & 1 \end{bmatrix}$$

$$= \begin{bmatrix} -8/26 & 5/26 \\ -16/26 & 10/26 \\ 2/26 & 2/26 \\ 12/26 & -1/26 \end{bmatrix}$$

21.3 Find the generalized inverse of

$$\mathbf{A} = \begin{bmatrix} -3 & 1 \\ -2 & 1 \\ -1 & 1 \\ 0 & 1 \\ 1 & 1 \\ 2 & 1 \\ 3 & 1 \end{bmatrix}$$

Since the columns of **A** are linearly independent, we use (*21.2*). Then

$$\mathbf{A}^H\mathbf{A} = \begin{bmatrix} 28 & 0 \\ 0 & 7 \end{bmatrix} \quad \text{and} \quad (\mathbf{A}^H\mathbf{A})^{-1} = \begin{bmatrix} 1/28 & 0 \\ 0 & 1/7 \end{bmatrix}$$

and
$$\mathbf{A}^+ = (\mathbf{A}^H\mathbf{A})^{-1}\mathbf{A}^H = \begin{bmatrix} -3/28 & -2/28 & -1/28 & 0 & 1/28 & 2/28 & 3/28 \\ 1/7 & 1/7 & 1/7 & 1/7 & 1/7 & 1/7 & 1/7 \end{bmatrix}$$

21.4 Construct a singular-value decomposition for the matrix in Problem 21.1.

We use Steps 21.6 through 21.11.

STEP 21.6:
$$\mathbf{A}^H\mathbf{A} = \begin{bmatrix} 12 & 12 & -20 \\ 12 & 12 & -20 \\ -20 & -20 & 44 \end{bmatrix}$$

which has eigenvalues 64, 4, and 0 with corresponding orthonormal eigenvectors

$$\mathbf{X}_1 = \begin{bmatrix} -1/\sqrt{6} \\ -1/\sqrt{6} \\ 2/\sqrt{6} \end{bmatrix} \quad \mathbf{X}_2 = \begin{bmatrix} 1/\sqrt{3} \\ 1/\sqrt{3} \\ 1/\sqrt{3} \end{bmatrix} \quad \mathbf{X}_3 = \begin{bmatrix} -1/\sqrt{2} \\ 1/\sqrt{2} \\ 0 \end{bmatrix}$$

STEP 21.7:

$$\mathbf{D}=\begin{bmatrix}\sqrt{64} & 0\\ 0 & \sqrt{4}\end{bmatrix}=\begin{bmatrix}8 & 0\\ 0 & 2\end{bmatrix}$$

STEP 21.8:

$$\mathbf{V}_1\begin{bmatrix}-1/\sqrt{6} & 1/\sqrt{3}\\ -1/\sqrt{6} & 1/\sqrt{3}\\ 2/\sqrt{6} & 1/\sqrt{3}\end{bmatrix} \quad\text{and}\quad \mathbf{V}_2=\begin{bmatrix}-1/\sqrt{2}\\ 1/\sqrt{2}\\ 0\end{bmatrix} \quad\text{with}\quad \mathbf{V}=[\mathbf{V}_1\,|\,\mathbf{V}_2]$$

STEP 21.9:

$$\mathbf{U}_1=\mathbf{A}\mathbf{V}_1\mathbf{D}^{-1}=\begin{bmatrix}2 & 2 & -2\\ 2 & 2 & -2\\ -2 & -2 & 6\end{bmatrix}\begin{bmatrix}-1/\sqrt{6} & 1/\sqrt{3}\\ -1/\sqrt{6} & 1/\sqrt{3}\\ 2/\sqrt{6} & 1/\sqrt{3}\end{bmatrix}\begin{bmatrix}1/8 & 0\\ 0 & 1/2\end{bmatrix}=\begin{bmatrix}-1/\sqrt{6} & 1/\sqrt{3}\\ -1/\sqrt{6} & 1/\sqrt{3}\\ 2/\sqrt{6} & 1/\sqrt{3}\end{bmatrix}$$

STEP 21.10: Augmenting the 3×3 identity matrix onto $\mathbf{U}_1$, we generate

$$\begin{bmatrix}-1/\sqrt{6} & 1/\sqrt{3} & 1 & 0 & 0\\ -1/\sqrt{6} & 1/\sqrt{3} & 0 & 1 & 0\\ 2/\sqrt{6} & 1/\sqrt{3} & 0 & 0 & 1\end{bmatrix}$$

STEP 21.11: The first three columns of this matrix form a maximal set of linearly independent column vectors. Discarding the last two columns and applying the modified Gram-Schmidt process to the first three columns, we obtain

$$\mathbf{U}=\begin{bmatrix}-1/\sqrt{6} & 1/\sqrt{3} & -1/\sqrt{2}\\ -1/\sqrt{6} & 1/\sqrt{3} & 1/\sqrt{2}\\ 2/\sqrt{6} & 1/\sqrt{3} & 0\end{bmatrix}$$

which, in this case, is identical to **V**. A direct calculation shows that

$$\mathbf{A}=\mathbf{V}\begin{bmatrix}8 & 0 & 0\\ 0 & 2 & 0\\ 0 & 0 & 0\end{bmatrix}\mathbf{U}^H$$

21.5 Construct a singular-value decomposition for the matrix in Problem 21.3.

Using Steps 21.6 through 21.11, we first form

$$\mathbf{A}^H\mathbf{A}=\begin{bmatrix}28 & 0\\ 0 & 7\end{bmatrix}$$

which has eigenvalues 28 and 7, with corresponding orthonormal eigenvectors $\mathbf{X}_1=[1,0]^T$ and $\mathbf{X}_2=[0,1]^T$. Then

$$\mathbf{D}=\begin{bmatrix}\sqrt{28} & 0\\ 0 & \sqrt{7}\end{bmatrix} \quad\text{and}\quad \mathbf{V}_1=\begin{bmatrix}1 & 0\\ 0 & 1\end{bmatrix}$$

$\mathbf{V}_2$ is empty, so $\mathbf{V}=\mathbf{V}_1$, and

$$\mathbf{U}_1=\mathbf{A}\mathbf{V}_1\mathbf{D}^{-1}=\begin{bmatrix}-3 & 1\\ -2 & 1\\ -1 & 1\\ 0 & 1\\ 1 & 1\\ 2 & 1\\ 3 & 1\end{bmatrix}\begin{bmatrix}1 & 0\\ 0 & 1\end{bmatrix}\begin{bmatrix}1/\sqrt{28} & 0\\ 0 & 1/\sqrt{7}\end{bmatrix}=\begin{bmatrix}-3/\sqrt{28} & 1/\sqrt{7}\\ -2/\sqrt{28} & 1/\sqrt{7}\\ -1/\sqrt{28} & 1/\sqrt{7}\\ 0 & 1/\sqrt{7}\\ 1/\sqrt{28} & 1/\sqrt{7}\\ 2/\sqrt{28} & 1/\sqrt{7}\\ 2/\sqrt{28} & 1/\sqrt{7}\end{bmatrix}$$

We augment $\mathbf{U}_1$ on the right by the 7×7 identity matrix and then determine that the first seven columns of the augmented matrix form a maximal set of linearly independent column vectors. We discard the last two columns of the augmented identity matrix and apply the modified Gram-Schmidt process to what remains, generating

$$\mathbf{U} = \begin{bmatrix} -3/\sqrt{28} & 1/\sqrt{7} & 15/\sqrt{420} & 0 & 0 & 0 & 0 \\ -2/\sqrt{28} & 1/\sqrt{7} & -10/\sqrt{420} & 10/\sqrt{210} & 0 & 0 & 0 \\ -1/\sqrt{28} & 1/\sqrt{7} & -7/\sqrt{420} & -8/\sqrt{210} & 2/\sqrt{10} & 0 & 0 \\ 0 & 1/\sqrt{7} & -4/\sqrt{420} & -5/\sqrt{210} & -2/\sqrt{10} & 3/\sqrt{30} & 0 \\ 1/\sqrt{28} & 1/\sqrt{7} & -1/\sqrt{420} & -2/\sqrt{210} & -1/\sqrt{10} & -4/\sqrt{30} & 1/\sqrt{6} \\ 2/\sqrt{28} & 1/\sqrt{7} & 2/\sqrt{420} & 1/\sqrt{210} & 0 & -1/\sqrt{30} & -2/\sqrt{6} \\ 3/\sqrt{28} & 1/\sqrt{7} & 5/\sqrt{420} & 4/\sqrt{210} & 1/\sqrt{10} & 2/\sqrt{30} & 1/\sqrt{6} \end{bmatrix}$$

A direct calculation shows that

$$\mathbf{A} = \mathbf{U} \begin{bmatrix} \sqrt{28} & 0 \\ 0 & \sqrt{7} \\ 0 & 0 \\ 0 & 0 \\ 0 & 0 \\ 0 & 0 \\ 0 & 0 \end{bmatrix} \mathbf{V}^H$$

21.6 Use (*21.4*) to calculate the generalized inverse of the matrix in Problem 21.1.

Using what we have already found in Problem 21.4, we compute

$$\mathbf{A}^+ = \mathbf{V}_1\mathbf{D}^{-1}\mathbf{U}_1^H = \begin{bmatrix} -1/\sqrt{6} & 1/\sqrt{3} \\ -1/\sqrt{6} & 1/\sqrt{3} \\ 2/\sqrt{6} & 1/\sqrt{3} \end{bmatrix} \begin{bmatrix} 1/8 & 0 \\ 0 & 1/2 \end{bmatrix} \begin{bmatrix} -1/\sqrt{6} & -1/\sqrt{6} & 2/\sqrt{6} \\ 1/\sqrt{3} & 1/\sqrt{3} & 1/\sqrt{3} \end{bmatrix} = \begin{bmatrix} 3/16 & 3/16 & 1/8 \\ 3/16 & 3/16 & 1/8 \\ 1/8 & 1/8 & 1/4 \end{bmatrix}$$

21.7 Use (*21.4*) to calculate the generalized inverse of the matrix in Problem 21.3.

Using what we have already found in Problem 21.5, we compute

$$\mathbf{A}^+ = \mathbf{V}_1\mathbf{D}^{-1}\mathbf{U}_1^H$$

$$= \begin{bmatrix} 1 & 0 \\ 0 & 1 \end{bmatrix} \begin{bmatrix} 1/\sqrt{28} & 0 \\ 0 & 1/\sqrt{7} \end{bmatrix} \begin{bmatrix} -3/\sqrt{28} & -2/\sqrt{28} & -1/\sqrt{28} & 0 & 1/\sqrt{28} & 2/\sqrt{28} & 3/\sqrt{28} \\ 1/\sqrt{7} & 1/\sqrt{7} & 1/\sqrt{7} & 1/\sqrt{7} & 1/\sqrt{7} & 1/\sqrt{7} & 1/\sqrt{7} \end{bmatrix}$$

$$= \begin{bmatrix} -3/28 & -2/28 & -1/28 & 0 & 1/28 & 2/28 & 3/28 \\ 1/7 & 1/7 & 1/7 & 1/7 & 1/7 & 1/7 & 1/7 \end{bmatrix}$$

21.8 Solve the following system of equations in the least-squares sense:

$$2x_1 + 2x_2 - 2x_3 = 1$$
$$2x_1 + 2x_2 - 2x_3 = 3$$
$$-2x_1 - 2x_2 + 6x_3 = 2$$

This system is inconsistent. Writing it in matrix form and then using (*21.5*) and the results of either Problem 21.1 or Problem 21.6, we obtain

$$\mathbf{X} = \begin{bmatrix} 3/16 & 3/16 & 1/8 \\ 3/16 & 3/16 & 1/8 \\ 1/8 & 1/8 & 1/4 \end{bmatrix} \begin{bmatrix} 1 \\ 3 \\ 2 \end{bmatrix} = \begin{bmatrix} 1 \\ 1 \\ 1 \end{bmatrix}$$

Therefore, $x_1 = x_2 = x_3 = 1$ is the solution in the least-squares sense.

21.9 Solve the following system of equations in the least-squares sense:

$$x_3 + 2x_4 = 1$$
$$x_1 + 2x_2 + 2x_3 + 3x_4 = 2$$

Writing this system in matrix form and then using (*21.5*) and the results of Problem 21.2, we obtain

$$\mathbf{X} = \begin{bmatrix} -8/26 & 5/26 \\ -16/26 & 10/26 \\ 2/26 & 2/26 \\ 12/26 & -1/26 \end{bmatrix} \begin{bmatrix} 1 \\ 2 \end{bmatrix} = \begin{bmatrix} 1/13 \\ 2/13 \\ 3/13 \\ 5/13 \end{bmatrix}$$

Thus, $x_1 = 1/13$, $x_2 = 2/13$, $x_3 = 3/13$, and $x_4 = 5/13$.

21.10 Verify that the solution obtained in Problem 21.9 is the solution of minimum Euclidean norm for the set of equations given in that problem.

Interchanging the order of the two equations, we obtain the system

$$\begin{aligned} x_1 + 2x_2 + 2x_3 + 3x_4 &= 2 \\ x_3 + 2x_4 &= 1 \end{aligned}$$

whose coefficient matrix is in row-echelon form. Using the techniques of Chapter 2, we determine the solution to be

$$\mathbf{X} = \begin{bmatrix} -2x_2 + x_4 \\ x_2 \\ -2x_4 + 1 \\ x_4 \end{bmatrix} \tag{1}$$

with x_2 and x_4 arbitrary. For this vector,

$$\|\mathbf{X}\|_2^2 = (-2x_2 + x_4)^2 + x_2^2 + (-2x_4 + 1)^2 + x_4^2 = 5x_2^2 + 6x_4^2 - 4x_2x_4 - 4x_4 + 1$$

The minimum of this function occurs at its critical points. Setting the first partial derivatives equal to zero, we get

$$\begin{aligned} 10x_2 - 4x_4 &= 0 \\ -4x_2 + 12x_4 &= 4 \end{aligned}$$

which has the unique solution $x_2 = 2/13$ and $x_4 = 5/13$. Substituting these values in (*1*) gives us the same solution as we obtained in Problem 21.9.

21.11 Sales data were taken over a seven-year period of time and coded so that the midpoint of the period coincides with time $t = 0$. The results are given in the following table (in hundreds of thousands of dollars):

Time	−3	−2	−1	0	1	2	3
Sales	10	15	19	27	28	34	42

A graph of these data reveals a near linear type of growth. Find the equation of the straight line that best fits the data in the least-squares sense.

A straight line here would satisfy the equation

$$S = at + b \tag{1}$$

where S denotes sales, t denotes time, and a and b are constants to be determined. Substituting each data pair into Eq. (*1*) yields the system:

$$\begin{aligned} -3a + b &= 10 \\ -2a + b &= 15 \\ -a + b &= 19 \\ b &= 27 \\ a + b &= 28 \\ 2a + b &= 34 \\ 3a + b &= 42 \end{aligned}$$

Writing this system in matrix form, and then using (*21.5*) and the results of either Problem 21.3 or Problem 21.7, we obtain

$$\mathbf{X} = \begin{bmatrix} a \\ b \end{bmatrix} = \begin{bmatrix} -3/28 & -2/28 & -1/28 & 0 & 1/28 & 2/28 & 3/28 \\ 1/7 & 1/7 & 1/7 & 1/7 & 1/7 & 1/7 & 1/7 \end{bmatrix} \begin{bmatrix} 10 \\ 15 \\ 19 \\ 27 \\ 28 \\ 34 \\ 42 \end{bmatrix} = \begin{bmatrix} 143/28 \\ 25 \end{bmatrix}$$

The equation of the line that best fits the data in the least-squares sense is $S = \frac{143}{28}t + 25$.

21.12 Working to four significant digits, show that (*21.2*) is numerically unstable when applied to

$$\mathbf{A} = \begin{bmatrix} 1 & 1 \\ 1 & 1 \\ 1 & 1.004 \end{bmatrix}$$

Rounding all stored (intermediate) numerical quantities to four significant digits, we calculate:

$$\mathbf{A}^H\mathbf{A} = \begin{bmatrix} 3.000 & 3.004 \\ 3.004 & 3.008 \end{bmatrix} \qquad \text{so} \qquad \det \mathbf{A}^H\mathbf{A} = -1.600 \times 10^{-5}$$

Then $$(\mathbf{A}^H\mathbf{A})^{-1} = \frac{1}{-1.600 \times 10^{-5}} \begin{bmatrix} 3.008 & -3.004 \\ -3.004 & 3.000 \end{bmatrix} = \begin{bmatrix} -1.880 \times 10^5 & 1.878 \times 10^5 \\ 1.878 \times 10^5 & -1.875 \times 10^5 \end{bmatrix}$$

and $$\mathbf{A}^+ = (\mathbf{A}^H\mathbf{A})^{-1}\mathbf{A}^H = \begin{bmatrix} -200 & -200 & 551.2 \\ 300 & 300 & -450 \end{bmatrix}$$

This compares unfavorably with the actual generalized inverse

$$\mathbf{A}^+ = \begin{bmatrix} 125.5 & 125.5 & -250 \\ -125 & -125 & 250 \end{bmatrix}$$

calculated without rounding off. Similarly poor results are obtained when results are rounded to k digits and the component 1.004 is replaced with the more general $1 + 4 \times 10^{-k+1}$.

21.13 Show that the generalized inverse is unique.

We assume that **D** and **E** are two generalized inverses for the same matrix **A** and then show that $\mathbf{D} = \mathbf{E}$. Since both **D** and **E** are assumed to satisfy conditions I1, I2, and I3 we know that **DA**, **AD**, **EA**, and **AE** are all Hermitian and that

$$\mathbf{ADA} = \mathbf{A} \qquad \mathbf{DAD} = \mathbf{D} \qquad \mathbf{AEA} = \mathbf{A} \qquad \mathbf{EAE} = \mathbf{E}$$

Multiplying the first of these equations on the right by **E**, we obtain $\mathbf{ADAE} = \mathbf{AE}$, from which we infer that

$$\mathbf{AE} = (\mathbf{AE})^H = \{(\mathbf{AD})(\mathbf{AE})\}^H = (\mathbf{AE})^H(\mathbf{AD})^H = (\mathbf{AE})(\mathbf{AD}) = (\mathbf{AEA})\mathbf{D} = \mathbf{AD}$$

Multiplying that same equation on the left by **E**, we obtain $\mathbf{EADA} = \mathbf{EA}$, from which we deduce that

$$\mathbf{EA} = (\mathbf{EA})^H = \{(\mathbf{EA})(\mathbf{DA})\}^H = (\mathbf{DA})^H(\mathbf{EA})^H = (\mathbf{DA})(\mathbf{EA}) = \mathbf{D}(\mathbf{AEA}) = \mathbf{DA}$$

Then $$\mathbf{E} = \mathbf{EAE} = (\mathbf{EA})\mathbf{E} = (\mathbf{DA})\mathbf{E} = \mathbf{D}(\mathbf{AE}) = \mathbf{DAD} = \mathbf{D}$$

21.14 Show that if **P** and **Q** are unitary matrices of appropriate order so that the product **PAQ** is defined, then $(\mathbf{PAQ})^+ = \mathbf{Q}^H\mathbf{A}^+\mathbf{P}^H$.

Let $\mathbf{G} = \mathbf{PAQ}$. We need to show that $\mathbf{G}^+ = \mathbf{Q}^H\mathbf{A}^+\mathbf{P}^H$ satisfies conditions I1, I2, and I3, given that **A** and $\mathbf{A}^+$ do.

$$\textbf{I1:}\quad \mathbf{GG^+ = (PAQ)(Q^HA^+P^H) = PA(QQ^H)A^+P^H = PAIA^+P^H = P(AA^+)P^H}$$
$$\mathbf{G^+G = (Q^HA^+P^H)(PAQ) = Q^HA^+(P^HP)AQ = Q^HA^+IAQ = Q^H(A^+A)Q}$$

Both are Hermitian since $\mathbf{AA^+}$ and $\mathbf{A^+A}$ are.

$$\textbf{I2:}\quad \mathbf{GG^+G = (PAQ)(Q^HA^+P^H)(PAQ) = PA(QQ^H)A^+(P^HP)AQ = PAIA^+IAQ = P(AA^+A)Q = PAQ = G}$$

$$\textbf{I3:}\quad \mathbf{G^+GG^+ = (Q^HA^+P^H)(PAQ)(Q^HA^+P^H) = Q^HA^+(P^HP)A(QQ^H)A^+P^H = Q^HA^+IAIA^+P^H}$$
$$\mathbf{= Q^H(A^+AA^+)P^H = Q^HA^+P^H = G^+}$$

21.15 Show that if $\mathbf{A}$ can be factored into the product $\mathbf{BC}$, where both $\mathbf{B^HB}$ and $\mathbf{CC^H}$ are invertible, then $\mathbf{A^+ = C^H(CC^H)^{-1}(B^HB)^{-1}B^H}$.

We need to show that $\mathbf{A^+}$ satisfies the three conditions required of a generalized inverse.

$$\textbf{I1:}\quad \mathbf{AA^+ = (BC)C^H(CC^H)^{-1}(B^HB)^{-1}B^H = B(CC^H)(CC^H)^{-1}(B^HB)^{-1}B^H = B(B^HB)^{-1}B^H}$$
$$\mathbf{A^+A = C^H(CC^H)^{-1}(B^HB)^{-1}B^H(BC) = C^H(CC^H)^{-1}(B^HB)^{-1}(B^HB)C = C^H(CC^H)^{-1}C}$$

Both are obviously Hermitian.

$$\textbf{I2:}\quad \mathbf{AA^+A = (BC)C^H(CC^H)^{-1}(B^HB)^{-1}B^H(BC) = B[(CC^H)(CC^H)^{-1}][(B^HB)^{-1}(B^HB)]C = BIIC = BC = A}$$

$$\textbf{I3:}\quad \mathbf{A^+AA^+ = C^H(CC^H)^{-1}(B^HB)^{-1}B^H(BC)C^H(CC^H)^{-1}(B^HB)^{-1}B^H}$$
$$\mathbf{= C^H(CC^H)^{-1}[(B^HB)^{-1}(B^HB)][(CC^H)(CC^H)^{-1}](B^HB)^{-1}B^H}$$
$$\mathbf{= C^H(CC^H)^{-1}II(B^HB)^{-1}B^H = C^H(CC^H)^{-1}(B^HB)^{-1}B^H = A^+}$$

21.16 Validate the algorithm given by Steps 21.1 through 21.5.

Steps 21.1 through 21.4 provide a procedure for factoring the matrix $\mathbf{PAQ}$ into the product $\mathbf{BC}$, where both $\mathbf{B^HB}$ and $\mathbf{CC^H}$ are nonsingular. Observe that the last $n - K$ columns of $\mathbf{PAQ}$ are linear combinations of the first K columns, so there must exist a matrix $\mathbf{F}$ such that both $\mathbf{A}_{22} = \mathbf{A}_{21}\mathbf{F}$ and $\mathbf{A}_{12} = \mathbf{A}_{11}\mathbf{F}$. Since $\mathbf{A}_{11}$ is invertible, $\mathbf{F} = \mathbf{A}_{11}^{-1}\mathbf{A}_{12}$ and $\mathbf{A}_{22} = \mathbf{A}_{21}\mathbf{A}_{11}^{-1}\mathbf{A}_{12}$. Now it follows from Problem 21.15 that $\mathbf{(PAQ)^+ = C^H(CC^H)^{-1}(B^HB)^{-1}B^H}$, and then from Problem 21.14 that

$$\mathbf{Q^HA^+P^H = C^H(CC^H)^{-1}(B^HB)^{-1}B^H}$$

The desired formula comes from multiplying both sides of this last equation by $\mathbf{Q}$ on the left and $\mathbf{P}$ on the right.

Although the factors $\mathbf{B}$ and $\mathbf{C}$ and the matrices $\mathbf{P}$ and $\mathbf{Q}$ are not unique, the product is, as a result of Problem 21.13.

21.17 Prove that $\mathbf{(A^H)^+ = (A^+)^H}$.

Set $\mathbf{G = A^H}$. We need to show that $\mathbf{G^+ = (A^+)^H}$ satisfies conditions I1, I2, and I3.

$$\textbf{I1:}\quad \mathbf{(GG^+)^H = \{A^H(A^+)^H\}^H = A^+A = (A^+A)^H = A^H(A^+)^H = GG^+}$$
$$\mathbf{(G^+G)^H = \{(A^+)^HA^H\}^H = AA^+ = (AA^+)^H = (A^+)^HA^H = G^+G}$$

Both are Hermitian by the definition (13.2).

$$\textbf{I2:}\quad \mathbf{G^+GG^+ = (A^+)^HA^H(A^+)^H = (A^+AA^+)^H = (A^+)^H = G^+}$$
$$\textbf{I3:}\quad \mathbf{GG^+G = A^H(A^+)^HA^H = (AA^+A)^H = A^H = G}$$

Thus, $\mathbf{G^+ = (A^+)^H}$ satisfies all the conditions for a generalized inverse for $\mathbf{A^H}$, and since the generalized inverse is unique, it follows that $\mathbf{(A^H)^+ = G^+ = (A^+)^H}$.

Supplementary Problems

In Problems 21.18 through 21.24, find the generalized inverse of the given matrix.

21.18 $\begin{bmatrix} 1 \\ 1 \end{bmatrix}$ **21.19** $\begin{bmatrix} 1 & 1 & 3 \\ 1 & 1 & 3 \end{bmatrix}$ **21.20** $\begin{bmatrix} 1 & 1 & 1 \\ 1 & 1 & 1 \\ 1 & 1 & 1 \end{bmatrix}$ **21.21** $\begin{bmatrix} 1 & 3 \\ 1 & 2 \end{bmatrix}$

21.22 $\begin{bmatrix} 1 & 1 \\ 2 & 0 \\ 0 & 1 \end{bmatrix}$ **21.23** $\begin{bmatrix} 1 & 2 & 3 & 0 & 1 \\ -1 & 0 & 2 & -2 & 3 \end{bmatrix}$ **21.24** $\begin{bmatrix} 1 & 1 & 2 \\ 1 & 1 & 1 \\ 1 & 2 & 3 \end{bmatrix}$

In Problems 21.25 through 21.38, find the least-squares solution to the given system of equations.

21.25
$$\begin{aligned} x_1 + x_2 + 3x_3 &= 1 \\ x_1 + x_2 + 3x_3 &= 2 \end{aligned}$$

21.26
$$\begin{aligned} x_1 + x_2 + x_3 &= 1 \\ x_1 + x_2 + x_3 &= 2 \\ x_1 + x_2 + x_3 &= 3 \end{aligned}$$

21.27
$$\begin{aligned} x_1 + x_2 &= 1 \\ 2x_1 \qquad &= 2 \\ x_2 &= 3 \end{aligned}$$

21.28
$$\begin{aligned} x_1 + 2x_2 + 3x_3 \qquad\qquad + x_5 &= 1 \\ -x_1 + 2 \qquad x_3 - 2x_4 + 3x_5 &= 1 \end{aligned}$$

21.29 Show that the least-squares solution of $\mathbf{AX} = \mathbf{B}$ must necessarily satisfy the system $\mathbf{A}^H\mathbf{AX} = \mathbf{A}^H\mathbf{B}$ (the *normal equations* for the original system).

21.30 Show that if $\mathbf{A} = \mathbf{QR}$ is a **QR** decomposition of **A**, then the normal equations can be written as $\mathbf{R}^H\mathbf{RX} = \mathbf{R}^H\mathbf{Q}^H\mathbf{B}$, which reduces to

$$\mathbf{RX} = \mathbf{Q}^H\mathbf{B} \tag{1}$$

when the columns of **A** are linearly independent.

21.31 A numerically stable procedure for determining the least-squares solution to the matrix system $\mathbf{AX} = \mathbf{B}$ when the columns of **A** are linearly independent is to first determine the **QR** decomposition for $\mathbf{A}^H\mathbf{A}$ and then solve (*1*) of Problem 21.30 for **X**. Use this procedure to solve Problem 21.11.

21.32 Use the procedure described in Problem 21.31 to solve Problem 21.27.

21.33 The following data appear to be quadratic when graphed:

x	0	1	2	3	4
y	10	14	18	32	49

Determine the normal equations for the quadratic equation, $y = ax^2 + bx + c$, that best fits these data in the least-squares sense (see Problem 21.29). Solve for a, b, and c using the procedure described in Problem 21.31.

21.34 Construct a singular-value decomposition for

$$\mathbf{A} = \begin{bmatrix} 1 & 1 & 3 \\ 1 & 1 & 3 \end{bmatrix}$$

21.35 Construct a singular-value decomposition for

$$\mathbf{A} = \begin{bmatrix} 7 & -9 \\ 4 & -6 \end{bmatrix}$$

21.36 Prove that $\mathbf{A}^+ = \mathbf{A}^{-1}$ when **A** is nonsingular.

21.37 Prove that $\mathbf{0}^+ = \mathbf{0}$.

21.38 Prove that $(\mathbf{A}^+)^+ = \mathbf{A}$.

21.39 Prove that $(k\mathbf{A})^+ = (1/k)\mathbf{A}^+$, provided $k \neq 0$.

21.40 Prove that if $\mathbf{A}$ is Hermitian, then so too is $\mathbf{A}^+$.

21.41 Prove that if $\mathbf{A}$ is Hermitian and idempotent, then $\mathbf{A}^+ = \mathbf{A}$.

21.42 Prove that $\mathbf{AA}^+$ and $\mathbf{A}^+\mathbf{A}$ are Hermitian and idempotent.

21.43 Show that if $\mathbf{A}$ has order $m \times n$ with $m \geq n$, then $\mathbf{A}$ can be factored into $\mathbf{A} = \mathbf{U\Sigma V}^H$, where $\mathbf{\Sigma}$ is an $n \times n$ diagonal matrix, $\mathbf{V}$ is an $n \times n$ unitary matrix, and $\mathbf{U}$ is an $m \times n$ matrix with orthonormal columns.

21.44 Using the matrices identified in Problem 21.43, show that $\mathbf{P} = \mathbf{V\Sigma V}^H$ is positive semidefinite, and $\mathbf{M} = \mathbf{UV}^H$ is a matrix with orthonormal columns.

21.45 Using the results of Problems 21.43 and 21.44, show that any $m \times n$ matrix $\mathbf{A}$ with $m \geq n$ can be factored into $\mathbf{A} = \mathbf{MP}$, where $\mathbf{M}$ has orthonormal columns and $\mathbf{P}$ is positive semidefinite. Such a factorization is called a *polar decomposition of* $\mathbf{A}$.

21.46 Find a polar decomposition for the matrix in Problem 21.1.

21.47 Find a polar decomposition for the matrix in Problem 21.3.

21.48 Show that the positive semidefinite matrix $\mathbf{P}$ defined in Problem 21.44 as part of the polar decomposition of $\mathbf{A}$ can be given by $\mathbf{P} = \sqrt{\mathbf{A}^H\mathbf{A}}$.

21.49 Show that if $\mathbf{A}$ is invertible, then the matrix $\mathbf{M}$ defined in Problem 21.44 as part of the polar decomposition of $\mathbf{A}$ reduces to $\mathbf{M} = (\mathbf{PA}^{-1})^H = (\mathbf{A}^H)^{-1}\mathbf{P}$.

21.50 Use the results of Problems 21.48 and 21.49 to determine a polar decomposition for

$$\mathbf{A} = \begin{bmatrix} 4 & 3 \\ 3 & -4 \end{bmatrix}$$

Answers to Supplementary Problems

CHAPTER 1

1.19 (*a*) $\begin{bmatrix} 1 & 4 \\ 4 & -2 \end{bmatrix}$ (*b*) $\begin{bmatrix} 3 & 6 \\ 9 & -12 \end{bmatrix}$ (*c*) $\begin{bmatrix} 2 & -2 \\ 3 & -14 \end{bmatrix}$ (*d*) $\begin{bmatrix} 1 & 3 & -1 \\ -6 & -7 & -1 \\ -5 & -4 & -2 \end{bmatrix}$ (*e*) undefined

1.20 (*a*) −10; (*b*) 23; (*c*) −1

1.21 (*a*) $\begin{bmatrix} 2 & 6 \\ -4 & -2 \end{bmatrix}$ (*b*) $\begin{bmatrix} 6 & -8 \\ 7 & -6 \end{bmatrix}$ (*c*) $\begin{bmatrix} 2 & -4 \\ 6 & -2 \end{bmatrix}$ (*d*) $\begin{bmatrix} 2 & -4 \\ 6 & -2 \end{bmatrix}$ (*e*) $\begin{bmatrix} 6 & 7 \\ -8 & -6 \end{bmatrix}$

1.22 (*a*) $\begin{bmatrix} 26 & 26 & 13 \\ -10 & -10 & -5 \\ 16 & 16 & 8 \end{bmatrix}$ (*b*) $\begin{bmatrix} 3 & 6 & 3 \\ 6 & 12 & 6 \\ 9 & 18 & 9 \end{bmatrix}$ **1.23** $\begin{bmatrix} 9 & 0 \\ -13 & 20 \end{bmatrix}$

1.24 (*a*) $\begin{bmatrix} 13 \\ -5 \\ 8 \end{bmatrix}$ (*b*) undefined **1.25** (*a*) undefined; (*b*) [2, 5, 5]

1.26 (*a*) $\begin{bmatrix} 1 & 2 & 3 \\ 2 & 4 & 6 \\ 3 & 6 & 9 \end{bmatrix}$ (*b*) [14] **1.27** $\begin{bmatrix} 1 & 2 \\ 0 & 1 \end{bmatrix}$

1.28 $\begin{bmatrix} 1 & 2 \\ 0 & 1 \end{bmatrix}$ **1.29** $\begin{bmatrix} 1 & 5/3 & 0 \\ 0 & 1 & 3 \\ 0 & 0 & 0 \end{bmatrix}$

1.30 $\begin{bmatrix} 1 & 1 & 1/2 \\ 0 & 0 & 0 \\ 0 & 0 & 0 \end{bmatrix}$ **1.31** $\begin{bmatrix} 1 \\ 0 \\ 0 \end{bmatrix}$

1.32 (*a*) 2; (*b*) 2; (*c*) 2; (*d*) 1; (*e*) 1

1.33 There are many examples; one is

$$\begin{bmatrix} 1 & 2 \\ 2 & 4 \end{bmatrix}\begin{bmatrix} 6 & -2 \\ -3 & 1 \end{bmatrix}$$

1.34 Total sales revenue for the flight

1.35 $\mathbf{ND} = [20700, 15300, 4900]^T$, a vector representing the money invested by each store in unsold computers of both brands

CHAPTER 2

2.15 (*b*) and (*c*) are solutions.

2.16 $\left[\begin{array}{ccc|c} 1 & 3 & 1 & 5 \\ 2 & 1 & -3 & 15 \\ 1 & 7 & 5 & 1 \end{array}\right]$ **2.17** $\left[\begin{array}{ccccc|c} 2 & -4 & 7 & 6 & -4 & 17 \\ 0 & 6 & -3 & -4 & -5 & 2 \\ 2 & 8 & 1 & -2 & -14 & 10 \end{array}\right]$

2.18 (*a*) $x_1 = 39, x_2 = 5, x_3 = -4$; (*b*) $x_1 = -7 - 3x_2$, x_2 arbitrary, $x_3 = 5$, $x_4 = 0$

2.19 $x_1 = 8 + 2x_3$, $x_2 = -1 - x_3$, x_3 arbitrary

2.20 The system is not consistent.

2.21 Only the trivial solution $x_1 = x_2 = x_3 = 0$ exists.

2.22 $x_1 = 4 + x_3$, $x_2 = -2x_3$, x_3 arbitrary

2.23 $x_1 = 8$, $x_2 = -2$, $x_3 = -3$

2.24 $x_1 = x_2 = x_3 = 2$, $x_4 = -1$

2.25 $x_1 = -560$, $x_2 = 4860$, $x_3 = -10\,920$, $x_4 = 7000$

2.26 $x_1 = -19\,999.302022$, $x_2 = -9999.298601$, $x_3 = 0.690113$, $x_4 = 10\,000.707204$. The system is inconsistent when results are continually rounded to four significant figures.

2.27 $x_1 = 0.49998$, $x_2 = 0.00001$, $x_3 = 0.5$

2.28 (*a*) Consistent only if $k = 7$, and then $x_1 = 2 - x_3$, $x_2 = 1 + x_3$, x_3 arbitrary; (*b*) consistent only if $k = 2$, and then $x_1 = x_2 = 2$.

2.29 There are infinitely many solutions to the system

$$\begin{aligned} 12c + 10d + 6r &= 440 \\ 5c + 3d + r &= 120 \end{aligned}$$

2.30
$$\begin{aligned} b - 0.03i + 0.03c + 0.03s &= 0 \\ 0.01i - c \qquad\quad &= 0 \\ 0.04i - 0.04c - s &= 0 \end{aligned}$$

2.31 $\mathbf{A}(\mathbf{Y} - \mathbf{Z}) = \mathbf{AY} - \mathbf{AZ} = \mathbf{B} - \mathbf{B} = \mathbf{0}$

2.32 $\mathbf{Y} - \mathbf{Z}$ is a solution of the homogeneous system; simply call that difference $\mathbf{H}$.

CHAPTER 3

3.13 (*a*) $\begin{bmatrix} 1 & 0 & 0 \\ 0 & 0 & 1 \\ 0 & 1 & 0 \end{bmatrix}$ (*b*) $\begin{bmatrix} 7 & 0 & 0 \\ 0 & 1 & 0 \\ 0 & 0 & 1 \end{bmatrix}$ (*c*) $\begin{bmatrix} 1 & 0 & 0 \\ -3 & 1 & 0 \\ 0 & 0 & 1 \end{bmatrix}$

3.14 (*a*) $\begin{bmatrix} 1 & 0 & 0 & 0 \\ 0 & 0 & 1 & 0 \\ 0 & 1 & 0 & 0 \\ 0 & 0 & 0 & 1 \end{bmatrix}$ (*b*) $\begin{bmatrix} 1 & 0 & 0 & 0 \\ 0 & 1 & 0 & 0 \\ 0 & 0 & 1 & 0 \\ -3 & 0 & 0 & 1 \end{bmatrix}$ (*c*) $\begin{bmatrix} 1 & 0 & 5 & 0 \\ 0 & 1 & 0 & 0 \\ 0 & 0 & 1 & 0 \\ 0 & 0 & 0 & 1 \end{bmatrix}$

3.15 (*a*) $\begin{bmatrix} 1 & 0 \\ 3/2 & -1/2 \end{bmatrix}$ (*b*) $\begin{bmatrix} -2 & 1 \\ 3/2 & -1/2 \end{bmatrix}$ **3.16** $\dfrac{1}{5}\begin{bmatrix} -1 & -4 & 2 \\ -6 & 1 & 2 \\ 3 & 2 & -1 \end{bmatrix}$

3.17 Let $\mathbf{A}$ and $\mathbf{B}$ be $n \times n$ lower triangular matrices. Then $a_{ik} = 0$ if $k > i$, and $b_{kj} = 0$ if $j > k$. Set $\mathbf{C} = \mathbf{AB}$. For $j > i$,

$$c_{ij} = \sum_{k=1}^{n} a_{ik}b_{kj} = \sum_{k=1}^{i} a_{ik}b_{kj} + \sum_{k=i+1}^{n} a_{ik}b_{kj} = \sum_{k=1}^{i} a_{ik}(0) + \sum_{k=i+1}^{n} (0)b_{kj} = 0$$

3.18 $\begin{bmatrix} 1 & 0 & 0 \\ 4 & -3 & 0 \\ 7 & -6 & 0 \end{bmatrix}\begin{bmatrix} 1 & 2 & 3 \\ 0 & 1 & 2 \\ 0 & 0 & 1 \end{bmatrix}$ **3.19** $\begin{bmatrix} 2 & 0 & 0 \\ 1 & 7/2 & 0 \\ -1 & 3/2 & 11/7 \end{bmatrix}\begin{bmatrix} 1 & -1/2 & 2 \\ 0 & 1 & 2/7 \\ 0 & 0 & 1 \end{bmatrix}$

3.20 $\begin{bmatrix} 1 & 0 & 0 & 0 \\ 2 & 1 & 0 & 0 \\ 0 & 2 & 2 & 0 \\ 0 & 3 & 0 & 10 \end{bmatrix}\begin{bmatrix} 1 & 0 & 0 & 2 \\ 0 & 1 & 0 & -3 \\ 0 & 0 & 1 & 3 \\ 0 & 0 & 0 & 1 \end{bmatrix}$

3.21 The factorization cannot be done.

3.22 $\begin{bmatrix} 1 & 0 & 0 & 0 \\ 2 & -6 & 0 & 0 \\ 1 & -5 & 41/6 & 0 \\ 2 & -2 & -20/3 & -155/41 \end{bmatrix}\begin{bmatrix} 1 & 2 & 3 & 4 \\ 0 & 1 & 7/6 & 7/6 \\ 0 & 0 & 1 & -13/41 \\ 0 & 0 & 0 & 1 \end{bmatrix}$

3.23 $x_1 = x_2 = 1,\ x_3 = 2,\ x_4 = -2$ **3.24** $x_1 = 4,\ x_2 = 0,\ x_3 = 1,\ x_4 = 0$

3.25 $x_1 = 0,\ x_2 = x_3 = 1,\ x_4 = -1$ **3.26** $x_1 = 4 + y_3,\ x_2 = -2y_3,\ x_3 = y_3 = y_3$; y_3 arbitrary

3.27 Cannot be solved; $\mathbf{LY} = \mathbf{B}$ is inconsistent. **3.28** $x_1 = 8,\ x_2 = -2,\ x_3 = -3$

3.29 $x_1 = x_2 = x_3 = 2,\ x_4 = -1$

3.30 $\mathbf{A}^2 = \begin{bmatrix} 7 & 10 \\ 15 & 22 \end{bmatrix}$ $\mathbf{A}^3 = \begin{bmatrix} 37 & 54 \\ 81 & 118 \end{bmatrix}$ **3.31** $\begin{bmatrix} 32 & 0 & 0 \\ 0 & 1 & 0 \\ 0 & 0 & -1 \end{bmatrix}$

3.32 Each diagonal element is raised to the pth power.

3.34 $(\mathbf{I} - \mathbf{A})^2 = (\mathbf{I} - \mathbf{A})(\mathbf{I} - \mathbf{A}) = \mathbf{I}^2 - \mathbf{IA} - \mathbf{AI} + \mathbf{A}^2 = \mathbf{I} - \mathbf{A} - \mathbf{A} + \mathbf{A} = \mathbf{I} - \mathbf{A}$

3.35 By induction: The proposition is obviously true for $p = 1$. If it is true for $p = k - 1$, then

$$(\mathbf{A}^k)^T = (\mathbf{A}^{k-1}\mathbf{A})^T = \mathbf{A}^T(\mathbf{A}^{k-1})^T = (\mathbf{A}^T)^1(\mathbf{A}^T)^{k-1} = (\mathbf{A}^T)^k$$

CHAPTER 4

4.15 These are elementary matrices: (*a*) the matrix is its own inverse; (*b*) change 7 to 1/7; (*c*) change 4 to −4; (*d*) change −3 to 3.

4.16 $\begin{bmatrix} 1/2 & 0 \\ -1/2 & 1 \end{bmatrix}$ **4.17** $\begin{bmatrix} -5 & 3 \\ 2 & -1 \end{bmatrix}$

4.18 No inverse **4.19** $\dfrac{1}{11}\begin{bmatrix} -1 & 4 \\ 2 & 3 \end{bmatrix}$

4.20 $\begin{bmatrix} 1 & 0 & 0 \\ -1 & 1/2 & 0 \\ 1/3 & 0 & 1/3 \end{bmatrix}$ **4.21** $\begin{bmatrix} 1 & -1 & 1/5 \\ 0 & -1 & 3/5 \\ 0 & 0 & 1/5 \end{bmatrix}$

4.22 No inverse

4.23 $\dfrac{1}{28}\begin{bmatrix} -1 & 4 & 7 \\ 6 & 4 & -14 \\ 8 & -4 & 0 \end{bmatrix}$

4.24 $\frac{1}{48}\begin{bmatrix} -6 & -16 & 17 \\ 0 & 16 & -8 \\ 18 & 16 & -11 \end{bmatrix}$ **4.25** $\begin{bmatrix} 1 & -3/2 & 1/2 \\ 1 & -1 & 0 \\ -2 & 7/2 & -1/2 \end{bmatrix}$

4.26 $\frac{1}{2}\begin{bmatrix} -26 & 11 & 7 & -1 \\ 0 & 1 & -1 & 1 \\ 16 & -7 & -3 & -1 \\ 2 & -1 & -1 & 1 \end{bmatrix}$

4.27 $x_1 = 37/28$, $x_2 = -26/28$, $x_3 = -44/28$

4.28 $x_1 = 3$, $x_2 = 0$, $x_3 = -3$ **4.29** $x_1 = -9/2$, $x_2 = 1/2$, $x_3 = 5/2$, $x_4 = 1/2$

4.30 $(\mathbf{A}^T)^{-1}$ is the inverse of $\mathbf{A}^T$. Now $\mathbf{A}^T(\mathbf{A}^{-1})^T = (\mathbf{A}^{-1}\mathbf{A})^T = I^T = \mathbf{I}$, so $(\mathbf{A}^{-1})^T$ is also the inverse of $\mathbf{A}^T$. Equality follows from the uniqueness of the inverse.

4.31 Each part follows from the uniqueness of the inverse: (*a*) $\mathbf{A}^{-1}\mathbf{B}^{-1}$ and $\mathbf{B}^{-1}\mathbf{A}^{-1}$ are both inverses of $\mathbf{AB}$; (*b*) $\mathbf{A}^{-1}\mathbf{B}$ and $\mathbf{BA}^{-1}$ are both inverses of $\mathbf{AB}^{-1}$; (*c*) $\mathbf{AB}^{-1}$ and $\mathbf{B}^{-1}\mathbf{A}$ are both inverses of $\mathbf{BA}^{-1}$.

CHAPTER 5

5.21 (*a*) -3; (*b*) -33; (*c*) $-2916 + 3015 = -3(-33) = 99$

5.22 (*a*) 0; (*b*) -28; (*c*) $0 = 0$

5.23 (*a*) -48; (*b*) $\det \mathbf{F}$ is undefined because $\mathbf{F}$ is not square.

5.24 (*a*) $-\frac{1}{3}\begin{bmatrix} 7 & -6 \\ -4 & 3 \end{bmatrix}$ (*b*) $-\frac{1}{33}\begin{bmatrix} -6 & 3 \\ -5 & 8 \end{bmatrix}$

5.25 $-\frac{1}{28}\begin{bmatrix} 1 & -4 & -7 \\ -6 & -4 & 14 \\ -8 & 4 & 0 \end{bmatrix}$ **5.26** $-\frac{1}{48}\begin{bmatrix} 6 & 16 & -17 \\ 0 & -16 & 8 \\ -18 & -16 & 11 \end{bmatrix}$

5.27 0 **5.28** 2

5.29 $\frac{1}{2}\begin{bmatrix} -26 & 11 & 7 & -1 \\ 0 & 1 & -1 & 1 \\ 16 & -7 & -3 & -1 \\ 2 & -1 & -1 & 1 \end{bmatrix}$

5.30 Denote the equal rows as rows i and j. Add -1 times row i to row j, and then use Properties 5.5 and 5.6.

5.31 Use Property 5.4 n times, once on each row of $\mathbf{A}$.

5.32 $\det \mathbf{A}^2 = \det \mathbf{AA} = \det \mathbf{A} \det \mathbf{A}$

5.33 $\det \mathbf{AB} = \det \mathbf{A} \det \mathbf{B} = \det \mathbf{B} \det \mathbf{A} = \det \mathbf{BA}$

5.34 $1 = \det \mathbf{I} = \det \mathbf{AA}^{-1} = \det \mathbf{A} \det \mathbf{A}^{-1}$; so $\det \mathbf{A}^{-1} = 1/\det \mathbf{A}$.

5.35 Note that $\det \mathbf{A} = \det \mathbf{LU} = \det \mathbf{L} \det \mathbf{U}$, and then use Property 5.2. In particular, $\det \mathbf{U} = 1$ since it has only unity elements on its main diagonal.

CHAPTER 6

6.15 Linearly dependent **6.16** Linearly dependent

6.17 Linearly independent **6.18** Linearly dependent

6.19 Linearly independent **6.20** Linearly dependent

6.21 No

6.22 (*a*) Yes, $[0, 0, 1] = 0[1, 1, 2] + 1[2, 2, 2] + (-1)[2, 2, 1]$; (*b*) no

6.23 (*a*) Yes, $[2, 1, 2, 1] = 2[2, 0, 1, 1] + (-1)[0, 1, 2, -1] + (-2)[1, -1, -1, 1] + 0[0, 0, 1, 2]$;
(*b*) Yes, $[0, 0, 0, 1] = (1/3)[2, 0, 1, 1] + (-2/3)[0, 1, 2, -1] + (-2/3)[1, -1, -1, 1] + (1/3)[0, 0, 1, 2]$

6.24 $[a, b, c] = \frac{4a - 2b + c}{5}[1, 0, 1] + \frac{a + 2b - c}{5}[1, 2, 0] + \frac{-2a + b + 2c}{5}[0, 1, 2]$

6.25 $\{[1, 2]^T\}$ **6.26** $\{[1, 1, 2], [2, 2, 2]\}$

6.27 $\{[1, 2, 1, -1], [1, 0, -1, 2], [0, 1, 1, 0]\}$

6.28 $[5/3, 5/6] = (1/2)[1, 1] + (1/3)[3, 0] + (1/6)[1, 2]$

6.29 $\begin{bmatrix} 0 \\ 7 \end{bmatrix} = \frac{2}{3}\begin{bmatrix} 3 \\ 6 \end{bmatrix} + \frac{1}{3}\begin{bmatrix} -6 \\ 9 \end{bmatrix} + 0\begin{bmatrix} 2 \\ 1 \end{bmatrix} + 0\begin{bmatrix} -1 \\ 1 \end{bmatrix}$

6.30 No vector in the set $\{\mathbf{V}_1, \mathbf{V}_2, \ldots, \mathbf{V}_r, \mathbf{V}\}$ can be written as a linear combination of vectors preceding it, not the first r vectors (as a result of Property 6.2), because they are linearly independent, and not the last vector in the set, because of the hypothesis. It then follows from Property 6.2 that the entire set is linearly independent.

6.31 Consider $\{\mathbf{V}_1, \mathbf{V}_2, \ldots, \mathbf{V}_n, \mathbf{0}\}$. Then $c_1 = c_2 = \cdots = c_n = 0$, $c_{n+1} = 1$ is a set of constants not all zero such that $c_1\mathbf{V}_1 + c_2\mathbf{V}_2 + \cdots + c_n\mathbf{V}_n + (1)(\mathbf{0}) = \mathbf{0}$.

6.32 Only $\begin{bmatrix} 0 \\ 0 \end{bmatrix}$ **6.33** $\begin{bmatrix} -3x_2 \\ x_2 \end{bmatrix} = x_2\begin{bmatrix} -3 \\ 1 \end{bmatrix}$ with x_2 arbitrary

6.34 $\begin{bmatrix} x_3 - 2x_4 \\ -x_3 \\ x_3 \\ x_4 \end{bmatrix} = x_3\begin{bmatrix} 1 \\ -1 \\ 1 \\ 0 \end{bmatrix} + x_4\begin{bmatrix} -2 \\ 0 \\ 0 \\ 1 \end{bmatrix}$ with x_3 and x_4 arbitrary

CHAPTER 7

All x_i in the solutions for this chapter denote arbitrary constants.

7.18 $x_2\begin{bmatrix} -3 \\ 1 \end{bmatrix}$ for $\lambda = 0$ and $x_2\begin{bmatrix} 1 \\ 1 \end{bmatrix}$ for $\lambda = 4$

7.19 $x_2\begin{bmatrix} 1 \\ 1 \end{bmatrix}$ for $\lambda = 1$, an eigenvalue of multiplicity two

7.20 $x_2\begin{bmatrix} -4 \\ 1 \end{bmatrix}$ for $\lambda = 3$ and $x_2\begin{bmatrix} 2 \\ 1 \end{bmatrix}$ for $\lambda = 9$

7.21 $x_2\begin{bmatrix}1/2\\1\end{bmatrix}$ for $\lambda = 6$ and $x_2\begin{bmatrix}7/2\\1\end{bmatrix}$ for $\lambda = -6$

7.22 $x_2\begin{bmatrix}(-3-\sqrt{5})/2\\1\end{bmatrix}$ for $\lambda = \sqrt{5}$ and $x_2\begin{bmatrix}(-3+\sqrt{5})/2\\1\end{bmatrix}$ for $\lambda = -\sqrt{5}$

7.23 $x_2\begin{bmatrix}-1-i2\\1\end{bmatrix}$ for $\lambda = 2+i2$ and $x_2\begin{bmatrix}-1+i2\\1\end{bmatrix}$ for $\lambda = 2-i2$

7.24 $x_1\begin{bmatrix}1\\0\\0\end{bmatrix}$ for $\lambda = 1$, $x_2\begin{bmatrix}1\\1\\0\end{bmatrix}$ for $\lambda = 2$, and $x_3\begin{bmatrix}0\\1\\1\end{bmatrix}$ for $\lambda = 3$

7.25 $x_1\begin{bmatrix}1\\0\\0\end{bmatrix}+x_3\begin{bmatrix}0\\1\\1\end{bmatrix}$ for $\lambda = 0$ (of multiplicity two) and $x_2\begin{bmatrix}-1\\1\\0\end{bmatrix}$ for $\lambda = -1$

7.26 $x_2\begin{bmatrix}-2\\1\\0\end{bmatrix}+x_3\begin{bmatrix}-3\\0\\1\end{bmatrix}$ for $\lambda = -3$ (of multiplicity two) and $x_3\begin{bmatrix}1/3\\2/3\\1\end{bmatrix}$ for $\lambda = 11$

7.27 $\begin{bmatrix}1\\0\\0\end{bmatrix}$ for $\lambda = 5$ (of multiplicity three)

7.28 $\begin{bmatrix}1\\0\\0\end{bmatrix}$ and $\begin{bmatrix}0\\0\\1\end{bmatrix}$ for $\lambda = 5$ (of multiplicity three)

7.29 $\begin{bmatrix}1\\0\\0\end{bmatrix}$ $\begin{bmatrix}0\\1\\0\end{bmatrix}$ and $\begin{bmatrix}0\\0\\1\end{bmatrix}$ for $\lambda = 5$ (of multiplicity three)

7.30 $\begin{bmatrix}1\\0\\0\\0\end{bmatrix}$ and $\begin{bmatrix}0\\0\\1\\0\end{bmatrix}$ for $\lambda = 3$ (of multiplicity four)

7.31 $\begin{bmatrix}1\\0\\0\\0\end{bmatrix}$ for $\lambda = 3$ (of multiplicity four)

7.32 $\begin{bmatrix}-1\\1\\0\\0\end{bmatrix}$ and $\begin{bmatrix}0\\0\\-1\\1\end{bmatrix}$ for $\lambda = 1$ (of multiplicity three) and $\begin{bmatrix}-3\\2\\0\\0\end{bmatrix}$ for $\lambda = 2$

7.33 $\begin{bmatrix}1\\0\\-1\end{bmatrix}$ $\begin{bmatrix}1\\-1\\1\end{bmatrix}$ and $\begin{bmatrix}1\\2\\1\end{bmatrix}$ for eigenvalues 2, 3, and 6, respectively

7.34 $\begin{bmatrix}1\\1\\-2\end{bmatrix}$ $\begin{bmatrix}1\\1\\1\end{bmatrix}$ and $\begin{bmatrix}1\\-1\\0\end{bmatrix}$ for eigenvalues 12, 6, and 4, respectively

7.35 $[-1, 1]$ and $[1, 3]$, corresponding to eigenvalues 0 and 4, respectively

7.36 $[-1, 1]$, corresponding to the eigenvalue 1 (of multiplicity two)

7.37 $[1, -2]$ and $[1, 4]$, corresponding to eigenvalues 3 and 9, respectively

7.38 $[1, -1, 0]$, $[1, 1, 2]$, and $[1, 1, -1]$, corresponding to eigenvalues 1, 3, and 6, respectively

7.39 $[-1, 1, 0]$, $[1, 0, 1]$, and $[1, 1, -1]$, corresponding to eigenvalues 0, 0, and 3, respectively

7.40 $[1, 1, 0]$, $[-1, 0, 1]$, and $[1, -1, 1]$, corresponding to eigenvalues 2, 2, and 5, respectively

7.42 $\mathbf{A}^2\mathbf{X} = \mathbf{A}(\mathbf{AX}) = \mathbf{A}(\lambda\mathbf{X}) = \lambda(\mathbf{AX}) = \lambda(\lambda\mathbf{X}) = \lambda^2\mathbf{X}$

7.43 $(\mathbf{A} - c\mathbf{I})\mathbf{X} = \mathbf{AX} - c\mathbf{IX} = \lambda\mathbf{X} - c\mathbf{X} = (\lambda - c)\mathbf{X}$

7.44 The proof is by induction on the order of the matrices. The proposition is certainly true for 1×1 matrices. Assume it is true for $k \times k$ matrices, and let $\mathbf{A}$ be an arbitrary $(k+1) \times (k+1)$ matrix. Designate as $\mathbf{A}'$ the matrix obtained from $\mathbf{A}$ by deleting its first row and column. Then $\mathbf{A}'$ has order $k \times k$, and the induction hypothesis can be used on it. Evaluating $\det(\mathbf{A} - \lambda\mathbf{I})$ by expansion of the first row, we get

$$\begin{aligned}\det(\mathbf{A} - \lambda\mathbf{I}) &= (a_{11} - \lambda)\det(\mathbf{A}' - \lambda\mathbf{I}') + O(\lambda^{k-2}) \\ &= (a_{11} - \lambda)(-1)^k\{\lambda^k - (\text{trace } \mathbf{A}')\lambda^{k-1} + O(\lambda^{k-2})\} + O(\lambda^{k-2}) \qquad \text{(by induction)} \\ &= (-1)^{k+1}\{\lambda^{k+1} - (a_{11} + \text{trace } \mathbf{A}')\lambda^k + O(\lambda^{k-1})\} \\ &= (-1)^{k+1}\{\lambda^{k+1} - (\text{trace } \mathbf{A})\lambda^k + O(\lambda^{k-1})\}\end{aligned}$$

7.45 Denote the eigenvalues of $\mathbf{A}$ as $\lambda_1, \lambda_2, \ldots, \lambda_n$. Then

$$\det(\mathbf{A} - \lambda\mathbf{I}) = (-1)^n(\lambda - \lambda_1)(\lambda - \lambda_2)\cdots(\lambda - \lambda_n) = (-1)^n\{\lambda^n - (\lambda_1 + \lambda_2 + \cdots + \lambda_n)\lambda^{n-1} + O(\lambda^{n-2})\}$$

But from Problem 7.44,

$$\det(\mathbf{A} - \lambda\mathbf{I}) = (-1)^n\{\lambda^n - (\text{trace } \mathbf{A})\lambda^{n-1} + O(\lambda^{n-2})\}$$

The result follows from equating the coefficients of λ^{n-1} in the two expressions for the characteristic polynomial.

7.46
$$\begin{aligned}\text{trace}(\mathbf{A} + \mathbf{B}) &= (a_{11} + b_{11}) + (a_{22} + b_{22}) + \cdots + (a_{nn} + b_{nn}) \\ &= (a_{11} + a_{22} + \cdots + a_{nn}) + (b_{11} + b_{22} + \cdots + a_{nn}) \\ &= \text{trace } \mathbf{A} + \text{trace } \mathbf{B}\end{aligned}$$

7.47
$$\text{trace } \mathbf{AB} = \sum_{i=1}^{n}\left(\sum_{k=1}^{n} a_{ik}b_{ki}\right) = \sum_{k=1}^{n}\left(\sum_{i=1}^{n} b_{ki}a_{ik}\right) = \text{trace } \mathbf{BA}$$

7.48 Using the results of Problem 7.47, we have

$$\text{trace } \mathbf{S}^{-1}\mathbf{AS} = \text{trace}\{\mathbf{S}^{-1}(\mathbf{AS})\} = \text{trace}\{(\mathbf{AS})\mathbf{S}^{-1}\} = \text{trace}\{\mathbf{A}(\mathbf{SS}^{-1})\} = \text{trace } \mathbf{AI} = \text{trace } \mathbf{A}$$

7.49 Denote the eigenvalues of $\mathbf{A}$ as $\lambda_1, \lambda_2, \ldots, \lambda_n$. Then $\det(\mathbf{A} - \lambda\mathbf{I}) = (-1)^n(\lambda - \lambda_1)(\lambda - \lambda_2)\cdots(\lambda - \lambda_n)$. Set $\lambda = 0$, and $\det \mathbf{A} = \lambda_1\lambda_2\cdots\lambda_n$.

7.50 The proof is by induction on the order of the matrix. When $\mathbf{C}$ has order $(k+1) \times (k+1)$, expand $\det(\mathbf{C} - \lambda\mathbf{I})$ along the first column, obtaining the sum of two determinants. Use the induction hypothesis on the cofactor matrix of the element $-\lambda$ in the (1,1) position. The second determinant is easy to evaluate because the cofactor matrix of $-a_0$ is lower triangular.

CHAPTER 8

8.19 $\lim_{k\to\infty} \mathbf{A}_k = \begin{bmatrix} 1 & \frac{2}{3} \\ 0 & 1 \end{bmatrix}$ $\quad \lim_{k\to\infty} \mathbf{B}_k = \begin{bmatrix} 1 & 0 & 0 \\ 2 & 1 & 0 \end{bmatrix}$

$\lim_{k\to\infty} \mathbf{C}_k$ does not exist because $\lim_{k\to\infty} \{(k - k^2)/(k+1)\} = -\infty$.

8.20 Every square matrix **A**

8.21 All eigenvalues must be less than 4 in absolute value.

8.22 (*a*) $\sin \mathbf{A} = \dfrac{1}{6}\begin{bmatrix} 4\sin 5 + 2\sin(-1) & 2\sin 5 - 2\sin(-1) \\ 4\sin 5 - 4\sin(-1) & 2\sin 5 + 4\sin(-1) \end{bmatrix} = \begin{bmatrix} -0.919773 & -0.039151 \\ -0.078302 & -0.880622 \end{bmatrix}$

(*b*) $e^{\mathbf{A}} = \dfrac{1}{6}\begin{bmatrix} 4e^5 + 2e^{-1} & 2e^5 - 2e^{-1} \\ 4e^5 - 4e^{-1} & 2e^5 + 4e^{-1} \end{bmatrix} = \begin{bmatrix} 99.0647 & 49.3484 \\ 98.6969 & 49.7163 \end{bmatrix}$

8.23 (*a*) $\cos \mathbf{A} = \dfrac{1}{2}\begin{bmatrix} 4\cos 1 - 2\cos(-1) & -2\cos 1 + 2\cos(-1) \\ 4\cos 1 - 4\cos(-1) & -2\cos 1 + 4\cos(-1) \end{bmatrix} = \begin{bmatrix} 0.540302 & 0 \\ 0 & 0.540302 \end{bmatrix}$

(*b*) $3\mathbf{A}^{47} + 2\mathbf{A}^{18} = \begin{bmatrix} 11 & -6 \\ 12 & -7 \end{bmatrix}$

8.24 (*a*) $\sin \mathbf{A} = \mathbf{0}$; (*b*) $\cos \mathbf{A} = \mathbf{I}$

8.25 $\begin{bmatrix} e^{-2t} & 0 \\ 0 & e^{-3t} \end{bmatrix}$ **8.26** $\begin{bmatrix} 3e^{t} - 2e^{-t} & 3e^{t} - 3e^{-t} \\ -2e^{t} + 2e^{-t} & -2e^{t} + 3e^{-t} \end{bmatrix}$

8.27 $\begin{bmatrix} e^{2t} & te^{2t} \\ 0 & e^{2t} \end{bmatrix}$ **8.28** $\begin{bmatrix} \cos 2t + 2\sin 2t & \frac{5}{2}\sin 2t \\ -2\sin 2t & \cos 2t - 2\sin 2t \end{bmatrix}$

8.29 $e^{2t}\begin{bmatrix} 1 & t & t^2/2 \\ 0 & 1 & t \\ 0 & 0 & 1 \end{bmatrix}$ **8.30** $e^{2t}\begin{bmatrix} 1 & 0 & 0 \\ 0 & 1 & t \\ 0 & 0 & 1 \end{bmatrix}$

8.31 $\dfrac{1}{9}\begin{bmatrix} 9e^{-t} & -3e^{-t} + 3e^{2t} & e^{-t} - e^{2t} + 3te^{2t} \\ 0 & 9e^{2t} & 9te^{2t} \\ 0 & 0 & 9e^{2t} \end{bmatrix}$ **8.32** $\begin{bmatrix} \sin 2t & 0 & 0 & 0 \\ 0 & \sin 2t & t\cos 2t & 0 \\ 0 & 0 & \sin 2t & 0 \\ 0 & 0 & 0 & \sin 3t \end{bmatrix}$

8.33 $\mathbf{X}(t) = \begin{bmatrix} \frac{11}{6}e^{2t} + \frac{1}{6}e^{-4t} \\ \frac{22}{6}e^{2t} - \frac{4}{6}e^{-4t} \end{bmatrix}$

8.34 $\mathbf{C} = [c_1, c_2]^T$ is arbitrary. The solution vector, in terms of **C**, is

$$\mathbf{X}(t) = \begin{bmatrix} \{(-3c_1 + c_2 + \frac{1}{9})t + (c_1 - \frac{2}{27})\}e^{3t} + \frac{1}{9}t + \frac{2}{27} \\ \{(-9c_1 + 3c_2 + \frac{1}{3})t + (c_2 - \frac{1}{9})\}e^{3t} + \frac{1}{9} \end{bmatrix}$$

8.35 $\mathbf{X}(t) = \begin{bmatrix} \frac{1}{6}e^{-4t} + \frac{1}{3}e^{2t} - \frac{1}{2} \\ -\frac{4}{6}e^{-4t} + \frac{2}{3}e^{2t} \end{bmatrix}$ **8.36** $\mathbf{X}(t) = \begin{bmatrix} \frac{1}{6}e^{-4t} + \frac{4}{3}e^{2t} - \frac{1}{2} \\ \frac{-4}{6}e^{-4t} + \frac{8}{3}e^{2t} \end{bmatrix}$ **8.37** $\mathbf{X} = \begin{bmatrix} -1 & -1 \\ -\frac{1}{2} & -\frac{1}{2} \end{bmatrix}$

CHAPTER 9

9.16 (*c*) and (*d*)

9.17 $\mathbf{X}_4$, $\mathbf{X}_3 = [-1, 0, 2, 1, 0]^T$, $\mathbf{X}_2 = [2, 0, -1, 0, 0]^T$, $\mathbf{X}_1 = [-1, 0, 0, 0, 0]^T$

9.18 $[0, 1, 0]^T$

9.19 $\mathbf{X}_3$, $\mathbf{X}_2 = [0, 1, 0]^T$, $\mathbf{X}_1 = [1, 0, 0]^T$

9.20 (*a*) Two chains of length 2; (*b*) one chain of length 2 and two chains of length 1; (*c*) one chain of length 3 and one chain of length 1; (*d*) two chains of length 3 and one chain of length 2; (*e*) one chain of length 3, one chain of length 2, and three chains of length 1; (*f*) three chains of length 2 and two chains of length 1; (*g*) the eigenvalue rank numbers as given are impossible.

9.21 (*a*) $N_4 = N_3 = N_2 = 1$ and $N_1 = 2$ for $\lambda = 1$; (*b*) the vectors found in Problem 9.17, along with $\mathbf{Y}_1 = [0, 1, 0, 0, 0]^T$

9.22 (*a*) $N_3 = N_2 = N_1 = 1$ for $\lambda = 5$; (*b*) the vectors found in Problem 9.19

9.23 (*a*) $N_2 = N_1 = 1$ for $\lambda = -2$; (*b*) $\mathbf{X}_2 = [0, 1]^T$, $\mathbf{X}_1 = [-3, 3]^T$

9.24 (*a*) $N_1 = 1$ for both $\lambda = 0$ and $\lambda = 4$; (*b*) $\mathbf{X}_1 = [-3, 1]^T$, $\mathbf{Y}_1 = [1, 1]^T$

9.25 (*a*) $N_2 = 1$, $N_1 = 2$ for $\lambda = 2$; (*b*) $\mathbf{X}_2 = [0, 0, 1]^T$, $\mathbf{X}_1 = [1, 0, 0]^T$, $\mathbf{Y}_1 = [0, -1, 2]^T$

9.26 (*a*) $N_1 = 2$ for $\lambda = 4$ and $N_1 = 1$ for $\lambda = -1$; (*b*) $\mathbf{X}_1 = [1, 0, 1]^T$, $\mathbf{Y}_1 = [1, 1, 0]^T$, $\mathbf{Z}_1 = [-2, 0, 3]^T$

9.27 (*a*) $N_2 = 1$ and $N_1 = 2$ for $\lambda = 1$; (*b*) $\mathbf{X}_2 = [0, 0, 1]^T$, $\mathbf{X}_1 = [1, 1, -2]^T$, $\mathbf{Y}_1 = [-2, 1, 1]^T$

9.28 (*a*) $N_2 = N_1 = 1$ for $\lambda = 2$ and $N_1 = 1$ for $\lambda = 4$; (*b*) $\mathbf{X}_2 = [0, 1, 1]^T$, $\mathbf{X}_1 = [4, 4, -8]^T$, $\mathbf{Y}_1 = [-1, 0, 1]^T$

9.29 (*a*) $N_2 = N_1 = 1$ for both $\lambda = 2$ and $\lambda = 3$; (*b*) $\mathbf{X}_2 = [0, 0, 1, 0]^T$, $\mathbf{X}_1 = [1, 0, -1, 1]^T$, $\mathbf{Y}_2 = [-5, -1, 2, 0]^T$, $\mathbf{Y}_1 = [0, 0, 0, 1]^T$

9.30 (*a*) $N_2 = N_1 = 1$ for $\lambda = 2$ and $N_1 = 2$ for $\lambda = 3$; (*b*) $\mathbf{X}_2 = [0, 0, 1, 0]^T$, $\mathbf{X}_1 = [1, 0, -1, 1]^T$, $\mathbf{Y}_1 = [0, 0, 0, 1]^T$, $\mathbf{Z}_1 = [-3, -1, 1, 0]^T$

9.31 (*a*) $N_3 = N_2 = 1$ and $N_1 = 2$ for $\lambda = 5$, $N_1 = 1$ for $\lambda = 2$; (*b*) $\mathbf{X}_3 = [0, 0, 1, 0, 0]^T$, $\mathbf{X}_2 = [2, -1, 0, 0, 0]^T$, $\mathbf{X}_1 = [-1, 0, 0, 0, 0]^T$, $\mathbf{Y}_1 = [0, 0, 0, 1, 1]^T$, $\mathbf{Z}_1 = [0, 0, 0, -2, 1]^T$

9.32 (*a*) $(\lambda - 2)^2$; (*b*) $(\lambda - 4)(\lambda + 1)$; (*c*) $(\lambda - 1)^2$; (*d*) $(\lambda - 2)^2(\lambda - 4)$; (*e*) $(\lambda - 2)^2(\lambda - 3)^2$; (*f*) $(\lambda - 2)^2(\lambda - 3)$

CHAPTER 10

10.20 None are similar.

10.21 $\begin{bmatrix} -3 & 0 \\ 3 & 1 \end{bmatrix}$

10.22 $\begin{bmatrix} -3 & 1 \\ 1 & 1 \end{bmatrix}$

10.23 $\begin{bmatrix} 0 & 1 & 0 \\ -1 & 0 & 0 \\ 2 & 0 & 1 \end{bmatrix}$

10.24 $\begin{bmatrix} 1 & 1 & -2 \\ 0 & 1 & 0 \\ 1 & 0 & 3 \end{bmatrix}$

10.25 $\begin{bmatrix} 1 & 0 & 0 & -5 \\ 0 & 0 & 0 & -1 \\ -1 & 1 & 0 & 2 \\ 1 & 0 & 1 & 0 \end{bmatrix}$

10.26 (a) $\begin{bmatrix} 2 & 1 & 0 & 0 \\ 0 & 2 & 0 & 0 \\ 0 & 0 & 2 & 1 \\ 0 & 0 & 0 & 2 \end{bmatrix}$ (b) $\begin{bmatrix} 2 & 0 & 0 & 0 \\ 0 & 2 & 0 & 0 \\ 0 & 0 & 2 & 1 \\ 0 & 0 & 0 & 2 \end{bmatrix}$

(c) $\begin{bmatrix} 2 & 0 & 0 & 0 \\ 0 & 2 & 1 & 0 \\ 0 & 0 & 2 & 1 \\ 0 & 0 & 0 & 2 \end{bmatrix}$ (d) $\begin{bmatrix} 2 & 1 & 0 & 0 & 0 & 0 & 0 & 0 \\ 0 & 2 & 1 & 0 & 0 & 0 & 0 & 0 \\ 0 & 0 & 2 & 0 & 0 & 0 & 0 & 0 \\ 0 & 0 & 0 & 2 & 1 & 0 & 0 & 0 \\ 0 & 0 & 0 & 0 & 2 & 1 & 0 & 0 \\ 0 & 0 & 0 & 0 & 0 & 2 & 0 & 0 \\ 0 & 0 & 0 & 0 & 0 & 0 & 2 & 1 \\ 0 & 0 & 0 & 0 & 0 & 0 & 0 & 2 \end{bmatrix}$

(e) $\begin{bmatrix} 2 & 0 & 0 & 0 & 0 & 0 & 0 & 0 \\ 0 & 2 & 0 & 0 & 0 & 0 & 0 & 0 \\ 0 & 0 & 2 & 0 & 0 & 0 & 0 & 0 \\ 0 & 0 & 0 & 2 & 1 & 0 & 0 & 0 \\ 0 & 0 & 0 & 0 & 2 & 1 & 0 & 0 \\ 0 & 0 & 0 & 0 & 0 & 2 & 0 & 0 \\ 0 & 0 & 0 & 0 & 0 & 0 & 2 & 1 \\ 0 & 0 & 0 & 0 & 0 & 0 & 0 & 2 \end{bmatrix}$ (f) $\begin{bmatrix} 2 & 0 & 0 & 0 & 0 & 0 & 0 & 0 \\ 0 & 2 & 0 & 0 & 0 & 0 & 0 & 0 \\ 0 & 0 & 2 & 1 & 0 & 0 & 0 & 0 \\ 0 & 0 & 0 & 2 & 0 & 0 & 0 & 0 \\ 0 & 0 & 0 & 0 & 2 & 1 & 0 & 0 \\ 0 & 0 & 0 & 0 & 0 & 2 & 0 & 0 \\ 0 & 0 & 0 & 0 & 0 & 0 & 2 & 1 \\ 0 & 0 & 0 & 0 & 0 & 0 & 0 & 2 \end{bmatrix}$

10.27 $\begin{bmatrix} -2 & 1 \\ 0 & -2 \end{bmatrix}$ **10.28** $\begin{bmatrix} 0 & 0 \\ 0 & 4 \end{bmatrix}$

10.29 $\begin{bmatrix} 2 & 0 & 0 \\ 0 & 2 & 1 \\ 0 & 0 & 2 \end{bmatrix}$ **10.30** $\begin{bmatrix} 4 & 0 & 0 \\ 0 & 4 & 0 \\ 0 & 0 & -1 \end{bmatrix}$

10.31 $\begin{bmatrix} 2 & 1 & 0 & 0 \\ 0 & 2 & 0 & 0 \\ 0 & 0 & 3 & 1 \\ 0 & 0 & 0 & 3 \end{bmatrix}$ **10.32** $\begin{bmatrix} 3 & 0 & 0 & 0 \\ 0 & 3 & 0 & 0 \\ 0 & 0 & 2 & 1 \\ 0 & 0 & 0 & 2 \end{bmatrix}$

10.33 $\begin{bmatrix} 5 & 0 & 0 & 0 & 0 \\ 0 & 2 & 0 & 0 & 0 \\ 0 & 0 & 5 & 1 & 0 \\ 0 & 0 & 0 & 5 & 1 \\ 0 & 0 & 0 & 0 & 5 \end{bmatrix}$

10.34 (a) $\begin{bmatrix} \cos 2 & 0 & 0 \\ 0 & \cos 2 & 0 \\ 0 & 0 & \cos 2 \end{bmatrix}$ (b) $\begin{bmatrix} \cos 2 & -\sin 2 & -\frac{1}{2}\cos 2 \\ 0 & \cos 2 & -\sin 2 \\ 0 & 0 & \cos 2 \end{bmatrix}$

(c) $\begin{bmatrix} \cos 2 & -\sin 2 & 0 & 0 \\ 0 & \cos 2 & 0 & 0 \\ 0 & 0 & \cos 2 & -\sin 2 \\ 0 & 0 & 0 & \cos 2 \end{bmatrix}$ (d) $\begin{bmatrix} \cos 2 & 0 & 0 & 0 & 0 & 0 \\ 0 & \cos 2 & 0 & 0 & 0 & 0 \\ 0 & 0 & \cos 2 & -\sin 2 & -\frac{1}{2}\cos 2 & \frac{1}{6}\sin 2 \\ 0 & 0 & 0 & \cos 2 & -\sin 2 & -\frac{1}{2}\cos 2 \\ 0 & 0 & 0 & 0 & \cos 2 & -\sin 2 \\ 0 & 0 & 0 & 0 & 0 & \cos 2 \end{bmatrix}$

10.35 $e^{-2}\begin{bmatrix} -2 & -3 \\ 3 & 4 \end{bmatrix}$ **10.36** $e^{2}\begin{bmatrix} 1 & 2 & 1 \\ 0 & 1 & 0 \\ 0 & 0 & 1 \end{bmatrix}$

10.37 $2^{30}\begin{bmatrix} 2 & 62 & 31 \\ 0 & 2 & 0 \\ 0 & 0 & 2 \end{bmatrix}$ **10.38** $\begin{bmatrix} e^{2t} & te^{2t} \\ 0 & e^{2t} \end{bmatrix}$

10.39 $\begin{bmatrix} 3e^{t} - 2e^{-t} & 3e^{t} - 3e^{-t} \\ -2e^{t} + 2e^{-t} & -2e^{t} + 3e^{-t} \end{bmatrix}$ **10.40** $e^{2t}\begin{bmatrix} 1 & 0 & 0 \\ 0 & 1 & t \\ 0 & 0 & 1 \end{bmatrix}$

10.41 $\dfrac{1}{9}\begin{bmatrix} 9e^{-t} & -3e^{-t} + 3e^{2t} & e^{-t} - e^{2t} + 3te^{2t} \\ 0 & 9e^{2t} & 9te^{2t} \\ 0 & 0 & 9e^{2t} \end{bmatrix}$

10.42 Premultiply (*10.1*) on the left by **S**; then postmultiply on the right by $\mathbf{S}^{-1}$ and set $\mathbf{T} = \mathbf{S}^{-1}$.

10.43 Premultiply (*10.1*) on the left by **S** to obtain **SA** = **BS**. Then

$$\mathbf{BY} = \mathbf{B(SX)} = \mathbf{(BS)X} = \mathbf{(SA)X} = \mathbf{S(AX)} = \mathbf{S}(\lambda\mathbf{X}) = \lambda(\mathbf{SX}) = \lambda\mathbf{Y}$$

CHAPTER 11

11.13 (*a*) 0; (*b*) 0; (*c*) 3; (*d*) 2; (*e*) 14; (*f*) −6; (*g*) 2

11.14 (*a*) $1 + i$; (*b*) $1 - i$; (*c*) $4 - i2$; (*d*) $-1 - i$; (*e*) $i5$; (*f*) $50 - i25$

11.15 $\mathbf{Q}_1 = (1/\sqrt{2})[1, 1]^T, \mathbf{Q}_2 = (1\sqrt{2})[-1, 1]^T$

11.16 $\mathbf{Q}_1 = (1/\sqrt{13})[3, 2]^T, \mathbf{Q}_2 = (1/\sqrt{6{,}877})[46, -69]^T$

11.17 The given vectors are not linearly independent. The Gram-Schmidt process produces $\mathbf{Y}_2 = \mathbf{0}$, on which Step 11.4 cannot be performed.

11.18 $\mathbf{Q}_1 = (1/\sqrt{2})[i, 1]^T, \mathbf{Q}_2 = (1/2)[1 - i, 1 + i]^T$

11.19 $\mathbf{Q}_1 = (1/\sqrt{2})[1, 1, 0]^T$, $\mathbf{Q}_2 = (1/\sqrt{6})[-1, 1, 2]^T$, $\mathbf{Q}_3 = (1/\sqrt{3})[1, -1, 1]^T$

11.20 $\mathbf{Q}_1 = (1/\sqrt{6})[2, 1, -1]^T$, $\mathbf{Q}_2 = (1/\sqrt{291})[-1, 13, 11]^T$, $\mathbf{Q}_3 = (1/\sqrt{194})[8, -7, 9]^T$

11.21 $\mathbf{Q}_1 = (1/\sqrt{3})[1, 1, 1]^T$, $\mathbf{Q}_2 = (1/\sqrt{6})[1, 1, -2]^T$, $\mathbf{Q}_3 = (1/\sqrt{2})[-1, 1, 0]^T$

11.22 $\mathbf{Q}_1 = (1/\sqrt{2})[1, 0, i]^T$, $\mathbf{Q}_2 = (1/\sqrt{14})[1 + i2, 2, 2 - i]^T$, $\mathbf{Q}_3 = (1/\sqrt{35})[-2 - i, 4 - i3, -1 + i2]^T$

11.23 $\mathbf{Q}_1 = (1/\sqrt{2})[1, 1, 0, 0]^T$, $\mathbf{Q}_2 = (1/\sqrt{6})[-1, 1, -2, 0]^T$, $\mathbf{Q}_3 = (1/\sqrt{3})[1, -1, -1, 0]^T$, $\mathbf{Q}_4 = [0, 0, 0, -1]^T$

11.24 $\mathbf{Q}_1 = (1/\sqrt{2})[1, i, 0, 0]^T$, $\mathbf{Q}_2 = (1/\sqrt{18})[i, 1, 4, 0]^T$, $\mathbf{Q}_3 = (1/\sqrt{117})[i4, 4, -2, 9]^T$, $\mathbf{Q}_4 = (1/\sqrt{13})[2, -i2, i, i2]^T$

11.25 $\mathbf{Q}_1 = (1/\sqrt{10})[1, 0]^T, \mathbf{Q}_2 = (1/\sqrt{10})[-7, 5]^T$

11.26 $\mathbf{Q}_1 = (1/\sqrt{8})[1, 1, 0]^T$, $\mathbf{Q}_2 = (1/\sqrt{24})[-1, 1, 2]^T$, $\mathbf{Q}_3 = (1/\sqrt{3})[1, -1, 1]^T$

11.27 Consider the equation $c_1\mathbf{Q}_1 + c_2\mathbf{Q}_2 + \cdots + c_n\mathbf{Q}_n = \mathbf{0}$, where the $\mathbf{Q}_i$ vectors form an orthogonal set. For each i ($i = 1, 2, \ldots, n$),

$$0=\langle \mathbf{0},\mathbf{Q}_i\rangle_{\mathbf{W}}=\left\langle \sum_{j=1}^{n} c_j\mathbf{Q}_j,\mathbf{Q}_i\right\rangle_{\mathbf{W}}=\sum_{j=1}^{n} c_j\langle \mathbf{Q}_j,\mathbf{Q}_i\rangle_{\mathbf{W}}=c_i\langle \mathbf{Q}_i,\mathbf{Q}_i\rangle_{\mathbf{W}}$$

Since $\mathbf{Q}_i$ is nonzero, so too is $\langle \mathbf{Q}_i,\mathbf{Q}_i\rangle_{\mathbf{W}}$; thus, $c_i=0$.

11.28 Set $\mathbf{WX}=\mathbf{Y}=[y_1,y_2,\ldots,y_n]^T$. Then $\langle \mathbf{X},\mathbf{X}\rangle_{\mathbf{W}}=(\mathbf{WX})\cdot(\overline{\mathbf{WX}})=\mathbf{Y}\cdot\overline{\mathbf{Y}}=\Sigma_{i=1}^{n}|y_i|^2$. Therefore, $\langle \mathbf{X},\mathbf{X}\rangle_{\mathbf{W}}\ge 0$. Furthermore, if $\mathbf{X}\ne\mathbf{0}$, then $\mathbf{Y}=\mathbf{W}^{-1}\mathbf{X}\ne\mathbf{0}$, and $\Sigma_{i=1}^{n}|y_i|^2$ is positive.

11.29 If we continue on Problem 11.28, it follows that $\langle \mathbf{X},\mathbf{X}\rangle_{\mathbf{W}}=0$ if and only if $\mathbf{Y}=\mathbf{0}$ and that is the case if and only if $\mathbf{X}=\mathbf{W}^{-1}\mathbf{Y}=\mathbf{0}$.

CHAPTER 12

12.19 (*a*) 1; (*b*) $\sqrt{14}$; (*c*) $\sqrt{21}$; (*d*) $\sqrt{65}$; (*e*) $\sqrt{48}$

12.20 (*a*) 1; (*b*) 6; (*c*) 7; (*d*) 9; (*e*) 12

12.21 (*a*) 1; (*b*) 3; (*c*) 4; (*d*) 8; (*e*) 4

12.22 (*a*) $\sqrt{10}$; (*b*) $\sqrt{46}$; (*c*) $\sqrt{29}$; (*d*) $\sqrt{298}$; (*e*) $\sqrt{464}$

12.23 (*a*) $3^{1/4}$; (*b*) $\sqrt{3}$; (*c*) 3; (*d*) $(155)^{1/3}$; (*e*) 11; (*f*) 4

12.24 (*a*) 1; (*b*) $\sqrt{29}$; (*c*) 8; (*d*) $\sqrt{30}$

12.25 (*a*) 1; (*b*) 7; (*c*) 8; (*d*) $5+\sqrt{5}$

12.26 (*a*) 1; (*b*) 5; (*c*) 8; (*d*) 5

12.27 (*a*) 1; (*b*) $133^{1/3}$; (*c*) 8; (*d*) 5.1448

12.28 (*a*) $\sqrt{30}$; (*b*) $\sqrt{79}$; (*c*) $\sqrt{66}$; (*d*) $\sqrt{22}$; (*e*) $\sqrt{75}$

12.29 (*a*) 6; (*b*) 12; (*c*) 9; (*d*) 5; (*e*) 8.9443

12.30 (*a*) 7; (*b*) 8; (*c*) 9; (*d*) 5; (*e*) 9.4721

12.31 (*a*) 5.4650; (*b*) 8.6099; (*c*) 8.1231; (*d*) 4.2992; (*e*) 7.1517

12.32 For any induced norm, $\|\mathbf{I}\|=\max_{\|\mathbf{X}\|=1}(\|\mathbf{IX}\|)=\max_{\|\mathbf{X}\|=1}(\|\mathbf{X}\|)=1$.

12.33 Denote the ith row of $\mathbf{A}$ as $\mathbf{A}_i$, and the jth column of $\mathbf{B}$ as $\mathbf{B}_j$. Then

$$\|\mathbf{AB}\|_F^2=\sum_{i=1}^{n}\sum_{j=1}^{n}\left|\sum_{k=1}^{n} a_{ik}b_{kj}\right|^2=\sum_{i=1}^{n}\sum_{j=1}^{n}|\langle \mathbf{A}_i^T,\bar{\mathbf{B}}_j\rangle|^2$$

and, by the Schwarz inequality,

$$\|\mathbf{AB}\|_F^2\le\sum_{i=1}^{n}\sum_{j=1}^{n}\langle \mathbf{A}_i^T,\mathbf{A}_i^T\rangle\langle \bar{\mathbf{B}}_j,\bar{\mathbf{B}}_j\rangle$$

$$=\sum_{i=1}^{n}\sum_{j=1}^{n}\left(\sum_{k=1}^{n}|a_{ik}|^2\sum_{m=1}^{n}|b_{mj}|^2\right)=\left(\sum_{i=1}^{n}\sum_{k=1}^{n}|a_{ik}|^2\right)\left(\sum_{j=1}^{n}\sum_{m=1}^{n}|b_{mj}|^2\right)=\|\mathbf{A}\|_F^2\|\mathbf{B}\|_F^2$$

12.34 $\|\mathbf{X}+\mathbf{Y}\|_{\mathbf{W}}^2 = \langle \mathbf{X}+\mathbf{Y}, \mathbf{X}+\mathbf{Y}\rangle_{\mathbf{W}} = \langle \mathbf{X}, \mathbf{X}\rangle_{\mathbf{W}} + \langle \mathbf{X}, \mathbf{Y}\rangle_{\mathbf{W}} + \langle \mathbf{Y}, \mathbf{X}\rangle_{\mathbf{W}} + \langle \mathbf{Y}, \mathbf{Y}\rangle_{\mathbf{W}}$
$= \langle \mathbf{X}, \mathbf{X}\rangle_{\mathbf{W}} + \langle \mathbf{Y}, \mathbf{Y}\rangle_{\mathbf{W}} = \|\mathbf{X}\|_{\mathbf{W}}^2 + \|\mathbf{Y}\|_{\mathbf{W}}^2$

12.35 (*a*) $\sqrt{30}$; (*b*) 8; (*c*) $\sqrt{66}$; (*d*) $\sqrt{22}$; (*e*) $\sqrt{75}$

12.36 (*a*) 5.3723; (*b*) 6.7958; (*c*) 8.1231

12.37 4.8990

12.38 The eigenvalues of $\mathbf{A}$ and $\mathbf{A}^T$ are identical.

12.39 (*a*) 15; (*b*) 4.158; (*c*) 66; (*d*) 2.729; (*e*) 2.147

12.40 $\mathbf{I}^{-1} = \mathbf{I}$ and $\|\mathbf{I}\| = 1$ from Problem 12.31.

12.41 For nonsingular $\mathbf{A}$, $1 = \|\mathbf{I}\| = \|\mathbf{A}\mathbf{A}^{-1}\| \le \|\mathbf{A}\|\|\mathbf{A}^{-1}\| = c(\mathbf{A})$.

CHAPTER 13

13.20 $\mathbf{C}$, $\mathbf{E}$, and $\mathbf{F}$

13.21 $\mathbf{B}$, $\mathbf{C}$, $\mathbf{E}$, $\mathbf{F}$, and $\mathbf{H}$.

13.22 $\mathbf{Q}_1 = [1/\sqrt{2}, 1/\sqrt{2}, 0]^T$, $\mathbf{Q}_2 = [-1/\sqrt{6}, 1/\sqrt{6}, 2/\sqrt{6}]^T$, and $\mathbf{Q}_3 = [1/\sqrt{3}, -1/\sqrt{3}, 1/\sqrt{3}]^T$, corresponding to eigenvalues -2, -2, and 1, respectively.

13.23 $\mathbf{Q}_1 = [1, 0, 0, 0, 0]^T$, $\mathbf{Q}_2 = [0, 1/\sqrt{2}, 1/\sqrt{2}, 0, 0]^T$, $\mathbf{Q}_3 = [0, -1/\sqrt{6}, 1/\sqrt{6}, 2/\sqrt{6}, 0]^T$, $\mathbf{Q}_4 = [0, 0, 0, 0, 1]^T$, and $\mathbf{Q}_5 = [0, 1/\sqrt{3}, -1/\sqrt{3}, 1/\sqrt{3}, 0]^T$, corresponding to eigenvalues 3, 3, 3, 0 and 0, respectively.

13.24 $\mathbf{Q}_1 = [i/\sqrt{3}, -1/\sqrt{3}, 1/\sqrt{3}, 0]^T$, $\mathbf{Q}_2 = [i/\sqrt{3}, 0, -1/\sqrt{3}, 1/\sqrt{3}]^T$, $\mathbf{Q}_3 = [i/\sqrt{6}, 2/\sqrt{6}, 1/\sqrt{6}, 0]^T$, and $\mathbf{Q}_4 = [-i/\sqrt{6}, 0, 1/\sqrt{6}, 2/\sqrt{6}]^T$, corresponding to eigenvalues 1, 1, 4, and 4, respectively.

13.25 (*a*) $\mathbf{F}$ cannot be reduced using only $E3$; (*b*) three positive values and two zeros; (*c*) four positive values

13.26 (*a*) $\begin{bmatrix} 2 & 3 \\ -3 & 4 \end{bmatrix}$ (*b*) $\begin{bmatrix} 9 & 11 \\ -4 & -3 \end{bmatrix}$

13.27 (*a*) $\mathbf{E}$ itself (*b*) $\dfrac{1}{16}\begin{bmatrix} -1 & 4-i2 & 2+i11 \\ 5+i16 & 64-i90 & 72-i145 \\ -2-i16 & i100 & -16+i90 \end{bmatrix}$

13.28 $\mathbf{A}^* = \begin{bmatrix} 3 & -i \\ -i & 2 \end{bmatrix}$ $\mathbf{B}^* = \begin{bmatrix} -i & -2 \\ 2 & i \end{bmatrix}$ $\mathbf{G}^* = \begin{bmatrix} 1 & 2 & 0 \\ 2 & -2 & 0 \end{bmatrix}$

13.29 $(\mathbf{A}+\mathbf{B})^H = (\overline{\mathbf{A}+\mathbf{B}})^T = (\bar{\mathbf{A}}+\bar{\mathbf{B}})^T = \bar{\mathbf{A}}^T + \bar{\mathbf{B}}^T = \mathbf{A}^H + \mathbf{B}^H = \mathbf{A}+\mathbf{B}$

13.30 $(\mathbf{AB})^H = (\overline{\mathbf{AB}})^T = (\overline{\mathbf{BA}})^T = (\bar{\mathbf{B}}\bar{\mathbf{A}})^T = \bar{\mathbf{A}}^T\bar{\mathbf{B}}^T = \mathbf{A}^H\mathbf{B}^H = \mathbf{AB}$, implying that powers of Hermitian matrices are Hermitian, because $\mathbf{A}$ commutes with itself.

13.31 Construct the vector $\mathbf{E}_k$ so that it has a 1 as its kth component and all other components are zero. Then $\langle \mathbf{AE}_k, \mathbf{E}_k\rangle = a_{kk}$, which is real by Property 13.5.

13.32 $\mathbf{AA}^H = \mathbf{A}(-\mathbf{A}) = (-\mathbf{A})\mathbf{A} = \mathbf{A}^H\mathbf{A}$

13.33 $(i\mathbf{A})^H = (\overline{i\mathbf{A}})^T = (\bar{i}\bar{\mathbf{A}})^T = (-i\bar{\mathbf{A}})^T = -i\bar{\mathbf{A}}^T = -i\mathbf{A}^H = i(-\mathbf{A}^H) = i\mathbf{A}$

13.34 $\langle \mathbf{AX}, \mathbf{X}\rangle = \langle \mathbf{X}, \mathbf{A}^*\mathbf{X}\rangle = \langle \mathbf{X}, \mathbf{A}^H\mathbf{X}\rangle = \langle \mathbf{X}, -\mathbf{AX}\rangle = -\langle \mathbf{X}, \mathbf{AX}\rangle = -\overline{\langle \mathbf{AX}, \mathbf{X}\rangle}$

13.35 Let λ be an eigenvalue of **A** corresponding to the eigenvector **X**. Then

$$\lambda\langle \mathbf{X}, \mathbf{X}\rangle = \langle \lambda\mathbf{X}, \mathbf{X}\rangle = \langle \mathbf{AX}, \mathbf{X}\rangle = \langle \mathbf{X}, \mathbf{A}^*\mathbf{X}\rangle = \langle \mathbf{X}, \mathbf{A}^H\mathbf{X}\rangle = \langle \mathbf{X}, -\mathbf{AX}\rangle = \langle \mathbf{X}, -\lambda\mathbf{X}\rangle = -\bar{\lambda}\langle \mathbf{X}, \mathbf{X}\rangle$$

Thus, $\lambda = -\bar{\lambda}$, and λ is pure imaginary.

13.36 $-\mathbf{A}^H = -\bar{\mathbf{A}}^T = -\mathbf{A}^T = \mathbf{A}$

13.37 $\frac{1}{2}(\mathbf{A} + \mathbf{A}^H)$ is Hermitian and $\frac{1}{2}(\mathbf{A} - \mathbf{A}^H)$ is skew-Hermitian for any matrix **A**. For real **A**, these matrices are symmetric and skew-symmetric, respectively.

13.38 According to (*8.1*), $f(\mathbf{A})$ can be written as an $(n-1)$-degree polynomial in **A**. Since the eigenvalues of **A** are real, so are the coefficients of such a polynomial. The result then follows from Problems 13.29 and 13.30.

CHAPTER 14

14.17 **A**, **E** and **G** are positive definite; **B** and **D** are positive semidefinite.

14.18 $\mathbf{A}^{1/2} = \begin{bmatrix} 1.6927 & 0.27849 & -0.23915 \\ 0.27849 & 1.6927 & -0.23915 \\ -0.23915 & -0.23915 & 2.2103 \end{bmatrix}$ **14.19** $\mathbf{B}^{1/2} = \frac{1}{3}\begin{bmatrix} 2 & 2 & -1 \\ 2 & 2 & -1 \\ -1 & -1 & 5 \end{bmatrix}$

14.20 $\mathbf{K}^{1/2} = \begin{bmatrix} 4 & i3 \\ -i3 & 4 \end{bmatrix}$ **14.21** $\mathbf{L} = \begin{bmatrix} \sqrt{3} & 0 & 0 \\ \sqrt{1/3} & \sqrt{8/3} & 0 \\ -\sqrt{1/3} & -\sqrt{1/6} & \sqrt{9/2} \end{bmatrix}$

14.22 $\mathbf{L} = \begin{bmatrix} 2 & 0 & 0 \\ -i & 3 & 0 \\ 1 & 1/3 & \sqrt{71}/3 \end{bmatrix}$ **14.23** $\mathbf{L} = \begin{bmatrix} 3 & 0 & 0 & 0 \\ -1 & 2.2361 & 0 & 0 \\ 0 & 1.3416 & 2.6833 & 0 \\ -1 & -0.44721 & -0.89443 & 2 \end{bmatrix}$

14.24 $\langle (\mathbf{A}+\mathbf{B})\mathbf{X}, \mathbf{X}\rangle = \langle \mathbf{AX}, \mathbf{X}\rangle + \langle \mathbf{BX}, \mathbf{X}\rangle > 0$

14.25 $\langle \mathbf{A}^H\mathbf{X}, \mathbf{X}\rangle = \langle \mathbf{X}, \mathbf{AX}\rangle = \overline{\langle \mathbf{AX}, \mathbf{X}\rangle} = \langle \mathbf{AX}, \mathbf{X}\rangle > 0$

14.26 The eigenvalues of $\mathbf{A}^{-1}$ are the reciprocals of the eigenvalues of **A** (Property 7.4) and are, therefore, positive. Furthermore, $(\mathbf{A}^{-1})^H = (\mathbf{A}^H)^{-1} = \mathbf{A}^{-1}$, so $\mathbf{A}^{-1}$ is Hermitian. A Hermitian matrix with positive eigenvalues is positive definite.

14.27 For any $\mathbf{X} \neq \mathbf{0}$, set $\mathbf{Y} = \mathbf{CX}$. Then $\mathbf{Y} \neq \mathbf{0}$ because $\mathbf{X} = \mathbf{C}^{-1}\mathbf{Y}$, and $\langle \mathbf{BX}, \mathbf{X}\rangle = \langle \mathbf{C}^H\mathbf{ACX}, \mathbf{X}\rangle = \langle \mathbf{ACX}, \mathbf{CX}\rangle = \langle \mathbf{AY}, \mathbf{Y}\rangle > 0$.

14.28 $\mathbf{B} = \mathbf{A}t$ is Hermitian, so $f(\mathbf{B})$ is Hermitian for any function (Problem 13.38); in particular, $f(\mathbf{B}) = e^{\mathbf{B}}$. If λ is an eigenvalue of **A**, then λt is an eigenvalue of **B**, and $e^{\lambda t} > 0$ is an eigenvalue of $e^{\mathbf{B}}$ (Property 10.3). A Hermitian matrix with positive eigenvalues is positive definite.

14.29 If $\langle \mathbf{AX}, \mathbf{X}\rangle$ is positive for all complex-valued vectors **X**, then it is also real for such vectors. It follows from Property 13.5 that **A** is Hermitian.

14.30 If

$$\mathbf{A} = \begin{bmatrix} 1 & 2 \\ 0 & 1 \end{bmatrix} \quad \text{and} \quad \mathbf{X} = [x_1, x_2]^T$$

with **X** real and nonzero, then $\langle \mathbf{AX}, \mathbf{X} \rangle = x_1^2 + 2x_1x_2 + x_2^2 = (x_1 + x_2)^2 > 0$.

CHAPTER 15

15.16 C and E

15.17 (*a*) $\begin{bmatrix} 9/\sqrt{97} & 4/\sqrt{97} \\ 4/\sqrt{97} & -9/\sqrt{97} \end{bmatrix}$ (*b*) $\begin{bmatrix} 0 & 1 & 0 \\ -1/\sqrt{2} & 0 & 1/\sqrt{2} \\ 1/\sqrt{2} & 0 & 1/\sqrt{2} \end{bmatrix}$ (*c*) $\begin{bmatrix} 1/\sqrt{3} & 2/\sqrt{6} & 0 \\ -1/\sqrt{3} & 1/\sqrt{6} & 1/\sqrt{2} \\ 1/\sqrt{3} & -1/\sqrt{6} & 1/\sqrt{2} \end{bmatrix}$

15.18 (*a*) With $\mathbf{U} = \begin{bmatrix} 9/\sqrt{97} & 4/\sqrt{97} \\ 4/\sqrt{97} & -9/\sqrt{97} \end{bmatrix}$ we have $\mathbf{U}^H\mathbf{AU} = \begin{bmatrix} 3 & 13 \\ 0 & -2 \end{bmatrix}$

(*b*) With $\mathbf{U} = \begin{bmatrix} 0 & 1 & 0 \\ -1/\sqrt{2} & 0 & 1/\sqrt{2} \\ 1/\sqrt{2} & 0 & 1/\sqrt{2} \end{bmatrix}\begin{bmatrix} 1 & 0 & 0 \\ 0 & -\sqrt{2/3} & \sqrt{1/3} \\ 0 & \sqrt{1/3} & \sqrt{2/3} \end{bmatrix}$ we have $\mathbf{U}^H\mathbf{BU} = \begin{bmatrix} 3 & -\sqrt{3} & -\sqrt{6} \\ 0 & 2 & \sqrt{2} \\ 0 & 0 & 4 \end{bmatrix}$

(*c*) With $\mathbf{U} = \begin{bmatrix} 1/\sqrt{3} & 2/\sqrt{6} & 0 \\ -1/\sqrt{3} & 1/\sqrt{6} & 1/\sqrt{2} \\ 1/\sqrt{3} & -1/\sqrt{6} & 1/\sqrt{2} \end{bmatrix}\begin{bmatrix} 1 & 0 & 0 \\ 0 & -\sqrt{3}/2 & 1/2 \\ 0 & 1/2 & \sqrt{3}/2 \end{bmatrix}$ we have $\mathbf{U}^H\mathbf{CU} = \begin{bmatrix} 3 & 0 & 0 \\ 0 & 2 & 0 \\ 0 & 0 & 6 \end{bmatrix}$

15.19 (*a*) $\begin{bmatrix} -3/5 & -4/5 \\ -4/5 & 3/5 \end{bmatrix}$ (*b*) $\begin{bmatrix} 0 & 0 & 1 \\ 0 & 1 & 0 \\ 1 & 0 & 0 \end{bmatrix}$ (*c*) $\dfrac{1}{3}\begin{bmatrix} -1 & -2\sqrt{2} \\ -2\sqrt{2} & 1 \end{bmatrix}$

15.20 $\|\mathbf{UX}\|_2^2 = \langle \mathbf{UX}, \mathbf{UX} \rangle = \langle \mathbf{X}, \mathbf{X} \rangle = \|\mathbf{X}\|_2^2$

15.21 $\langle \mathbf{UX}, \mathbf{UY} \rangle / \|\mathbf{UX}\|_2 \|\mathbf{UY}\|_2 = \langle \mathbf{X}, \mathbf{Y} \rangle / \|\mathbf{X}\|_2 \|\mathbf{Y}\|_2$

15.22 Since **R** is both real symmetric and orthogonal, $\mathbf{R}^2 = \mathbf{RR} = \mathbf{R}^T\mathbf{R} = \mathbf{R}^{-1}\mathbf{R} = \mathbf{I}$.

15.23 The eigenvalues are nonnegative (the matrix is Hermitian) and have absolute value 1 (the matrix is unitary), so all eigenvalues must be 1.

15.24 Simply combine Properties 7.8 and 15.4.

15.25 $\mathbf{R}_{23} = \begin{bmatrix} 1 & 0 & 0 & 0 & 0 \\ 0 & \cos\theta & \sin\theta & 0 & 0 \\ 0 & -\sin\theta & \cos\theta & 0 & 0 \\ 0 & 0 & 0 & 1 & 0 \\ 0 & 0 & 0 & 0 & 1 \end{bmatrix}$ $\mathbf{R}_{42} = \begin{bmatrix} 1 & 0 & 0 & 0 & 0 \\ 0 & \cos\theta & 0 & -\sin\theta & 0 \\ 0 & 0 & 1 & 0 & 0 \\ 0 & \sin\theta & 0 & \cos\theta & 0 \\ 0 & 0 & 0 & 0 & 1 \end{bmatrix}$

15.26 Direct multiplication yields $\mathbf{R}_{pq}^T(\theta)\mathbf{R}_{pq}(\theta) = \mathbf{I}$.

15.27 The (k, p) element of the product is $a_{kp}\cos\theta - a_{kq}\sin\theta$. Choose θ to make this quantity equal to zero.

CHAPTER 16

16.11 (*a*) $\begin{bmatrix} 3 & 1 & -1 \\ 1 & 3 & -1 \\ -1 & -1 & 5 \end{bmatrix}$ (*b*) $\begin{bmatrix} 2 & 2 & -1 \\ 2 & 2 & -1 \\ -1 & -1 & 5 \end{bmatrix}$

(*c*) $\begin{bmatrix} 9 & -3 & 0 & -3 \\ -3 & 6 & 3 & 0 \\ 0 & 3 & 9 & -3 \\ -3 & 0 & -3 & 6 \end{bmatrix}$ (*d*) $\begin{bmatrix} 1 & -1 & 2 & -1 \\ -1 & 3 & 4 & 2 \\ 2 & 4 & 3 & 1 \\ -1 & 2 & 1 & 1 \end{bmatrix}$

16.12 (*a*) and (*c*) are positive definite; (*b*) is positive semidefinite.

16.13 (*a*) $\begin{bmatrix} 1 & 0 & 0 \\ 0 & 1 & 0 \\ 0 & 0 & 1 \end{bmatrix}$ (*b*) $\begin{bmatrix} 1 & 0 & 0 \\ 0 & 1 & 0 \\ 0 & 0 & 0 \end{bmatrix}$ (*c*) $\begin{bmatrix} 1 & 0 & 0 & 0 \\ 0 & 1 & 0 & 0 \\ 0 & 0 & 1 & 0 \\ 0 & 0 & 0 & 1 \end{bmatrix}$ (*d*) $\begin{bmatrix} 1 & 0 & 0 & 0 \\ 0 & 1 & 0 & 0 \\ 0 & 0 & -1 & 0 \\ 0 & 0 & 0 & -1 \end{bmatrix}$

16.14 They are not congruent because they do not have the same inertia matrix.

16.15 (*a*) Three positive eigenvalues; (*b*) two positive eigenvalues and one zero eigenvalue; (c) four positive eigenvalues; (*d*) two positive and two negative eigenvalues

16.16 An identity matrix

16.17 (*a*) $\begin{bmatrix} 0 & 1 & 0 \\ 2/\sqrt{17} & -2/\sqrt{17} & 1/\sqrt{17} \\ 1/2 & -1 & 0 \end{bmatrix}$ (*b*) $\begin{bmatrix} 1 & 0 & 0 \\ -3/2 & 0 & 1/2 \\ -2 & 1 & 0 \end{bmatrix}$ (*c*) $\begin{bmatrix} 1 & 0 & 0 \\ -8/9 & 1/9 & 1 \\ -1 & 0 & 1 \end{bmatrix}$

16.18 Inertia matrix $= \begin{bmatrix} 1 & 0 & 0 \\ 0 & 1 & 0 \\ 0 & 0 & -1 \end{bmatrix}$ $\mathbf{P} = \begin{bmatrix} 1 & 0 & 0 \\ -(1+i2) & 1 & 0 \\ (7+i10)/5 & -(5-i)/5 & 1/5 \end{bmatrix}$

16.19 Given $\mathbf{A} = \mathbf{P}_1\mathbf{B}\mathbf{P}_1^T$ and $\mathbf{B} = \mathbf{P}_2\mathbf{C}\mathbf{P}_2^T$. Set $\mathbf{P}_3 = \mathbf{P}_1\mathbf{P}_2$. Then $\mathbf{A} = \mathbf{P}_1\mathbf{B}\mathbf{P}_1^T = \mathbf{P}_1(\mathbf{P}_2\mathbf{C}\mathbf{P}_2^T)\mathbf{P}_1^T = (\mathbf{P}_1\mathbf{P}_2)\mathbf{C}(\mathbf{P}_1\mathbf{P}_2)^T = \mathbf{P}_3\mathbf{C}\mathbf{P}_3^T$.

16.20 If $\mathbf{A} = \mathbf{P}\mathbf{B}\mathbf{P}^T$ then $\mathbf{B} = \mathbf{Q}\mathbf{A}\mathbf{Q}^T$, where $\mathbf{Q} = \mathbf{P}^{-1}$.

16.21 Denote as **N** an inertia matrix congruent to **A**. Then **A** is also congruent to **N** (Problem 16.20). (*a*) If **B** is congruent to **A** and **A** is congruent to **N**, then **B** is congruent to **N** (Problem 16.19). (*b*) If both **A** and **B** are congruent to **N**, then **A** is congruent to **N** and **N** is congruent to **B** so **A** is congruent to **B**.

16.22 The eigenvalues of $\mathbf{A}^{-1}$ are the reciprocals of the eigenvalues of **A**, and therefore have the same signs. It follows from Property 16.1 that **A** and $\mathbf{A}^{-1}$ are congruent to the same inertia matrix and to themselves.

CHAPTER 17

17.20 Reducible, not primitive, not stochastic, $3 \le \sigma \le 4$

17.21 Irreducible, primitive, not stochastic, $3 \le \sigma \le 4$

17.22 Irreducible, not primitive, not stochastic, $1 \le \sigma \le 2$

17.23 Irreducible, primitive, not stochastic, $\sigma = 4$

17.24 Irreducible, primitive, not stochastic, $\sigma = 7$

17.25 Irreducible, primitive, stochastic, ergodic, $\sigma = 1$,

$$\mathbf{L} = \begin{bmatrix} 19/45 & 17/45 & 9/45 \\ 19/45 & 17/45 & 9/45 \\ 19/45 & 17/45 & 9/45 \end{bmatrix}$$

17.26 Reducible, not primitive, stochastic, ergodic, $\sigma = 1$,

$$\mathbf{L} = \begin{bmatrix} 1 & 0 & 0 \\ 1 & 0 & 0 \\ 1 & 0 & 0 \end{bmatrix}$$

17.27 Reducible, not primitive, stochastic, ergodic, $\sigma = 1$,

$$\mathbf{L} = \begin{bmatrix} 3/8 & 0 & 5/8 \\ 0 & 1 & 0 \\ 3/8 & 0 & 5/8 \end{bmatrix}$$

17.28 Reducible, not primitive, stochastic, not ergodic, $\sigma = 1$, no limit exists.

17.29 Irreducible, primitive, doubly stochastic, ergodic, $\sigma = 1$,

$$\mathbf{L} = \begin{bmatrix} 1/3 & 1/3 & 1/3 \\ 1/3 & 1/3 & 1/3 \\ 1/3 & 1/3 & 1/3 \end{bmatrix}$$

17.30 (*a*) 56.48 percent; (*b*) 55.56 percent

17.31 Probabilities are 0.2, 0.14, 0.154, 0.151, and 0.15162.

17.32 (*a*) Approximately 34 discharged, 31 ambulatory, 18 bedridden, and 17 dead; (*b*) approximately 65 discharged and 35 dead

CHAPTER 18

18.12 (*a*) Band matrix of width three, tridiagonal, Toeplitz, and Hessenberg; (*b*) circulant and Toeplitz; (*c*) Hessenberg; (*d*) circulant and Toeplitz; (*e*) tridiagonal, Toeplitz, and Hessenberg; (*f*) none; (*g*) tridiagonal and Hessenberg

18.13 $\lambda_1 = \lambda_2 = 2$, $\lambda_3 = \lambda_4 = 0$, with eigenvectors $[1, 1, 1, 1]^T$, $[1, -1, 1, -1]^T$, $[1, i, -1, -i]^T$, and $[1, -i, -1, i]^T$, respectively

18.14 $\lambda_1 = 6$, $\lambda_2 = (-3 - i\sqrt{3})/2$, and $\lambda_3 = (-3 + i\sqrt{3})/2$, with eigenvectors $[1, 1, 1]^T$, $[1, (-1 + i\sqrt{3})/2, (-1 - i\sqrt{3})/2]^T$, and $[1, (-1 - i\sqrt{3})/2, (-1 + i\sqrt{3})/2]^T$, respectively

18.15 4.46410, 1, −2.46410

18.16 $\begin{bmatrix} 1 & 0 & 0 \\ 3 & -5 & 0 \\ 0 & 3 & 11/5 \end{bmatrix}\begin{bmatrix} 1 & 2 & 0 \\ 0 & 1 & -2/5 \\ 0 & 0 & 1 \end{bmatrix}$

18.17 $\begin{bmatrix} -1 & 0 & 0 & 0 \\ 2 & 5 & 0 & 0 \\ 0 & 2 & -12/5 & 0 \\ 0 & 0 & 1 & -7/12 \end{bmatrix}\begin{bmatrix} 1 & -2 & 0 & 0 \\ 0 & 1 & 1/5 & 0 \\ 0 & 0 & 1 & -5/12 \\ 0 & 0 & 0 & 1 \end{bmatrix}$

18.18 With $\mathbf{U} = \begin{bmatrix} 1 & 0 & 0 \\ 0 & -1/\sqrt{2} & -1/\sqrt{2} \\ 0 & -1/\sqrt{2} & 1/\sqrt{2} \end{bmatrix}$ we have $\mathbf{U}^T\mathbf{B}\mathbf{U} = \begin{bmatrix} 2 & 0 & 0 \\ -\sqrt{2} & 4 & 0 \\ 0 & 3 & 3 \end{bmatrix}$

18.19 With **U** as in Problem 18.18, we have

$$\mathbf{U}^T\mathbf{C}\mathbf{U} = \begin{bmatrix} 3 & -\sqrt{2} & 0 \\ -\sqrt{2} & 5 & 1 \\ 0 & 1 & 3 \end{bmatrix}$$

18.20 With $\mathbf{U} = \begin{bmatrix} 1 & 0 & 0 & 0 \\ 0 & -1/\sqrt{2} & 0 & 1/\sqrt{2} \\ 0 & 0 & 1 & 0 \\ 0 & 1/\sqrt{2} & 0 & 1/\sqrt{2} \end{bmatrix}\begin{bmatrix} 1 & 0 & 0 & 0 \\ 0 & 1 & 0 & 0 \\ 0 & 0 & 0 & 1 \\ 0 & 0 & 1 & 0 \end{bmatrix}$ we have $\mathbf{U}^T\mathbf{A}\mathbf{U} = \begin{bmatrix} 1 & -\sqrt{2} & 0 & 0 \\ -\sqrt{2} & 1/2 & -1/2 & 0 \\ 0 & -1/2 & 1/2 & \sqrt{2} \\ 0 & 0 & \sqrt{2} & -1 \end{bmatrix}$

18.21 With $\mathbf{R}_{12}(0.927295) = \begin{bmatrix} 0.6 & 0.8 & 0 \\ -0.8 & 0.6 & 0 \\ 0 & 0 & 1 \end{bmatrix}$ we have $\mathbf{R}_{12}^T\mathbf{A}\mathbf{R}_{12} = \begin{bmatrix} 2.32 & 2.76 & 1.20 \\ -1.24 & 0.68 & -1.60 \\ 0 & 5.00 & 1.00 \end{bmatrix}$

18.22 **First iteration** ($k = 4$, $i = 1$, $j = 2$):

$$\mathbf{R}_{12}(-0.982794) = \begin{bmatrix} 0.554700 & -0.832050 & 0 & 0 \\ 0.832050 & 0.554700 & 0 & 0 \\ 0 & 0 & 1 & 0 \\ 0 & 0 & 0 & 1 \end{bmatrix}$$ and **A** becomes

$$\begin{bmatrix} 2.153846 & -1.230769 & 4.160251 & 0 \\ -1.230769 & -1.153846 & -0.832050 & 7.211103 \\ 4.160251 & -0.832050 & 2 & 1 \\ 0 & 7.211103 & 1 & -1 \end{bmatrix}$$

Second iteration ($k = 4$, $i = 2$, $j = 3$):

$$\mathbf{R}_{23}(1.433000) = \begin{bmatrix} 1 & 0 & 0 & 0 \\ 0 & 0.137361 & 0.990521 & 0 \\ 0 & -0.990521 & 0.137361 & 0 \\ 0 & 0 & 0 & 1 \end{bmatrix}$$ and **A** becomes

$$\begin{bmatrix} 2.153846 & -4.289876 & -0.647648 & 0 \\ -4.289876 & 2.166909 & 0.371544 & 0 \\ -0.647648 & 0.371544 & -1.320755 & 7.280110 \\ 0 & 0 & 7.280110 & -1 \end{bmatrix}$$

Third iteration ($k = 3$, $i = 1$, $j = 2$):

$$\mathbf{R}_{12}(-1.049953) = \begin{bmatrix} 0.497612 & -0.867400 & 0 & 0 \\ 0.867400 & 0.497612 & 0 & 0 \\ 0 & 0 & 1 & 0 \\ 0 & 0 & 0 & 1 \end{bmatrix}$$ and **A** becomes

$$\begin{bmatrix} -1.539591 & 2.171016 & 0 & 0 \\ 2.171016 & 5.860346 & 0.746655 & 0 \\ 0 & 0.746655 & -1.320755 & 7.280110 \\ 0 & 0 & 7.280110 & -1 \end{bmatrix}$$

18.23 $\mathbf{C} \otimes \mathbf{D} = \begin{bmatrix} 5 & -4 & 15 & -12 \\ 4 & -3 & 12 & -9 \\ 5 & -4 & 15 & -12 \\ 4 & -3 & 12 & -9 \end{bmatrix}$

18.24 No solution **18.25** $\mathbf{X} = \begin{bmatrix} -1 & 1 \\ 1 & -1 \end{bmatrix}$ **18.26** $\mathbf{X} = [1, -1, 0]^T$

CHAPTER 19

19.13

Iteration	Eigenvector components			Eigenvalue
0	1.0000	1.0000	1.0000	
1	1.0000	0.7778	0.7778	9.0000
2	1.0000	0.7183	0.7746	7.8889
3	1.0000	0.6887	0.7746	7.8732
4	1.0000	0.6737	0.7746	7.8730
5	1.0000	0.6661	0.7746	7.8730

19.14

Iteration	Eigenvector components			Eigenvalue
0	1.0000	1.0000	1.0000	
1	−0.7143	−0.2857	1.0000	7.0000
2	−0.7143	−0.2857	1.0000	−11.0000
3	−0.7143	−0.2857	1.0000	−11.0000

The eigenvalue is $14 + (-11) = 3$.

19.15

Iteration	Eigenvector components			Eigenvalues
0	1.0000	1.0000	1.0000	
1	1.0000	−0.5313	0.2188	32.0000
2	1.0000	0.2960	0.1038	30.4063
3	1.0000	−0.2102	0.1613	29.1192
4	1.0000	0.1040	0.1253	29.8552

19.16

Iteration	Eigenvector components			Eigenvalue
0	1.0000	1.0000	1.0000	
1	1.0000	−0.5000	1.0000	−2.0000
2	1.0000	0.2500	1.0000	−2.0000

19.17 The eigenvalues are −4, −2, and 1, with corresponding eigenvectors $\mathbf{V}_1 = [1, 0, -1]^T$, $\mathbf{V}_2 = [1, 0, 1]^T$, and $\mathbf{V}_3 = [0, 1, 0]^T$, respectively. With $\mathbf{X}_0 = [1, 1, 1]^T$, we have $\mathbf{X}_0 = 0\mathbf{V}_1 + 1\mathbf{V}_2 + 1\mathbf{V}_3$; thus, $\mathbf{X}_0$ is a linear combination of $\mathbf{V}_2$ and $\mathbf{V}_3$, with no component that is influenced by $\mathbf{V}_1$, the eigenvector corresponding to the dominant eigenvalue.

19.18 There is no convergence, implying that the dominant eigenvalue is complex.

19.19

Iteration	Eigenvector components				Eigenvalue
0	1.0000	1.0000	1.0000	1.0000	
1	0.9697	0.6970	1.0000	0.9394	33.0000
2	0.9591	0.6899	1.0000	0.9432	30.3939
3	0.9578	0.6891	1.0000	0.9437	30.3011
4	0.9577	0.6890	1.0000	0.9438	30.2902
5	0.9576	0.6889	1.0000	0.9438	30.2889
6	0.9576	0.6889	1.0000	0.9438	30.2887

19.20

Iteration	Eigenvector components			Eigenvalue
0	0.5774	0.5774	0.5774	
1	0.6448	−0.0586	0.7621	7.6667
2	0.7216	0.0147	0.6922	11.8454
3	0.7034	−0.0037	0.7108	11.9902
4	0.7080	0.0009	0.7062	11.9994
5	0.7069	−0.0002	0.7073	12.0000
6	0.7072	0.0001	0.7070	12.0000

19.21

Iteration	Eigenvector components			Eigenvalue
0	0.5774	0.5774	0.5774	
1	−0.0765	−0.9941	0.0765	−4.3333
2	0.1541	0.9760	−0.1541	−13.3158
3	−0.2282	−0.9465	0.2282	−13.6491

There is no convergence yet.

19.22

Iteration	Eigenvector components			Eigenvalue
0	1.0000	1.0000	1.0000	
1	1.0000	0.0000	−1.0000	−1.0000
2	1.0000	0.1389	−0.7778	9.0000
3	1.0000	0.1417	−0.7746	7.8889
4	1.0000	0.1418	−0.7746	7.8732

The eigenvalue is $1/7.8732 = 0.1270$.

19.23 There is no unique solution to $\mathbf{LZ} = \mathbf{X}$; hence $\lambda = 0$ is an eigenvalue.

19.24

Iteration	Eigenvector components				Eigenvalue
0	1.0000	1.0000	1.0000	1.0000	
1	−0.1765	1.0000	−0.6353	−0.7059	−0.1828
2	0.3041	1.0000	0.1596	0.0851	−0.4681
3	0.1405	1.0000	−0.0661	−0.0968	−0.3323

19.25 $\lambda = 9 + 1/3.08114 = 9.3246$

19.26 $\lambda = 2.5 + 1/39.5709 = 2.5253$

19.27 **First iteration:** shift = 7.66667; eigenvector = $[1, -0.486452, 0.89935]^T$
Second iteration: shift = 10.5092; eigenvector = $[0.976710, 0.067597, 1]^T$
Third iteration: shift = 11.969339; eigenvector = $[1, -0.000168, 0.999924]^T$
Fourth iteration: shift = 12.0000; eigenvector = $[1, 0, 1]^T$

CHAPTER 20

20.14 (*a*) $\mathbf{Q} = \begin{bmatrix} -0.6667 & 0.2357 & 0.7071 \\ 0.6667 & -0.2357 & 0.7071 \\ 0.3333 & 0.9428 & 0 \end{bmatrix} \qquad \mathbf{R} = \begin{bmatrix} 6 & -5 & -0.6667 \\ 0 & 2.8284 & 0.2357 \\ 0 & 0 & 2.1213 \end{bmatrix}$

(*b*) $\mathbf{Q} = \begin{bmatrix} 0.7428 & 0.6442 & -0.1826 \\ 0.5571 & -0.7459 & -0.3651 \\ -0.3714 & 0.1695 & -0.9129 \end{bmatrix} \qquad \mathbf{R} = \begin{bmatrix} 5.3852 & 3.1568 & -2.0426 \\ 0 & 1.0171 & -0.5425 \\ 0 & 0 & 0.7303 \end{bmatrix}$

(*c*) The third column is a linear combination of the previous two, so Step 20.1′ must be employed for **Q**:

$$\mathbf{Q} = \begin{bmatrix} 0 & -1/\sqrt{2} & 1/\sqrt{2} \\ -1 & 0 & 0 \\ 0 & -1/\sqrt{2} & -1/\sqrt{2} \end{bmatrix} \qquad \mathbf{R} = \begin{bmatrix} 1 & 0 & 1 \\ 0 & \sqrt{2} & 0 \\ 0 & 0 & 0 \end{bmatrix}$$

(*d*) $\mathbf{Q} = \begin{bmatrix} 0.9057 & -0.2959 & -0.3037 \\ 0.3985 & 0.8387 & 0.3712 \\ -0.1449 & 0.4572 & -0.8775 \end{bmatrix} \qquad \mathbf{R} = \begin{bmatrix} 27.6043 & -139.0361 & -85.0591 \\ 0 & 17.1747 & 10.3479 \\ 0 & 0 & 19.4391 \end{bmatrix}$

(*e*) $\mathbf{Q} = \begin{bmatrix} 0 & -\sqrt{2}/2 & 0 & \sqrt{2}/2 \\ -1 & 0 & 0 & 0 \\ 0 & -\sqrt{2}/2 & 0 & -\sqrt{2}/2 \\ 0 & 0 & -1 & 0 \end{bmatrix} \qquad \mathbf{R} = \begin{bmatrix} 1 & 0 & 1 & 0 \\ 0 & \sqrt{2} & 0 & \sqrt{2}/2 \\ 0 & 0 & 1 & 0 \\ 0 & 0 & 0 & \sqrt{2}/2 \end{bmatrix}$

(*f*) $\mathbf{Q} = \begin{bmatrix} 0.6644 & -0.3332 & 0.6606 & -0.1055 \\ 0.7474 & 0.2962 & -0.5872 & 0.0938 \\ 0 & -0.0254 & 0.1452 & 0.9891 \\ 0 & -0.8948 & -0.4446 & 0.0423 \end{bmatrix}$

$$\mathbf{R} = \begin{bmatrix} 1{,}324.5754 & 493.2902 & 904.3653 & 246.6451 \\ 0 & 983.5089 & 7{,}413.9901 & 98.3798 \\ 0 & 0 & 4{,}328.6070 & -197.4178 \\ 0 & 0 & 0 & 6.2313 \end{bmatrix}$$

20.15 10.1461, 5.92049, −0.0665892 **20.16** 10.6056, 3.39445, 3

20.17 3.41421, 2, 0.585786 **20.18** 1, $-17+i24$, $-17-i24$

20.19 3.61803, 2.61803, 1.38197, 0.381966 **20.20** 990, 660, 440, 330

20.21 9, 9, 9, 9 **20.22** $-1 \pm i\sqrt{3}, \pm i$

20.23 The **QR** algorithm does not converge. **20.24** 232.275, 79.6707, 63.8284, 24.2261

CHAPTER 21

21.18 [1/2, 1/2] **21.19** $\frac{1}{22}\begin{bmatrix} 1 & 1 \\ 1 & 1 \\ 3 & 3 \end{bmatrix}$ **21.20** $\frac{1}{9}\begin{bmatrix} 1 & 1 & 1 \\ 1 & 1 & 1 \\ 1 & 1 & 1 \end{bmatrix}$ **21.21** $\begin{bmatrix} -2 & 3 \\ 1 & -1 \end{bmatrix}$

21.22 $\frac{1}{9}\begin{bmatrix} 1 & 4 & -1 \\ 4 & -2 & 5 \end{bmatrix}$ **21.23** $\frac{1}{206}\begin{bmatrix} 26 & -23 \\ 36 & -16 \\ 38 & 6 \\ 16 & -30 \\ -6 & 37 \end{bmatrix}$ **21.24** $\begin{bmatrix} 1 & 1 & -1 \\ -2 & 1 & 1 \\ 1 & -1 & 0 \end{bmatrix}$

21.25 $x_1 = 3/22,\ x_2 = 3/22,\ x_3 = 9/22$ **21.26** $x_1 = x_2 = x_3 = 2/3$

21.27 $x_1 = 2/3,\ x_2 = 5/3$ **21.28** $x_1 = 3/206,\ x_2 = 20/206,\ x_3 = 44/206,\ x_4 = -14/206,\ x_5 = 31/206$

21.29 Form $\|\mathbf{AX} - \mathbf{B}\|_2^2$, and then set the first partial derivatives with respect to each component of **X** equal to zero.

21.30 $\mathbf{A}^H\mathbf{AX} = (\mathbf{QR})^H(\mathbf{QR})\mathbf{X} = \mathbf{R}^H(\mathbf{Q}^H\mathbf{Q})\mathbf{RX} = \mathbf{R}^H\mathbf{IRX} = \mathbf{R}^H\mathbf{RX}$
and $\mathbf{A}^H\mathbf{B} = (\mathbf{QR})^H\mathbf{B} = \mathbf{R}^H\mathbf{Q}^H\mathbf{B}$

When the columns of **A** are linearly independent, the diagonal elements of **R** are all nonzero, and $(\mathbf{R}^H)^{-1}$ exists.

21.31 Solve $\mathbf{RX} = \mathbf{Q}^H\mathbf{B}$ with

$$\mathbf{Q} = \begin{bmatrix} -0.566947 & 0.377964 \\ -0.377964 & 0.377964 \\ -0.188982 & 0.377964 \\ 0 & 0.377964 \\ 0.188982 & 0.377964 \\ 0.377964 & 0.377964 \\ 0.566947 & 0.377964 \end{bmatrix} \qquad \mathbf{R} = \begin{bmatrix} 5.291503 & 0 \\ 0 & 2.645751 \end{bmatrix}$$

21.32 Solve $\mathbf{RX} = \mathbf{Q}^H\mathbf{B}$ with

$$\mathbf{Q} = \begin{bmatrix} 0.447214 & 0.596285 \\ 0.894427 & -0.298142 \\ 0 & 0.745356 \end{bmatrix} \qquad \mathbf{R} = \begin{bmatrix} 2.236068 & 0.447214 \\ 0 & 1.341641 \end{bmatrix}$$

21.33 The normal equations are

$$\begin{aligned} 354a + 100b + 30c &= 1158 \\ 100a + 30b + 10c &= 342 \\ 30a + 10b + 5c &= 122 \end{aligned}$$

and the solution is $a = 2.57$, $b = -0.69$, $c = 10.54$.

21.34 $\mathbf{V} = \begin{bmatrix} 1/\sqrt{11} & 1/\sqrt{2} & 3/\sqrt{22} \\ 1/\sqrt{11} & -1/\sqrt{2} & 3/\sqrt{22} \\ 3/\sqrt{11} & 0 & -2/\sqrt{22} \end{bmatrix}$ $\mathbf{U} = \begin{bmatrix} 1/\sqrt{2} & 1/\sqrt{2} \\ 1/\sqrt{2} & -1/\sqrt{2} \end{bmatrix}$ $\mathbf{\Sigma} = \begin{bmatrix} \sqrt{22} & 0 & 0 \\ 0 & 0 & 0 \end{bmatrix}$

21.35 $\mathbf{V} = \mathbf{V}_1 = \begin{bmatrix} -0.597354 & 0.801978 \\ 0.801978 & 0.597354 \end{bmatrix}$ $\mathbf{U} = \mathbf{U}_1 = \begin{bmatrix} -0.845435 & 0.534078 \\ -0.534078 & -0.845435 \end{bmatrix}$ $\mathbf{D} = \begin{bmatrix} 13.4834 & 0 \\ 0 & 0.444992 \end{bmatrix}$

21.36 $\mathbf{A}^{-1}$ satisfies Properties I1 through I3 because

$$(\mathbf{A}\mathbf{A}^{-1})^H = \mathbf{I}^H = \mathbf{I} = \mathbf{A}\mathbf{A}^{-1}$$
$$\mathbf{A}\mathbf{A}^{-1}\mathbf{A} = \mathbf{A}(\mathbf{A}^{-1}\mathbf{A}) = \mathbf{A}\mathbf{I} = \mathbf{A}$$

and

$$\mathbf{A}^{-1}\mathbf{A}\mathbf{A}^{-1} = (\mathbf{A}^{-1}\mathbf{A})\mathbf{A}^{-1} = \mathbf{I}\mathbf{A}^{-1} = \mathbf{A}^{-1}$$

The result then follows from Property 21.1.

21.37 $\mathbf{A}^+ = \mathbf{0}$ satisfies conditions I1 through I3 when $\mathbf{A} = \mathbf{0}$.

21.38 Conditions I1 through I3 are symmetric with respect to $\mathbf{A}$ and $\mathbf{A}^+$. Thus, if $\mathbf{A}^+$ is the generalized inverse of $\mathbf{A}$, then $\mathbf{A}$ is also the generalized inverse of $\mathbf{A}^+$. That is, $\mathbf{A} = (\mathbf{A}^+)^+$.

21.39 Show that conditions I1 through I3 are satisfied.

21.40 It follows from Property 21.5 that $(\mathbf{A}^+)^H = (\mathbf{A}^H)^+ = \mathbf{A}^+$.

21.41 Conditions I1 through I3 are satisfied because

$$(\mathbf{A}^+\mathbf{A})^H = (\mathbf{A}\mathbf{A})^H = \mathbf{A}^H = \mathbf{A} = \mathbf{A}\mathbf{A} = \mathbf{A}^+\mathbf{A}$$
$$\mathbf{A}^+\mathbf{A}\mathbf{A} = \mathbf{A}\mathbf{A}\mathbf{A} = \mathbf{A}(\mathbf{A}\mathbf{A}) = \mathbf{A}\mathbf{A} = \mathbf{A}$$

and

$$\mathbf{A}^+\mathbf{A}\mathbf{A}^+ = \mathbf{A}\mathbf{A}\mathbf{A} = \mathbf{A}(\mathbf{A}\mathbf{A}) = \mathbf{A}\mathbf{A} = \mathbf{A} = \mathbf{A}^+$$

21.42 $\mathbf{A}\mathbf{A}^+$ and $\mathbf{A}^+\mathbf{A}$ are Hermitian from condition I1. Also,

$$(\mathbf{A}\mathbf{A}^+)(\mathbf{A}\mathbf{A}^+) = (\mathbf{A}\mathbf{A}^+\mathbf{A})\mathbf{A}^+ = \mathbf{A}\mathbf{A}^+ \quad \text{and} \quad (\mathbf{A}^+\mathbf{A})(\mathbf{A}^+\mathbf{A}) = (\mathbf{A}^+\mathbf{A}\mathbf{A}^+)\mathbf{A} = \mathbf{A}^+\mathbf{A}$$

21.43 Take $\mathbf{\Sigma}$ to be the $n \times n$ diagonal matrix containing all the singular values of $\mathbf{A}$, including zeros if they arise. Construct $\mathbf{D}$, $\mathbf{V}$, and $\mathbf{U}_1$ exactly as described in Steps 21.8 and 21.9, and then construct $\mathbf{U}$ by first following Step 21.10 but then keeping only the first n linearly independent columns and orthonormalizing them.

21.44 $\mathbf{P}$ is Hermitian and similar to $\mathbf{\Sigma}$, which has nonnegative eigenvalues. Since the columns of $\mathbf{U}$ are orthonormal, $\mathbf{U}^H\mathbf{U} = \mathbf{I}$, and, with $\mathbf{V}$ being unitary,

$$\mathbf{M}^H\mathbf{M} = (\mathbf{U}\mathbf{V}^H)^H(\mathbf{U}\mathbf{V}^H) = \mathbf{V}\mathbf{U}^H\mathbf{U}\mathbf{V}^H = \mathbf{V}\mathbf{I}\mathbf{V}^H = \mathbf{V}\mathbf{V}^H = \mathbf{I}$$

21.45 $\mathbf{M}\mathbf{P} = (\mathbf{U}\mathbf{V}^H)(\mathbf{V}\mathbf{\Sigma}\mathbf{V}^H) = \mathbf{U}(\mathbf{V}^H\mathbf{V})\mathbf{\Sigma}\mathbf{V}^H = \mathbf{U}\mathbf{\Sigma}\mathbf{V}^H = \mathbf{A}$

21.46 With $\mathbf{V}$ and $\mathbf{U}$ as given in Problem 21.4 and

$$\mathbf{\Sigma} = \begin{bmatrix} 8 & 0 & 0 \\ 0 & 2 & 0 \\ 0 & 0 & 0 \end{bmatrix}$$

we have $\mathbf{P} = \mathbf{V}\mathbf{\Sigma}\mathbf{V}^H = \mathbf{A}$ and $\mathbf{M} = \mathbf{U}\mathbf{V}^H = \mathbf{I}$.

21.47 With $\mathbf{D}$, $\mathbf{V}$, and $\mathbf{U}_1$ as given in Problem 21.5, we have $\mathbf{\Sigma} = \mathbf{D}$ and $\mathbf{U} = \mathbf{U}_1$. Then $\mathbf{P} = \mathbf{V}\mathbf{\Sigma}\mathbf{V}^H = \mathbf{\Sigma} = \mathbf{D}$ and $\mathbf{M} = \mathbf{U}\mathbf{V}^H = \mathbf{U} = \mathbf{U}_1$.

Index